Streamflow Measurement

Streamflow Measurement

Third edition

Reginald W. Herschy

Chairman British Standards Institution Technical Committee
on Hydrometry

LONDON AND NEW YORK

First published 1985 by Taylor & Francis

Second edition published 1995
Third edition published 2009
by Taylor & Francis

2 Park Square, Milton Park, Abingdon, Oxfordshire OX14 4RN
52 Vanderbilt Avenue, New York, NY 10017

*Taylor & Francis is an imprint of the Taylor & Francis Group,
an informa business*

First issued in paperback 2019

Typeset in Sabon by
RefineCatch Limited, Bungay, Suffolk

Library of Congress Cataloging in Publication Data
Herschy, Reginald W.
    Streamflow measurement / Reginald W. Herschy.—3rd ed.
        p. cm.
    Includes bibliographical references and index.
    ISBN 978–0–415–41342–8 (hardback : alk. paper)—
    ISBN 978–0–203–92129–4 (ebook)
        1. Stream measurements.   I. Title.
    GB1203.2.H47 2008
    551.48′30287—dc22   2008022423

ISBN13: 978–0–415–41342–8 (hbk)
ISBN13: 978–0–367–86518–4 (pbk)

Water, water, every where,
Nor any drop to drink.
                    Samuel Taylor Coleridge (1772–1834)

Contents

Tables

Figures

Acknowledgements

In the preparation of this third edition of the book, there are many friends and colleagues who have kindly supplied information and generously offered suggestions. In this connection, I am particularly indebted to Mr T. M. Marsh of the Centre for Ecology and Hydrology (CEH) UK and Mr S. C. Child of Hydro-Logic UK.

There are many colleagues both in the United Kingdom and overseas who have kindly helped me directly or indirectly and I am grateful to the British Standards Institution (BSI), the International Organization for Standardization (ISO), the European Committee for Standardisation (CEN) and the World Meteorological Organization (WMO) for permitting my past participation as chairman of their relevant committees on Hydrometry or Hydrology and for their permission to reproduce figures from their Standards.

I am indebted to my colleagues on BSI, ISO, CEN and WMO Committees for their advice and their encouragement.

I am grateful to the United Nations Development Programme (UNDP) and the World Meteorological Organization (WMO) for inviting me to undertake streamflow missions for them in China, India, Kenya and Lesotho.

Acknowledgement is kindly made to the United States Geological Survey, the Ministry of Water Conservancy of the People's Republic of China and the Central Water and Power Research Station in India for their permission to reproduce figures from their publications.

Acknowledgement is also made to John Wiley and Sons for permission to reproduce figures from my *Hydrometry: Principles and Practices* (first and second editions).

Throughout the text, I have tried to make specific acknowledgement in the Further Reading references regarding the source of material and any failure to do so is an unintentional oversight.

Thanks are recorded to the following for kindly permitting the reproduction of figures and tables in this book.

Bonacci, O. (IAHS), Fig. 2.29
Central Water and Power Research Station (CWPRS), Pune, India, Fig. 2.8

Diptone, Fig. 3.8

Environment Agency, Thames, Fig. 3.21(d)

Environment Canada, Fig. 3.15

HR Ltd., Figs. 10.14, 15.9

ISO *www.iso.org*, Figs. 2.26, 2.27, 2.28, 3.4, 10.4, 10.5, 10.7, 10.11, 10.12, 10.13, 10.15, 10.16, 10.17, 10.18, 11.4, 11.6, 11.7, 11.8, 11.9, 11.10, Tables 10.1, 10.4, 10.6, 10.7, 10.8, 10.12, 10.14, 10.15

John Wiley & Sons Ltd., Figs. 3.6, 3.25, 5.20. 7.1, 15.1(a), 15.2, 15.3, 15.4, 15.5, 15.6(a), 15.7

Leupold and Stevens, Fig. 3.13

Littlewood, I. G., Fig. 11.14

Ministry of Water Resources, Bureau of Hydrology, China, Figs. 2.16, 2.20, 3.2, 3.3, 3.21(a)(b)(c), 3.27, 3.30, 3.31, 4.5

Lucseva, A. A., Figs. 7.4, 7.5

Jones, R. C., Fig. 3.23

IAHS, Fig. 3.29

Neyrtec, Figs. 2.17, 2.21

ORE, Figs. 12.2, 15.7

Ott, Figs. 2.6(a)(c), 2.7(a), 3.10, 3.11

Republic of South Africa, Figs. 6.5, 6.6, 6.7, 6.10, 6.12

Rijkswaterstaat, Fig. 12.10

Sarasota, Figs. 12.4, 12.8, 12.9

Sargent, D. M., Figs 7.7, 7.8, 7.9

SonTek *www.sontek.com*, Figs. 5.19, 6.3(a)(b), 6.4, 6.8, 6.9, 6.11(b), 6.13, 6.15

Strangeways, I.C., Figs. 3.35, 3.36

Tilrem, O. A., Figs. 2.10, 2.18, 2.23, 2.29, 3.29, 3.32, 3.33, 3.34, 4.1, 4.2, 4.3, 4.8, 4.11, 5.1(a), 5.9, 5.11, 5.12, 5.13, 5.14, 5.15, 5.16, 9.1, 9.2, 9.3

United States Geological Survey *www.usgs.gov*, Figs. 2.5(a), 2.13, 2.19, 2.22, 2.24, 2.25, 3.7, 3.18, 3.24, 3.28, 4.4, 5.4, 5.5, 5.6, 5.7, 5.10, 5.17, 5.18, 6.3(c), 6.11(a), 6.14(a)

University of Dundee *www.dundee.ac.uk*, Fig. 5.21

University of Wales *www.sos.bangor.ac.uk*, Figs. 2.35, 2.36

Valeport, Figs. 2.6(b), 2.6(d), 2.7(b)

Preface

Since the publication of the second edition of the book in 1995, significant advances have taken place worldwide in streamflow measurement in both instrumentation and methodology.

The scope of this third edition of the book has therefore been extended to include new methods and instrumentation and a new Chapter 6 on the acoustic Doppler current profiler (ADCP) has been included to meet the growing need worldwide for this method.

The moving boat and electromagnetic methods are now included in a reduced form in Chapter 2 as is a new method now under research – the seismic flowmeter.

There are hardly any large floods which have been measured directly by standard methods and a section on indirect methods of flood peak estimation has therefore been included under Chapter 5 – Special problems.

The opportunity has been taken to rewrite certain sections of the book in view of technical developments and new or revised international standards. Chapters 9 and 15 come into this category. However, care has been taken to retain existing instrumentation, such as chart recorders, which are still used efficiently in many countries.

The book has been written in a global context as streamflow measurements are carried out to similar standards worldwide. It is indeed true to say that a gauging is being made by someone somewhere in the world at any moment in time.

The need for better and extended streamflow measurements is more necessary today than ever before especially with the need to address climate change and its effects on streamflow.

The international monitoring of streamflow measurement will become crucial in a changing hydrological world where our precious water resources are required to be audited and carefully managed.

Reginald W. Herschy
Reading

Chapter 1

Introduction

1.1 Global water

The management of our global water resources will require renewed effort by all concerned in the water industry and measurement of the world's rivers will play a large part in the distribution of these resources. In the world today over one quarter of the population still do not have safe drinking water.

The amount of fresh water available is small. Only about 0.6% of the global water is available for use and of that only 1% is in rivers; the rest is groundwater.

However, this 1% is only 0.006% of the total global water and it is this water that requires to be measured as streamflow; in fact about 70,000 km^3.

With tens of millions of people worldwide relying on fresh water, the importance of global streamflow measurement is crucial.

Figure 1.1 shows diagrammatically the components of the global water cycle.

1.2 Climate change

Considerable research has been undertaken into climate change by the IPCC (Intergovernment Panel on Climate Change). Two important elements for hydrometry to address are rainfall and streamflow. Climate change can be expected to lead to changes in precipitation and streamflow. Water supply in arid and semi-arid regions is very sensitive to small changes in rainfall and evaporation although the latter may be reduced because of increased CO_2 concentrations. In the runoff scenario, it is believed that a doubling of CO_2 might increase river flows by between 40 and 80% in certain parts of the world. Such estimates of runoff extremes, if confirmed, would require careful streamflow monitoring as well as considerable modification in the design of water related structures including flood control works and conveyance capacity.

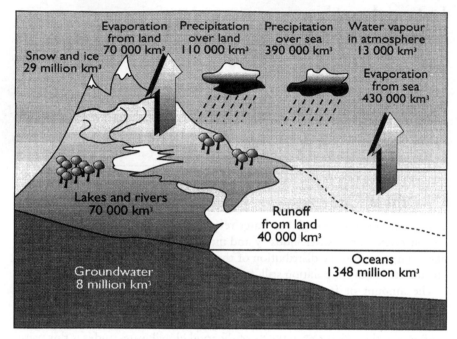

Figure 1.1 The hydrological (water) cycle with major reservoirs.

Note: The fluxes of evaporation, precipitation and annual run off are in km³ year⁻¹

1.3 Field measurements

Streamflow is the combined result of all climatological and geographical factors that operate in a drainage basin. It is the only phase of the hydrological cycle in which the water is confined in well-defined channels which permit accurate measurements to be made of the quantities involved. Other measurements of the hydrological cycle are point measurements for which the uncertainties, on an areal basis, are difficult, if not impossible, to estimate.

Good water management is founded on reliable streamflow information and the final reliability of the information depends on the initial field measurements. The hydrologist making these measurements has therefore the responsibility of ensuring raw data of acceptable quality are collected. The successful processing and publication of the data depend largely on the quality of the field measurements.

This book is therefore for field hydrologists and for students of hydrology in universities and colleges.

Objectives of a streamflow programme

There are many different uses of streamflow data within the broad context of water management, such as water supply, pollution control, irrigation, flood control, energy generation and industrial water use. The importance placed on any one of these purposes may vary from country to country. In India and China, for example, emphasis may be placed upon irrigation and flood control whereas in the United Kingdom water supply may be given priority. The emphasis for any one need may also change over short or longer periods of time. What appears to be axiomatic, however, is that none of these needs can be met without reliable streamflow data being available at the right time, the right place and the right quality.

Categories of streamflow data

The type of streamflow information required may be classified into two distinct categories. The first is that required for planning and design while the second is that required for current use, i.e. operational management.

Data for planning and design may not necessarily have an immediate use but are valuable in the long term for civil engineering works of various types and for flood forecasting and control. Planning and design data are also used to examine long-term trends as are data on the stream environment.

Current use data have an immediate high return value since they are invariably required initially for operation and control. Current use streamflow stations are operated for as long as the need remains.

Designers of water control and water-related facilities increasingly use the statistical characteristics of streamflow rather than flow over specific historic periods. The probability that the historical sequence of flow history at a given site will occur again is remote. Indeed, when a hydrologist makes just one measurement of discharge it is probable that the exact conditions under which the discharge occurred may rarely happen again.

It is often desirable to consider the future, not in terms of specific events, but in terms of probability of occurrence over a span of years. For example, many highway bridges are designed on the basis of the flood that will be exceeded on the average only once in 50 years. Storage reservoirs are designed on the basis of the probability of failure of a particular capacity to sustain a given draft rate. The water available for irrigation, dilution of waste or other purposes may be stated in terms of the mean flow, or probability of flow magnitudes, for periods of a year, season, month, week or day. In addition there is a trend towards flow simulation based on statistical characteristics, such as the mean, standard deviation and skew. To define statistical characteristics, a record of at least 30 years is desirable for reliable results and a study of Section 13.9 (Chapter 13) suggests caution in estimating trends or probabilities from short-term periods.

1.4 Cost effectiveness

In most countries the cost effectiveness of streamflow data collection is an important consideration; this is particularly the case where streamflow is included in the budget for water management. Cost effectiveness may be measured by the benefit:cost ratio, but to estimate this ratio for streamflow is difficult, mainly due to the problems associated with assessing the benefits accruing. This problem sometimes leaves the hydrological service at a disadvantage in bidding for funds.

The wide variety of uses of streamflow data also makes the estimation of national benefits difficult. The question of marginal gains through network changes is therefore not straightforward. It is, however, a fact that costs have risen sharply in providing gauging stations and in data capture and publication. The gains on the other hand are not easily quantified and each use of stream-flow data may demand different and perhaps sophisticated analysis before benefits of streamflow data collection can be realised. This, however, is not usually the case in developing countries where the gain from a flood control scheme or an irrigation scheme may be enough to cover the cost of the entire network many times over. Benefit:cost ratios in these circumstances may be as high as 50:1 or more.

In other countries, however, a period of years may elapse before a useful record is generated to quantify the benefits of current data and even then any satisfactory assessment is complicated.

The first objective therefore is to develop a suitable method to identify potential quantifiable benefits to various types of data user. Such benefits are usually to be found in data required for reservoir design, water abstraction, flood warning, flood control including flood proofing, irrigation, highway bridge design, hydroelectric power generation, river pollution control, sewage purification and so on.

The costs of providing these services are quantified over a defined period and the benefits accruing from streamflow data are calculated for each. More often the benefits may have to be determined from an agreed percentage of the cost of the services.

Flood proofing, for example, reduces the cost of flood damage and this figure can be conveniently quantified from flood damage records. If no flood damage records exist, a percentage benefit based on the cost of the scheme can usually be calculated. The benefits are calculated for each use to which the streamflow data are put and totalled. This total is divided by the cost of obtaining the streamflow data. This is usually the cost of the operation of the gauging stations and processing the data or, more conveniently, the sum of capital and staff costs.

In a benefit–cost study of the UK streamflow network carried out for the Department of the Environment (DOE) in 1989, the annual benefits were found to be in the range US$16.5–90 million depending on how these were quantified, the best estimate being US$31.5 million. The annual cost of operating the network, including overheads, management, data processing, etc. was US$13.5 million. The benefit:cost ratio was therefore in the range 1.2–7 with a best estimate of 2.3. It was concluded, therefore, that even at the lowest level of benefit:cost ratio, the UK streamflow network represents a sound economic investment. The ratio would have been higher if some of the intangibles (e.g. consents for discharge effluents) could have been quantified. Not surprisingly only small annual economic benefits could be quantified from flood forecasting, flood warning or flood alleviation.

1.5 International standards in stream gauging

Water in a stream in a specific locality knows no jurisdictional boundaries, local or national. That same water may eventually move to any other part of the earth through the hydrological cycle. Streamflow data are therefore needed from all parts of the earth to enable hydrologists to discover the quantity of the earth's water resources on a comprehensive and continuous basis. Streamflow records that have been gathered by non-standard methods may be suspect. For this and other reasons, the International Organization for Standardization (ISO) set up in 1956, a technical committee on streamflow measurement. This committee, known as TC113, has produced a number of international standards on streamflow which are now used worldwide. Of the 104 ISO member countries, some 37 are members of TC113.

The methods described in this book generally follow the principles and recommendations of the ISO Standards.

In addition, the World Meteorological Organization (WMO), publishes guides and technical reports on stream gauging and selected ISO Standards, in the form of technical regulations which are circulated to some 187 WMO member countries.

Standardisation activity at the European level is the responsibility of CEN (European Committee for Standardization – Comité Européen de Normalisation) and CENELEC (European Committee for Electrotechnical Standardization). Together these bodies make up the Joint European Standards Institution (ESI). The aim of European standardisation is the harmonisation of standards on a Euro-wide basis in order to facilitate the exchange of goods and services by eliminating barriers to trade which might result from requirements of a technical nature. The national standardisation institutes of 27 countries support CEN. In addition, other European countries have affiliate status. Streamflow is under TC318 'Hydrometry', formed in 1994.

1.6 Summary of methods

A summary of the methods of streamflow measurement follows together with a reference to the chapters in which each method is discussed.

Velocity–area method (Chapter 2)

The discharge is derived from the sum of the products of stream velocity, depth and distance between verticals (Fig. 1.2), the stream velocity usually being obtained by a current meter. For a continuous record of discharge in a stable prismatic open channel with no variable backwater effects, a unique relation exists between water level (stage) and discharge. Once established, this stage–discharge relation is used to derive discharge values from recordings of stage. With the exception of the dilution method, which is a direct method, it could be inferred that all methods of streamflow measurement are based as the velocity–area principle.

The stage–discharge relation is covered in Chapter 4 and since the measurement of stage is one of the most important factors in all methods, a separate chapter is devoted to it (Chapter 3). Special problems, associated with velocity–area stations in particular, such as corrections for soundings from cableways, stilling well lag and draw-down, rapidly changing discharge and measurements under ice cover, are presented in Chapter 5.

Acoustic Doppler Current Profiler (ADCP) (Chapter 6)

The ADCP instrument is mounted on a motorised boat that moves across the river perpendicular to the current. Velocities are measured when the ADCP transmits acoustic pulses along three or four beams at a constant frequency. These beams are positioned at precise horizontal angles from each other and

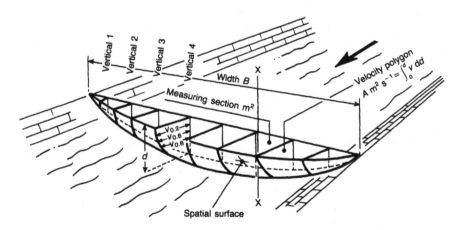

Figure 1.2 The measuring section. The volume of water is bounded by the measuring section, the water surface, the bed and the spatial surface as shown schematically. At any section XX, the area of the velocity polygon is the integral vdd (with limits from 0 to d) and equal to A m^{2}s^{-1}. The volume of water passing per second is then found from the integral of Adb (with limits from 0 to b) and equal to the integral of vdd db (with limits from 0 to d and 0 to b) which is equal to the total flow Q in m^{3}s^{-1}.

directed at a known angle from the vertical, typically 20° or 30° The instrument processes echoes throughout the water column along each beam. The difference in frequency (Doppler shift) between transmitted pulses and received echoes (Doppler effect) can be used to measure the relative velocity between the instrument and the suspended material in the water that reflects the pulses back to the instrument (backscattering). The ADCP uses the Doppler effect to compute a velocity component along each beam and the system software calculates velocity in three dimensions using trigonometric relations. The ADCP may therefore be regarded as a velocity–area method giving a single value of discharge, usually to provide a point on the stage–discharge curve.

Float gauging (Chapter 7)

The water velocity is measured by recording the time taken for a float to travel a known distance along the channel. Observations are made using floats at different positions across the channel and discharge is derived from the sum of the products of velocity, width and depth.

Generally, this method is used only when the flow is either too fast or too slow to use a current meter or where ice floes would cause damage to the meter.

Slope–area method (Chapter 8)

The discharge is derived from measurements of the slope of the water surface and the cross-section of the channel over a fairly straight reach, assuming a roughness coefficient for the channel boundaries.

Stage–fall–discharge method (Chapter 9)

In a stable open channel affected by backwater, a relation is established between fall (slope) and discharge.

Weirs and flumes (Chapter 10)

The relation between stage (or head) and discharge over a weir or through a flume is established from laboratory (or field) calibration. The discharge is subsequently derived from this rating equation.

Dilution method (Chapter 11)

A tracer liquid is injected into the channel and the water is sampled at a point further downstream where turbulence has mixed the tracer uniformly throughout the cross-section. The change in concentration between the solution injected and the water at the sampling station is converted into a measure of the discharge.

Moving boat method (Chapter 12)

A current meter is suspended from a boat which traverses the channel normal to the streamflow. The component of the velocity in the direction of the stream is computed from the resultant velocity and the angle of this resultant. The discharge is the sum of the products of the stream velocity, depth and distance between observation points.

Ultrasonic method (Chapter 12)

The velocity of flow is measured by transmitting an ultrasonic pulse diagonally across the channel in both directions simultaneously. The difference in time transits is a measure of the velocity which has to be multiplied by the cross-sectional area to derive discharge. The ultrasonic method therefore also follows the principles of velocity–area measurements.

Electromagnetic method (Chapter 12)

The discharge is found by measuring the electromotive force (emf) produced by a moving conductor (the flowing water) through a magnetic field produced by a coil placed either below or above the open channel. The emf is proportional to the discharge.

Accuracy (Chapter 13)

Considerable research into uncertainties in streamflow measurement over recent years has led to the publication of several international standards and this chapter has been updated from the first edition to address the latest methods of the assessment of uncertainties.

Hydrometric data processing (Chapter 14)

In view of the advances made in solid-state recording and data processing of streamflow measurements, this chapter has been completely rewritten but autographic chart recording methods have been retained.

Flow in pipes (Chapter 15)

Because of today's need for hydrologists to address all types of flow measurement, this chapter has been added to the present edition and includes flow in closed conduits under pressure and flow in partially filled pipes using both existing theory and practice and modern concepts.

1.7 Selection of method

Velocity–area method

Generally, consideration is given first to the possibility of installing a velocity–area station especially if it is known that a relation can be established between stage and discharge. Discharge measurements may then be carried out using a current meter by wading (when the depth and velocity permit), by cableway (when the span permits the installation of a cableway, and the river is too deep to wade), by boat (if the river is too wide for a cableway installation), by moving boat (if the river is wide enough), by floats (if the velocity is too low or too high to use a current meter or there are ice floes in the river), by slope–area (if no other method is suitable during floods) or from bridges (if these are considered suitable).

ADCP

The ADCP is now used in many countries giving very good results in the measurement of a single measurement of discharge. Systems now available are able to measure both large and small rivers and deep or shallow rivers when mounted on motor launches or small remote-controlled or tethered rafts or catamarans. A great advantage of the method is its speed whereby an ADCP measurement may be as much as ten-times quicker than a conventional measurement. The equipment now is considerably reduced in size and weight and units available now may be only a few kilograms in weight and less than 30 centimetres in height.

Weirs and flumes

In small rivers (under 100 m in width) a measuring structure may be considered, particularly if backwater conditions prevail. The main factors to be assessed for a measuring structure are cost, head loss (afflux) available, Froude number and bed conditions.

Flumes are normally only considered in smaller channels and especially in wastewater treatment works of which there are literally thousands in the UK alone.

Ultrasonic and electromagnetic methods

The ultrasonic and electromagnetic methods provide a continuous measurement of discharge for all designed stages of flow and continue to do so under backwater conditions even if the flow actually reverses due, for example, to tidal influence.

The main restrictions for the ultrasonic method are that a source of electrical power should be available, the river should not be more than about 300 m wide with suitable minimum depth and should have no weed growth or significant sediment transport.

The electromagnetic method also requires a source of power and is restricted to rivers about 40 m wide but continues to measure under weed conditions or heavy sediment load.

Dilution techniques

Dilution gauging is not in such general use as other methods because the technique requires specially trained staff. Nevertheless it is the most suitable method available for discharge measurements in turbulent mountain streams. It is used mainly for spot measurements especially in the calibration of other methods, for example measuring structures, but in certain situations it may be the only suitable method. It is also the only fully direct method for the measurement of discharge since the velocity, depth or area does not enter into the computation.

Stage–fall–discharge and slope–area methods

These methods are indirect methods of measurement, but have their place under conditions where the above methods are not suitable or are unavailable. The stage–fall–discharge method is particularly useful under backwater conditions especially in large rivers, when it may be the only suitable method. The slope–area method is useful in the measurement of floods, either current or historical, the latter from flood marks.

The stage–fall–discharge method may take the form of a permanent station; the slope–area method is used for measurements and may be employed at a permanent velocity–area station for measuring the highest flows. The latter method depends, however, on Manning's 'n' or Chezy's 'C' roughness co-efficients and, unless these are established on site from measurements, the methods may have a large current or historical uncertainty.

Further reading

Ackers, P., White, W. R., Perkins, J. A. and Harrison, A. J. M., *Weirs and Flumes for Flow Measurement*. John Wiley and Sons, Chichester. 1978.

Bos, M. G., *Discharge Measurement Structures*. Publication No. 161 Delft Hydraulics Laboratory, Delft. 1976.

Department of the Environment. *The Benefit Cost of Hydrometric Data: River Flow Gauging*. Report by CNS, Reading, UK. The Foundation for Water Research, UK. 1989.

Herschy, R.W., The analyses of uncertainties in the stage discharge relation in Flow Meas. Instrum. 4 (3). Butterworth-Heinemann Oxford 1994.

Herschy, R.W., General purpose flow measurement equations for flumes and thin plate weirs. 1995.

Herschy, R.W., Hydrometry: Principles and Practices. 2nd Edition, John Wiley & Sons Chichester 1998.

Herschy, R.W., editorial to: Open channel flow measurement. *Flow Meas, and Instr. 13 189–190* 2002.

Herschy, R.W. and Fairbridge, R.W., Encyclopedia of Lakes and Reservoirs, Springer, Dordrecht (in press).

Herschy, R.W. and Fairbridge, R.W., Encyclopedia of Hydrology and Water Resources, Kluwer, Dordrecht 1998.

ISO Guide to the expression of uncertainty (GUM) 1995.

ISO 1070 Slope area method 1992.

ISO 1088 Collection of and processing of data for determination of uncertainties in flow measurement 2007.

ISO 1100/1 Establishment and operation of a gauging station 1997.

ISO 1100/2 Stage discharge relation 1998.

ISO 1438/1 Thin plate weirs 2008.

ISO 6416 Ultrasonic method 1992.

ISO 748 Velocity area methods 2008.

ISO 772 Hydrometry vocabulary and symbols 2008.

ISO 9213 The electromagnetic method 2004.

ISOCEN 25377 Hydrometric uncertainty guide 2007.

Thomas, F., Open channel flow measurement using international standards: introducing a standards programme and selecting a standard. *Flow Meas. and Instr. 13 303–307* 2000.

Yorke, T.H. and Oberg, K.A. Measuring river velocity and discharge with acoustic Doppler profilers. *Flow Meas. and Instr. 191–195* 2000.

Chapter 2

The velocity–area method of streamflow measurement

2.1 General

The velocity–area method for the determination of discharge in open channels consists of measurements of stream velocity, depth of flow and distance across the channel between observation verticals. The velocity is measured at one or more points in each vertical by a current meter and an average velocity determined in each vertical. The discharge is derived from the sum of the product of mean velocity, depth and width between verticals. The discharge so obtained is normally used to establish a relation between water level (stage) and streamflow. Once established this stage–discharge relation is used to derive discharge values from records of stages at the gauging station.

Not all current meter measurements, however, are made to establish a stage–discharge relation and for many purposes individual determinations or 'spot measurements' are very often required for management functions. Such measurements may not require the measurement of stages but otherwise the method of measurement is the same. At some stations, however, a record of stages only may be required for purposes such as flood warning. At most gauging stations, however, both stages and discharges are measured to establish a relation between these two variables.

2.2 Spacing of verticals

In order to describe the bed shape and the horizontal and vertical velocity distributions completely, an infinite number of verticals would be necessary; for practical reasons, however, only a finite number is possible. In practice, therefore, the cross-section is divided into segments by spacing verticals at a sufficient number of locations across the channel to ensure an adequate sample of both velocity distribution and bed profile. The spacing and number of verticals are crucial for the accurate measurement of discharge and for this reason between 20 and 30 verticals are normally used. This practice applies to rivers of all widths except where the channel is so narrow that 20 or 30 verticals would be impracticable. We shall see in Chapter 13 that uncertainties in streamflow

measurement are expressed as percentages. The percentage uncertainty therefore for using, say, 20 verticals is of the same order for all widths of river
notwithstanding the width of the segments (in absolute terms the uncertainty
will increase as the width of segment increases).

Verticals may be spaced on the basis of the following criteria:

(a) equidistant;
(b) segments of equal flow;
(c) bed profile.

The choice will depend largely on the flow conditions, the geometry of the
cross-section and the width of river. For very wide rivers (over 300m), for
example, it is sometimes convenient to make the verticals equidistant; for
rivers having an asymmetrical horizontal velocity distribution, or a significant
variation in the horizontal velocity distribution, it is normally advisable to
space the verticals in such a manner so as to achieve segments of equal flow over
the required range; for rivers having abnormalities in the bed profile, the verticals are spaced so as to make allowance for depressions or obtrusions and
general irregularities of the bed. A general rule, however, for current meter
measurements is to make the width of segments less as the depth and velocities
become greater.

Irrespective of which criteria are followed, the spacing of the verticals is
arranged so that no segment contains more than, say, 10% of the total flow. The
best measurement is normally one having no segment with more than 5% of
the total flow.

2.3 Computation of current meter measurements

Mid-section method

In the mid-section method of computation it is assumed that the velocity
sampled at each vertical represents the mean velocity in a segment. The segment area extends laterally from half the distances from the preceding vertical
to half the distance to the next, and from the water surface to the sounded
depth as shown by the hatched area in Fig. 2.1. The segment discharge is then
computed for each segment and these are summed up to obtain the total
discharge. Referring to Fig. 2.1, which shows diagrammatically the cross-
section of a stream channel, the discharge passing through segment 5 is
computed as

$$q_5 = \bar{v}_5 \left(\frac{(b_5 - b_4) + (b_6 - b_5)}{2} \right) d_5 \qquad (2.1)$$

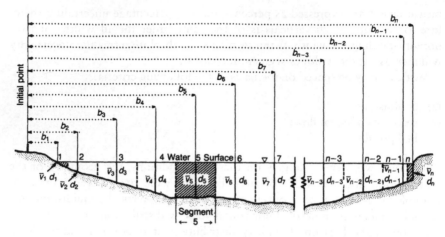

Figure 2.1 The mid-section method of computing current meter measurements. 1, 2, 3, . . . , n, number of vertical; $b_1, b_2, b_3, \ldots, b_n$, distance from initial point; $d_1, d_2, d_3 \ldots, d_n$, depth of flow at verticals; $\bar{v}$, average velocity in verticals.

$$= \bar{v}_5 \left(\frac{b_6 - b_4}{2} \right) d_5 \qquad\qquad (2.2)$$

where $q_5 =$ discharge through segment 5;
   $\bar{v}_5 =$ mean velocity in vertical 5;
   $b_4, b_5, b_6 =$ distance from an initial point on the bank to verticals 4, 5 and 6;
   $d_5 =$ depth of flow at vertical 5.

For the end segment, 1, shown hatched, the discharge may be computed as

$$q_1 = \bar{v}_1 \left(\frac{b_2 - b_1}{2} \right) d_1 \qquad\qquad (2.3)$$

and the end segment, n, as

$$q_n = \bar{v}_n \left(\frac{b_n - b_{n-1}}{2} \right) d_n. \qquad\qquad (2.4)$$

The preceding segment at the beginning of the cross-section is therefore considered coincident with vertical 1 and the next vertical at the end of the cross-section is considered coincident with vertical n.

In the example in Fig. 2.1, q_1 is zero because the depth at vertical 1 is zero. However, when the cross-section boundary is vertical at the edge of the water, the depth is not zero and the velocity at the end vertical may or may not be zero. The equations for q_1 and q_n are used whenever there is water on only one side of a vertical, such as piers, abutments and islands. It is usually necessary to estimate the velocity at the end segments as a percentage of the velocity on the adjacent vertical because it is not possible to locate the current meter close to a boundary. Alternatively, a current meter observation may be made as near the edge as possible and this velocity used in computing the discharge in the end segments. However, if the verticals 2 and $n - 1$ are placed as close as possible to the banks and the cross-section is wide, the discharge in in the end segments can normally be neglected. A typical computation of a current meter measurement employing the mid-section method is shown in Table 2.1. It will be noted in this example that 22 verticals have been used and that the discharge in any one segment does not exceed 10% of the total discharge.

Mean-section method

Segment discharges are computed between successive verticals. An example of one such segment is shown hatched in Fig. 2.2. The velocities and depths for successive verticals are each averaged, the segment discharge being the product of the two averages.

Referring to Fig. 2.2, the discharge passing through segment 5–6 is computed as

$$q_{5-6} = \left(\frac{\bar{v}_5 + \bar{v}_6}{2} \right) \left(\frac{d_5 + d_6}{2} \right) (b_6 - b_5) \tag{2.5}$$

where q_{5-6} = discharge through segment 5–6;
   $\bar{v}_5, \bar{v}_6$ = mean velocities in verticals 5 and 6;
   d_5, d_6 = depth of flow at verticals 5 and 6;
   b_5, b_6 = distance from an initial point on the bank to verticals 5 and 6.

It will be noted that the depth of flow at vertical 1 is zero and the problem of computing the flow in the end segments does not arise in this method nor does it arise when the bank is vertical and the velocity can be taken as approximately zero at the end vertical. The computation is therefore carried out for the end segments in exactly the same way as for the other segments. Nevertheless this facility does not give the mean-section method an overall advantage over the mid-section method, the latter being simpler to compute and therefore quicker if the calculations are being performed manually. There is little difference in time, however, if a pocket calculator is employed for the calculation.

Table 2.1 Typical computation for a current meter measurement by the mid-section method

(1) Verticals	(2) Distance from initial point (m)	(3) Depth (m)	(4) Meter position	(5) Revs	(6) Time (s)	(7) Velocity At point (m s⁻¹)	(8) Mean in vertical (m s⁻¹)	(9) Width (m)	(10) Area (m²)	(11) Discharge (m³ s⁻¹)
RB	4	0		0	0	0	0	0	0	0
1	5	0.31	0.6	40	60	0.193	0.193	1	0.31	0.060
2	6	0.40	0.6	45	59	0.219	0.219	1	0.40	0.089
3	7	0.51	0.6	51	61	0.238	0.238	1	0.51	0.121
4	8	0.85	0.6	52	61	0.243	0.243	1	0.85	0.206
5	9	1.23	0.2	55	60	0.260	0.235	1	1.23	0.289
			0.8	44	60	0.211				
6	10	1.58	0.2	58	62	0.265	0.240	1	1.58	0.379
			0.8	46	61	0.216				
7	11	1.69	0.2	60	61	0.278	0.251	1	1.69	0.424
			0.8	48	61	0.225				
8	12	1.71	0.2	65	62	0.295	0.274	1	1.71	0.468
			0.8	51	63	0.253				
9	13	1.87	0.2	70	62	0.317	0.287	1	1.87	0.537
			0.8	58	64	0.257				
10	14	1.84	0.2	69	62	0.313	0.287	1	1.84	0.528
			0.8	58	63	0.262				

11	15	1.71	0.278	—	1.71	0.475
12	16	1.65	0.262	—	1.65	0.432
13	17	1.50	0.258	—	1.50	0.387
14	18	1.36	0.241	—	1.36	0.328
15	19	1.19	0.228	—	1.19	0.271
16	20	1.17	0.211	—	1.17	0.247
17	21	0.92	0.216	—	0.92	0.199
18	22	0.81	0.188	—	0.81	0.152
19	23	0.70	0.184	—	0.70	0.129
20	24	0.63	0.167	—	0.63	0.105
21	25	0.55	0.150	—	0.55	0.082
22	26	0.48	0.125	—	0.36	0.045
LB	26.5	0	0	0	0	0
				Σ	24.54	5.953

0.2	66	61	0.305
0.8	55	62	0.252
0.2	62	61	0.287
0.8	52	62	0.238
0.2	60	61	0.278
0.8	50	60	0.238
0.2	58	62	0.265
0.8	47	62	0.217
0.2	55	61	0.257
0.8	42	63	0.193
0.2	51	62	0.235
0.8	39	60	0.188
0.6	46	61	0.216
0.6	41	63	0.188
0.6	39	61	0.184
0.6	36	63	0.167
0.6	31	61	0.150
0.6	26	64	0.125
	0	0	0

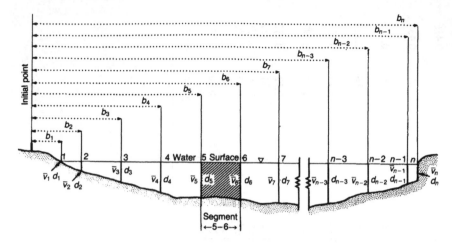

Figure 2.2 The mean-section method of computing current meter measurements. 1, 2, 3, . . . , n, number of vertical; $b_1, b_2, b_3, \ldots, b_n$, distance from initial point; $d_1, d_2, d_3, \ldots, d_n$, depth of flow at verticals; $\bar{v}$, average velocity in verticals.

Velocity–depth integration method

Whereas the previous two methods may be termed arithmetical methods of computing discharge, the velocity–depth integration method is a graphical method. If sufficient current meter observations have been made in the verticals, a curve of mean velocity × depth of flow (area of vertical velocity curve) may be drawn over the cross-section. The area of this curve represents the total discharge.

Referring to Fig. 2.3, the procedure is as follows:

(a) Draw the vertical velocity curve for each vertical by plotting the velocity observations against their corresponding depths of flow.
(b) Measure the area contained by each curve by planimeter.
(c) Plot these areas over the water surface line of the cross-section and draw a smooth curve through the points. The area enclosed between this curve and the water surface line represents the total discharge.

The areas contained by the curves are best measured by planimeter but if graph paper with millimetre divisions is used, the 10 mm squares can be counted to calculate the area with acceptable accuracy, making allowance for scale factors.

Velocity–contour method

This is also a graphical method and like the velocity–depth method described above requires a number of current meter observations in the verticals.

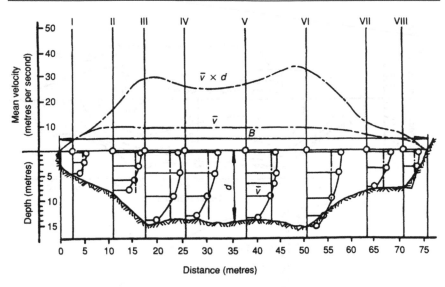

Figure 2.3 The velocity–depth integration method of computing current meter measurements. $Q = \Sigma_0^B \bar{v} d \, \Delta B.$

Referring to Fig. 2.4, the procedure is as follows:

(a) Vertical velocity distribution curves are drawn for each vertical.
(b) These curves are interpolated for convenient intervals of velocity (e.g. 0.25, 0.5, 0.75 m s^{-1}).
(c) Curves or contours of equal velocity (isovels) are drawn as shown in Fig. 2.4(a).
(d) Starting from the maximum, the areas enclosed by successive velocity contours are measured by planimeter and plotted on a diagram, as shown in Fig. 2.4(b), with the ordinate indicating velocity and the abscissa indicating the corresponding area enclosed by the respective velocity contour. The summation of the area enclosed by this curve represents the total discharge.

It can be seen from Fig. 2.4(b) that the maximum velocity plotted on the ordinate is 3.05 m s^{-1}, which in this example is found from the surface velocity distribution curve (Fig. 2.4(a)) and the maximum area plotted on the abscissa is about 1138 m^2, being the sum of the areas enclosed by each velocity contour and the water surface line.

Of the four methods of computation of discharge described above, the mid-section and mean-section methods are used almost universally. The two graphical methods are normally employed in special studies and in the investigation of velocity distribution. The use of graphical methods, however, does not relax the rule for the number or spacing of verticals.

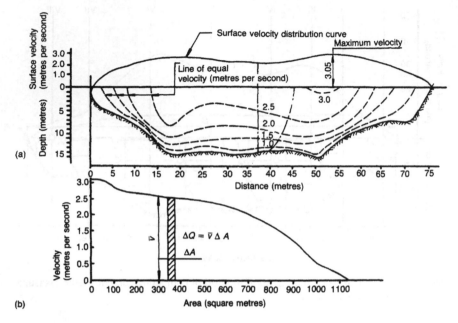

Figure 2.4 The velocity contour method of computing current meter measurements. $Q = \Sigma_0^A \bar{v}\, \Delta A$. (a) Velocity contours in a section, and (b) Total flow.

2.4 Measurement of velocity

The mean velocity in each vertical is determined by current meter observations by any of the following methods.

The velocity distribution method

In this method velocity observations are made in each vertical at a sufficient number of points distributed between the water surface and bed to define effectively the vertical velocity curve, the mean velocity being obtained by dividing the area between the curve and the plotting axes by the depth. The number of points required depends on the degree of curvature, particularly in the lower part of the curve, and usually varies between six and ten. Observations are normally made at 0.2, 0.6, and 0.8 of the depth from the surface, so that the results from the vertical velocity curve can be compared with various combinations of reduced points methods, and the highest and lowest points should be located as near to the water surface and bed as possible.

This method is the most accurate if done under ideal, steady-stage conditions but is not considered suitable for routine gauging due to the length of time required for the field observations and for the ensuing computation. It is used mainly for checking velocity distribution when the station is first established and for checking the accuracy of the reduced points methods.

The velocity curve may be extrapolated to the bed by the use of the following equation

$$v_x = v_a \left(\frac{x}{a}\right)^{1/c} \tag{2.6}$$

where v_x is the point velocity required in the extrapolated zone at distance x from the bed;

v_a is the velocity at the last measuring point on the velocity curve at distance a from the bed;

c is a constant varying from 5 for coarse beds to 7 for smooth beds and generally taken as 6.

Note: if $x = 0$ (bed level), v_x (at bed level) $= 0$.

An example of the use of equation (2.6) is as follows. In a velocity distribution measurement the lowest observation in the vertical was at a point 0.25 m from the bed. The value of the velocity at this point was 0.15 m s^{-1}. Find the approximate velocity at a point 0.1 m from the bed in order to complete the vertical velocity curve.

From equation (2.6)

$$v_x = 0.15 \left(\frac{0.10}{0.25}\right)^{1/6}$$

$$= 0.13 \text{ m s}^{-1}.$$

An alternative method of obtaining the velocity in the region beyond the last measuring point, and so to complete the vertical velocity curve, is based on the assumption that the velocity for some distance up from the bed may often be taken as being proportional to the logarithm of the distance x from the bed. If the observed values of velocities, therefore, are plotted against corresponding values of log x, the best-fitting straight line through these points can be extended to the bed. The required velocities close to the bed may then be read directly from the graph.

The 0.6 depth method

Velocity observations are made at a single point at 0.6 of the depth from the surface and the value obtained is accepted as the mean for the vertical. This assumption is based both on theory and on results of analysis of many vertical velocity curves, which showed that in the majority of cases the 0.6 method produced results of acceptable accuracy. The value of the method is its essential reliability, the ease and speed of setting the meter at a single point, and the reduced time necessary for completion of a gauging.

The 0.2 and 0.8 depth method

Velocity is observed at two points at 0.2 and 0.8 of the depth from the surface and the average of the two readings is taken as the mean for the vertical. Here again this assumption is based on theory and on the study of vertical velocity curves; experience has confirmed its essential accuracy. Generally the minimum depth of flow should be about 0.75 m when the 0.2 and 0.8 depth method is used.

Six-point method

Velocity observations are made by taking current meter readings on each vertical at 0.2, 0.4, 0.6 and 0.8 of the depth below the surface and as near as possible to the surface and bed. The mean velocity may be found by plotting in graphical form and using a planimeter, or from the equation

$$\bar{v} = 0.1 \, (v_{surface} + 2v_{0.2} + 2v_{0.4} + 2v_{0.6} + 2v_{0.8} + v_{bed}) \tag{2.7}$$

where v is the velocity.

Five-point method

Velocity observations are made by taking current meter readings on each vertical at 0.2, 0.6, and 0.8 of the depth below the surface and as near as possible to the surface and bed. The mean velocity may be found by plotting in graphical form and using a planimeter, or from the equation

$$\bar{v} = 0.1 \, (v_{surface} + 3v_{0.2} + 3v_{0.6} + 2v_{0.8} + v_{bed}). \tag{2.8}$$

Equations (2.7) and (2.8) are established from the area of a plane surface by a simple arithmetical procedure. In the six-point method, for example, the surface area of the curve ($\bar{v}D$) is approximately

$$(v_1 \times 0.1D + v_2 \times 0.2D + v_3 \times 0.2D + v_4 \times 0.2D + v_5 \\ \times 0.2D + v_6 \times 0.1D) \, \text{m}^2\text{s}^{-1}$$

and the average velocity is found by dividing the total depth by D, giving

$$0.1 \, (v_1 + 2v_2 + 2v_3 + 2v_4 + 2v_5 + v_6) \, \text{m s}^{-1}.$$

Similarly, equation (2.8) is established from the surface area of the curve giving

$$(v_1 \times 0.1D + v_2 \times 0.3D + v_3 \times 0.3D + v_4 \times 0.2D + v_5 \times 0.1D) \, \text{m}^2\text{s}^{-1}$$

and dividing by D gives the average velocity as

$$0.1\ (v_1 + 3v_2 + 3v_3 + 2v_4 + v_5)\ \text{m s}^{-1}.$$

Three-point method

Velocity observations are made by taking current meter readings on each vertical at 0.2, 0.6 and 0.8 of the depth below the surface. The average of the three values may be taken as the mean velocity in the vertical. Alternatively the 0.6 measurement may be weighted and the mean velocity obtained from the equation

$$\bar{v} = 0.25\ (v_{0.2} + 2v_{0.6} + v_{0.8}). \tag{2.9}$$

The origin of the average velocity occurring at 0.6 of the depth and also at the average of 0.2 and 0.8 of the depth from the surface is based essentially on the theoretical velocity distribution of velocity in an open channel. For the condition of turbulent flow over a rough boundary the vertical velocity curves have approximately the form of a parabola whose axis, coinciding with the filament of maximum velocity, is parallel with the surface and is in general situated between the surface and one-third of the depth of the water from the bed. As the depth and velocity increases, however, the curve approaches a vertical line in its limiting position (this fact is used to advantage in the moving boat method (p. 82) where the current meter is located at approximately 1 m from the surface).

The vertical distribution of velocity may be expressed approximately by the equation

$$v = \left(\frac{D-d}{a}\right)^{1/c} \tag{2.10}$$

where v is the velocity at depth d below the water surface, c is a coefficient usually having a value of 6 (equation (2.6)), D is the total depth of flow and a is a constant numerically equal to the distance above the bottom of the channel of a point at which the velocity has unit value.

Now integrating equation (2.10) for $\bar{v}$ (average velocity in the vertical)

$$\bar{v} = \frac{1}{D}\int_0^D v\ \mathrm{d}d$$

$$= \frac{1}{D}\int_0^D \left(\frac{D-d}{a}\right)^{1/c} \mathrm{d}d$$

$$= \frac{1}{D}\left[-\frac{ac}{c+1}\left(\frac{D-d}{a}\right)^{1/c+1}\right]_0^D \tag{2.11}$$

then

$$\bar{v} = \frac{c}{c+1}\left(\frac{D}{a}\right)^{1/c}. \tag{2.12}$$

Now making $v = \bar{v}$ in equation (2.10)

$$\frac{c}{c+1}\left(\frac{D}{a}\right)^{1/c} = \left(\frac{D-d}{a}\right)^{1/c} \tag{2.13}$$

and

$$\left(\frac{c}{c+1}\right)^c = \frac{D-d}{D}.$$

Hence

$$\frac{d}{D} = 1 - \left(\frac{c}{c+1}\right)^c \tag{2.14}$$

and substituting values of c between 5 and 8 in equation (2.14), d/D is approximately equal to 0.6.

Now if $v_{0.2}$ is the velocity at depth 0.2D, $v_{0.8}$ is the velocity at depth 0.8D and $\bar{v} = \frac{1}{2}(v_{0.2} + v_d)$, then from equations (2.10) and (2.12)

$$\frac{1}{2}\left[\left(\frac{D-0.2D}{a}\right)^{1/c} + \left(\frac{D-d}{a}\right)^{1/c}\right] = \frac{c}{c+1}\left(\frac{D}{a}\right)^{1/c}$$

so

$$\frac{D-d}{a} = \frac{D}{a}\left[\frac{2c}{c+1} - (0.8)^{1/c}\right]^c$$

and

$$d = D - D\left[\frac{2c}{c+1} - (0.8)^{1/c}\right]^c.$$
(2.15)

Substituting $c = 6$ in equation (2.15) gives d/D approximately equal to 0.82. Therefore

$$\bar{v} \doteqdot \tfrac{1}{2}(v_{0.2} + v_{0.8}).$$

Similarly if $\bar{v} = \tfrac{1}{2}(v_d + v_{0.8})$ then

$$d = D - D\left[\frac{2c}{c+1} - (0.2)^{1/c}\right]^c$$
(2.16)

and substituting $c = 6$ in equation (2.16) gives d/D approximately equal to 0.27. Therefore $\bar{v} \doteqdot \tfrac{1}{2}(v_{0.2} + v_{0.8})$, as before.

The foregoing theory is normally applicable to large rivers when the time of exposure of the current meter is sufficient to equalise pulsations but in general the 0.6 and 0.2 + 0.8 depth methods are almost universally used and give acceptable results. Also it will be noted that in Chapter 13 the uncertainty in the measurement of velocity due to the limited number of points taken in the vertical, X_p, is divided by m, the number of verticals. However, many rivers do not necessarily follow the theoretical parabolic velocity distribution even when the time of exposure of the meter is several minutes. In such situations and where sufficient depth is available, special gaugings are sometimes taken using the five-point or six-point method using a single rod with five or six meters attached to it and employing a special counter box to record the current meter observations.

2.5 Current meters

The current meter is still the most universally used instrument for velocity determination. The principle is based upon the relation between the speed of the water and the resulting angular velocity of the rotor. By placing a current meter at a point in a stream and counting the number of revolutions of the rotor during a measured time interval, the velocity of the water at that point can be determined. The number of revolutions of the rotor is obtained by various means depending on the design of the meter but normally this is achieved by an electric circuit through the contact chamber. Contact points or a reed switch in the chamber are designed to complete an electric circuit at selected frequencies of revolution, normally once per revolution, but also at frequencies of twice per revolution or once for five revolutions depending on design. In the case of the Braystoke propeller meter, a diametrically opposed pair of small cylindrical permanent magnets is inserted at the rear of the impeller boss. This operates an electrical reed switch enclosed in a glass envelope located in the current meter body. From the switch a twin-conductor cable carries the electric pulses generated by each revolution of the impeller to the counter device. In all types of design the electrical impulses produce either a signal which registers a unit on a counting device or an audible signal in a headphone. Intervals of time are measured by a stopwatch or by an automatic timing device. Latest developments in current meter design include the introduction by the United States Geological Survey of an optical head pick-up which improves low velocity response. This new pick-up system utilises a pivot bearing in the head and is actuated by a rotating fibre-optic bundle. The system generates four counts per revolution.

The SonTek Flow Tracker hand-held acoustic Doppler current meter for wading measurement can now be classed as a unique member of the current meter genre (Fig. 2.7c). Like the propeller and cup-type meters it is a point velocity meter but with the important innovation that it has an optional two or three dimensional velocity probe. It has a powerful software-friendly package with a keypad custom-designed for velocity and discharge measurements.

The electromagnetic current meter (Figure 2.6d and also Figure 2.12b) is a point velocity meter as distinct from the electromagnetic total flow meter described in Chapter 2. It has no moving parts and is of solid-state construction; it has the ability to measure very low velocities of the order of millimetres per second.

Another innovation is the solid-state current meter digitiser used in both the United States and the United Kingdom. This device counts the number of revolutions and at the end of the count period illuminates revolutions and seconds on a light emitting diode (LED) display. This value is held for a few seconds and then the velocity is displayed based on the time and counts recorded and the rating equations for the specific current meter in use.

Cup-type and propeller-type current meters

Current meters can be classified generally as those having vertical axis rotors (Fig. 2.5) and those having horizontal axis rotors (Fig. 2.6(a)(b)(c)), the former being known as cup-type current meters and the latter as propeller-type meters.

The cup-type current meter consists of a rotor revolving about a vertical shaft and hub assembly, bearings, main frame, a contact chamber containing the electrical contact, tail fin and means of attaching the instrument to rod or cable suspension equipment. The rotor is generally constructed of six conical cups fixed at equal angles on a ring mounted on the vertical shaft. This assembly is retained in the main frame by means of an upper shaft bearing and a lower pivot bearing.

(a)

(b)

Figure 2.5 (a) Cup-type current meter (Price) and pigmy (mini) Price. (b) Watts cup-type current meter (shown with sounding weight and hanger bar for cable suspension).

The propeller-type current meter consists of a propeller revolving about a horizontal shaft, ball-bearings in an oil chamber, the body containing the electrical contact, a tailpiece with or without a vane and a means of attaching the instrument to the suspension equipment. The meter may be supplied with

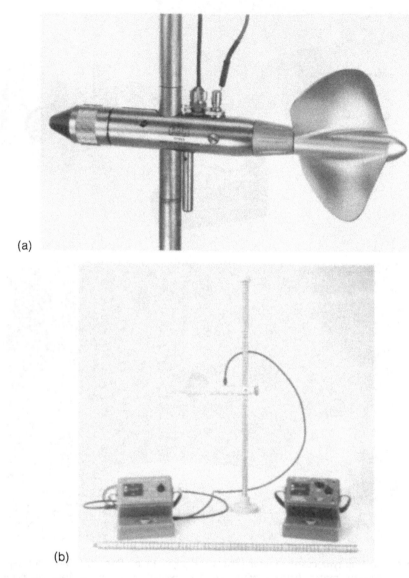

(a)

(b)

Figure 2.6 Propeller-type current meters: (a) Ott, and (b) Braystoke with alternative counter boxes, (c) Ott meter on cableway suspension, (d) ADS electromagnetic current meter ((c) and (d) opposite).

one or more propellers which differ in pitch and diameter and therefore may be used for various flow speeds.

Both types of current meter are available in miniature form (mini-meters) for use in very small depths of flow (Figure 2.7a and b). There is generally no

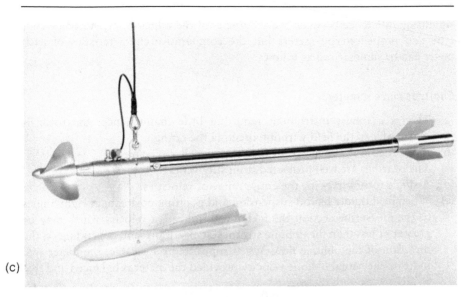

(c)

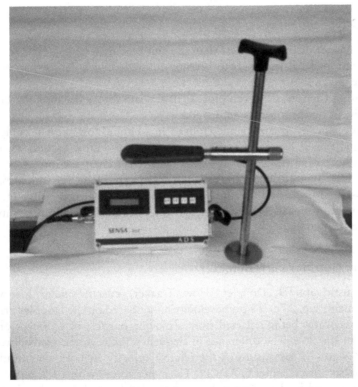

(d)

significant difference between the accuracies of the velocities registered by cup-type and propeller-type meters but the comparative characteristics of each meter can be summarised as follows:

Cup-type current meter

(a) This is a robust instrument requiring little maintenance; the rotor is replaceable in the field without affecting the rating.
(b) It operates at lower velocities than the propeller-type meter.
(c) The bearings are well protected from silty water.
(d) A single rotor serves for the entire range of velocities.
(e) When held rigidly by rod suspension and pointing upstream at right angles to the measuring section the meter will indicate a velocity which may be greater or less than the oblique stream velocity, if present, depending on the direction of the oblique flow. When supported on a cable, the meter will indicate the actual oblique velocity provided the meter is balanced and free to align itself with the stream.

Propeller-type current meter

(a) This meter disturbs flow less than the cup-type meter.
(b) The propeller is less likely to become entangled with debris than the cup-type meter.
(c) Bearing friction is less than for vertical shaft rotors because any bending moment on the rotor is eliminated.
(d) A propeller-type current meter used in oblique flow with a component propeller will register the velocity normal to the cross-section when held rigidly by road suspension at right angles to the cross-section.
(e) Propeller-type current meters are not so susceptible to vertical currents as cup-type meters and therefore give better results when used for measurements from boats. This advantage also relates to the moving boat method described later.

Acoustic Doppler current meter

The acoustic SonTek Doppler (Flow Tracker) current meter for wading measurement (Fig. 2.7c) is a point velocity meter based on the Doppler principle of acoustic waves being reflected from a moving particle of sediment or from bubbles in the flow. The difference in frequency between the transmitted and reflected waves is a measure of the relative velocity in both magnitude and direction (see also Chapter 6, ADCP). The meter comes with a 1.2 m two-piece, top-setting hand-held wading rod complete with keypad with the electronics and batteries contained in the controller mounted to the top of the wading rod. The controller is approximately 230 × 120 × 50 mm weighing about 1 kg. and is

(c)

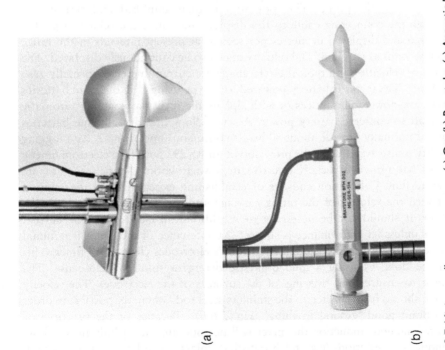

(a)   (b)

Figure 2.7 Mini-propeller current meters: (a) Ott, (b) Braystoke, (c) Acoustic Doppler current meter

waterproof (temporarily submersible without damage). The controller includes the ability to internally record data to be downloaded later by connection to a PC.

The minimum depth measurable is 20 mm and the velocity range of the meter is 0.003 to 4.00 m/s with a resolution of 0.0001 m/s.

The meter is custom-designed for both velocity and discharge measurements, the latter using the mid-section or mean-section methods. The meter's intelligent algorithm automatically prompts the user for the method of measurement.

The power supply requires 8 AA batteries with a typical battery life of 25 hours continuous operation (alkaline batteries). The operating temperature is −20° to 50° C. The manufacturer claims that calibration is not necessary.

Electromagnetic current meter

The electromagnetic current meter (Fig. 2.6(d)) employs the Faraday principle of electromagnetic induction whereby a magnetic field (in the velocity sensor) induces an electropotential in a moving conductor (the water). An alternating current is passed through a coil in the velocity probe which in turn sets up an alternating field in the probe head and surrounding water. Movement of water past the probe causes an electric potential in the water which is detected by two electrodes in the probe. This potential is then amplified and transmitted through the connecting cable to the display unit where the velocity signal is digitised and displayed in metres per second at present intervals in the range 2–60 seconds as required. The velocity may also be continuously displayed. The range of velocities of a typical electromagnetic current meter is normally zero to 4ms^{-1}. The meter is battery powered, one type using 10 No. C size batteries and a low-power microprocessor with alphanumeric display enabling automatic shut-off to conserve battery power. A set of 'long-life' dry alkaline batteries should normally provide about 30 hours continuous operation. A data logging facility is also available by the provision of an RS 232 port. The electromagnetic meter has proved especially effective in sewage works channels to measure sewage flow. Calibration consists of establishing experimentally the relation between the velocity of the rating carriage and the velocity indicated by the meter. It should not be necessary to recalibrate an electromagnetic current meter unless its performance is suspect. Maintenance of the meter is minimal and consists of keeping the velocity probe's electrodes clean at all times. This can be done by using a mild domestic detergent solution before use. The aim is to ensure good 'wetting' of the surfaces of the electrodes. The velocity probe should be connected to the stainless steel rods whenever possible in order to obtain good 'ground earthing' (Fig. 2.6(d)). Because of the principle of electromagnetic induction the meter will not operate successfully in very low-conductivity solutions (e.g. in laboratories) unless a small amount of common salt is added to the water (say 14 mg per litre). A simple test to check the meter is to immerse the meter in perfectly still water which is properly earthed. The display should read zero or a known small zero offset.

2.6 Rating of current meters

In order to determine the velocity of the water from the revolutions of the rotor of the rotating-element current meter, a relation is established between the angular speed of the rotor and the speed of the water which causes it to turn. This relation is known as the current meter rating.

The usual method of rating a current meter is to tow it through still water and observe the time of travel and the number of revolutions as the meter travels a given distance. The number of revolutions per second and the corresponding velocity are then computed. When these two quantities are plotted against each other on graph paper, a series of equations will usually be necessary to fit the points. A rating table is prepared by solving these equations.

The rating tank of the Central Water and Power Research Station in Pune, India is shown in Fig. 2.8. The tank is constructed of reinforced concrete and is 228 m long, 3.6 m wide and 2.13 m deep. The rating trolley is electrically driven with AC servo motors and driven with PLC (Programmable Logic Controlled) for precise speed control. The tank has a real-time PC-based data acquisition and processing system using specially developed software and conforms to ISO 3455. The speed range of the carriage is 0.01 m/s to 6.0 m/s. During calibration, the current meters are suspended from the mountings on the carriage (Fig. 2.8b) and drawn through still water at a number of steady speeds of the trolley. Simultaneous measurements of the trolley and the rate of the revolution of the rotor are made. The two sets of values are related by one or more equations. The calibration of each current meter is also provided in the form of calibration tables.

Rating tank laboratories may have different procedures but generally in calibrating a propeller-type meter, for example, the rate of revolutions of the rotor of the meter is expressed either as pulses or revolutions per second and the average speed of the carriage in metres per second. For each run the average speed is divided by the rate of revolution of the rotor to give the distance travelled per revolution.

For a propeller-type meter, this quotient is the effective pitch of the propeller in revolutions per second and is almost constant at speeds above 1 m/s for helical propellers. If it were strictly constant at all speeds the calibration equation for the propeller would be

$$v = kn \tag{2.17}$$

where v is the velocity in metres per second, n is the rate of revolution of the propeller in revolutions per second and k is a constant. Owing to frictional

(a)

(b)

Figure 2.8 The rating tank of the Central Water and Power Research Station in Pune, India showing (a) the carriage from the front and (b) the rear of the carriage showing current meter mountings.

causes, the relation between v and n is not linear near to the minimum speed of response of the meter. . The complete calibration of the current meter is expressed by one or more equations of the form

$$v = b + kn \qquad (2.18)$$

where b is the intercept on the velocity axis when $n = 0$.

Derivation of rating equations

There are several methods available for the determination of the rating equations and their limits of application.

From the time, distance travelled and number of propeller revolutions, the velocity in metres per second is computed for each run in the rating tank. An expanded method of plotting the data is used in order to magnify the normal scatter of the data points and to facilitate estimation of the rating curve. In the expanded method, an assumed asymptote to the actual rating curve $v = f(n)$ is drawn through the origin (Fig. 2.9).

The equation of the assumed asymptote is

$$v_0 = b_0 n \tag{2.19}$$

where b_0 is equal to the pitch of the propeller.

The expanded rating curve is drawn to show the difference Δv between the observed velocity v and the velocity determined by the assumed asymptote v_0 shown in Fig. 2.9. This difference is written

$$\Delta v = (v - v_0). \tag{2.20}$$

Substituting $b_0 n$ for v_0 gives

$$\Delta v = (v - b_0 n) \tag{2.21}$$

as the ordinate for the expanded rating curve shown in Fig. 2.10.

Typical observed data and run calculations in a rating tank for a propeller-type current meter are shown in Table 2.2. After the data are plotted on graph paper, the expanded rating curve is drawn through the plotted points as shown

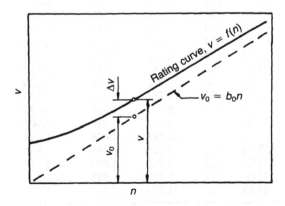

Figure 2.9 Schematic graph of current meter rating curve and assumed asymptote.

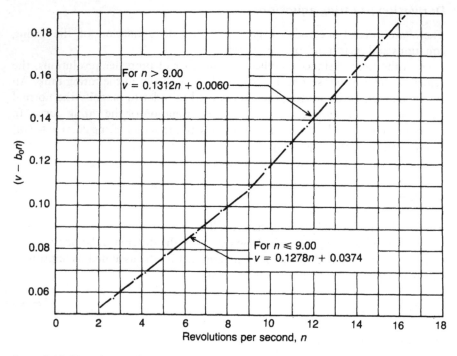

Figure 2.10 Plot of expanded current meter rating curve from data in Table 2.2 to establish the rating equations.

Table 2.2 Observed and calculated data of a current meter rating

Run	Distance (m)	Time (s)	Rev.	Rev. per second (n)	Metres per second (v)	$v_0 = b_0 n^a$	$\Delta v = (v - b_0 n)$
1	21.244	56.15	150	2.671	0.378	0.321	0.057
2	27.581	54.63	200	3.661	0.505	0.439	0.066
3	34.196	60.63	250	4.123	0.564	0.495	0.069
4	40.535	59.02	300	5.083	0.687	0.610	0.077
5	46.735	53.63	350	6.526	0.871	0.783	0.088
6	46.439	45.64	350	7.669	1.018	0.920	0.098
7	46.205	38.95	350	8.986	1.186	1.078	0.108
8	46.186	35.89	350	9.752	1.287	1.170	0.117
9	46.141	32.43	350	10.792	1.423	1.295	0.128
10	46.116	30.00	350	11.667	1.537	1.400	0.137
11	46.086	27.46	350	12.746	1.678	1.530	0.148
12	46.086	25.68	350	13.629	1.795	1.635	0.160
13	46.071	24.50	350	14.286	1.880	1.714	0.166
14	46.067	23.06	350	15.178	1.998	1.821	0.177
15	46.061	21.80	350	16.055	2.113	1.927	0.186

[a] Pitch of propeller $b_0 = 0.120$ m.

in Fig. 2.10. In most cases the rating curve may be estimated by one or more straight-line segments. In Fig. 2.10 two segments are satisfactory. Coordinates are selected from the straight-line segments (or extensions of these) for computation of the rating equations.

The form of the rating equation is based on the equation for a straight line (equation (2.18)). Substituting $(v_0 - b_0 n)$ for v, equation (2.18) becomes

$$v - b_0 n = kn + b. \tag{2.22}$$

Solving for v, the equation is simplified to

$$v = (b_0 + k)n + b \tag{2.23}$$

where v = velocity in ms^{-1};

   b_0 = pitch of propeller (m);

   k = slope of the plotted line;

   n = revolutions per second;

   b = intercept on the y-axis.

An example of the computation to determine the rating equations is shown in Table 2.3 In this table the two coordinates taken to compute the equations of the lower curve are

$$(v - 0.12n)_1 = 0.060, \quad n_1 = 2.900$$

and

$$(v - 0.12n)_2 = 0.170, \quad n_2 = 17.000.$$

Note that the lower curve has been extended upwards for the $(v - 0.12n)_2$ value of 0.170, $n_2 = 17.000$. Similarly the upper curve has been extended downwards for the $(v - 0.12n)_1$ value of 0.090, $n_1 = 7.500$.

The intersection of the two segments ($n = 9.0$) can be checked by solving the two rating equations. The equations are

$$v = 0.1278n + 0.0374 \quad (n < 9.00) \tag{2.24}$$

and

$$v = 0.1312n + 0.0060 \quad (n > 9.00). \tag{2.25}$$

Table 2.3 Computations to determine the current meter rating equations

(1)	(2)	(3)	(4)	(5)	(6)	(7)	(8)	(9)	(10)	(11)
$(v - 0.12n)_2$	$(v - 0.12n)_1$	n_2	n_1	Col. 1 − Col. 2	Col. 3 − Col. 4	$k =$ Col. 5 ÷ Col. 6	kn_1	$b =$ Col. 2 − Col. 8	$0.12 + k$	$v = (0.12 + k)n + b$
0.170	0.060	17.000	2.900	0.110	14.100	0.0078	0.0226	0.0374	0.1278	$v = 0.1278n + 0.0374$
0.185	0.090	15.950	7.500	0.095	8.450	0.0112	0.0840	0.0060	0.1312	$v = 0.1312n + 0.0060$

Figure 2.11 shows a variation in the above method for deriving the rating equations. In this example there are three segments but the procedure is similar in that the asymptote is drawn with equation $v_0 = b_0 n$ as before. For each measurement over the entire speed range, a notional value of flow speed v_c is calculated from the above relation, then the difference $(v_m - v_c)$ plotted to a convenient scale against the corresponding speed of rotation in Fig. 2.11. Straight lines are then fitted by inspection to the resulting curved plot, the slope of each line expressing the difference between the true hydraulic pitch in that range and the value adopted for the asymptote equation $v_0 = b_0 n$.

Also the intercepts of these lines on the axis of the plot indicate the value of the constant in each range. The equations are computed as follows

$$v = 0.2507n + 0.0218 \quad n < 0.5$$

$$v = 0.2735n + 0.0099 \quad 0.5 < n < 6$$

$$v = 0.2761n - 0.0050 \quad 6 < n < 11.$$

Calibration (rating) tables may be prepared manually from the rating equations or by computer. However, to employ a computer for the actual establishment of the rating equations requires a knowledge of the break points of the curve. This requires either a fairly comprehensive iteration programme or a preliminary graphical exercise. It is possible, however, that for some current meters a polynomial may be considered in the future (see Chapter 4 – The stage–discharge relation) which can be solved by a computer programme.

A typical group rating table for a Braystoke propeller-type current meter is shown in Table 2.4.

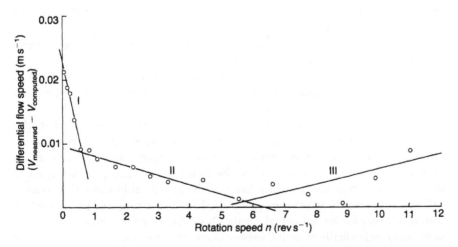

Figure 2.11 Alternative procedure for establishing the current meter rating equations.

Table 2.4 Group rating of the Braystoke Series 8011 current meter (propeller diameter 125 mm, geometric pitch 0.275 m): intermediate *n* values may be interpolated

n (*rev. s⁻¹*)	*v (m s⁻¹)*									
	0	*0.01*	*0.02*	*0.03*	*0.04*	*0.05*	*0.06*	*0.07*	*0.08*	*0.09*
0									0.033	0.036
0.5	0.141	0.144	0.147	0.149	0.152	0.155	0.157	0.160	0.163	0.165
1.0	0.275	0.277	0.280	0.283	0.285	0.288	0.291	0.293	0.296	0.299
2.0	0.541	0.544	0.547	0.549	0.552	0.555	0.557	0.560	0.563	0.565
3.0	0.808	0.811	0.813	0.816	0.819	0.821	0.824	0.827	0.829	0.832
4.0	1.075	1.077	1.080	1.083	1.085	1.088	1.091	1.093	1.096	1.099
5.0	1.342	1.344	1.347	1.350	1.352	1.355	1.358	1.360	1.363	1.366
6.0	1.608	1.611	1.614	1.616	1.619	1.622	1.624	1.627	1.630	1.632
7.0	1.875	1.878	1.880	1.883	1.886	1.888	1.891	1.894	1.896	1.899
8.0	2.142	2.144	2.147	2.150	2.152	2.155	2.158	2.160	2.163	2.166
9.0	2.408	2.411	2.414	2.416	2.419	2.422	2.424	2.427	2.430	2.432
10.0	2.675	2.678	2.680	2.683	2.686	2.688	2.691	2.694	2.696	2.699
11.0	2.942	2.944	2.947	2.950	2.952	2.955	2.958	2.960	2.963	2.966
11.2	2.995	2.998	3.000	3.003	3.006	3.008	3.011	3.014	3.016	–

Individual and group ratings

Current meters may be used with an individual rating or a group rating. Generally, meters on an individual rating require to be rerated each year. A group, or standard, rating on the other hand is established from a group of meters and although the confidence limits are wider than for an individual rating, the uncertainty is acceptable for routine flow measurement. The cost saving in using a group rating may be significant since only one or two of a batch requires rerating each year in order to check that the ratings are still within the original confidence limits.

Implicit in the use of a group rating is that identical meters are used, these meters being produced by rigid control in manufacture.

The group rating should be based on the calibration of a group of at least 10 current meters, the sample consisting of new current meters, well maintained used meters, or ratings from both.

Essentially, provided good care of the current meters is maintained, the meters are used with the group rating without further rerating. Any meter of the number may, however, be rerated from time to time if any suspicion arises from its performance, or if damage to the meter has occurred. However, if the new rating is within the tolerance band of the group calibration, the latter is still used. Further, the number of points used in the rerating may be reduced by about half that for an individual rating. Basically, therefore, instead of rating each meter individually after 100 uses or one year (whichever is less), the cost is reduced substantially since in any one year possibly only a few meters at the most should require a check rating; alternatively if a large number of meters are involved then a sample would be checked to ensure quality control. For authorities with only a few meters of the same type extra care is required in adopting a group rating although the same principle still holds good. With only a few meters, however, the detection of any deterioration in meter performance is more difficult, and care and maintenance are therefore fundamental. Newly purchased meters would of course be used on a group rating thus effecting a significant reduction in the initial cost of the meters. The evidence to date would appear to support the view that the adoption of group ratings would not lead to any deterioration in the accuracy of normal current meter measurements and would undoubtedly significantly reduce the present overall cost of ratings. However, strict control of manufacturer's tolerances is an essential requirement for the implementation of a group rating and, although these are explicitly stated in international standards, careful vigilance by the user is necessary. Of fundamental importance, however, is the care and attention of the current meter and where this is not up to the highest standards neither an individual rating nor a group rating will give acceptable results.

Minimum speed of response

The minimum speed of response of a rotating-element current meter is defined as the minimum speed at which the rotor of the current meter attains continuous and uniform angular motion. This speed is of the order of 0.03m s^{-1} for most current meters except mini-meters. The lower the minimum speed of response of a current meter, the lower the speed of flow which is measurable with confidence, always accepting that the uncertainty at this speed will be of the order of ±20% (see Chapter 13).

In order to ascertain the minimum speed of response in a rating tank the alignment of the meter with the tank and its horizontal balance are checked and adjusted if necessary. The carriage speed is set to a value lower than that at which the rotor is likely to start turning and the carriage is started. The speed of the carriage is gradually increased by turning the speed setting control until it is seen that the rotor of the meter starts to turn. It is possible that at this minimum speed the meter may not complete one revolution; friction or magnetic drag of the contact system may be sufficient to stop the rotor, or it may continue to rotate but have a pronounced hesitation at one position. The speed of the carriage is increased until the rotor is turning regularly. This may be judged partly by observation of a television monitor and partly by noting to the nearest second the reading of the timer as each pulse is counted. This is as near as an operator can get to ensuring that the rotor has attained uniform angular motion. Knowing the least speed at which the rotor turns evenly and the approximate interval between consecutive pulses, the pulse timers and counters are set, and the calibration measurements are made at the minimum speed.

The minimum speed of response is determined after a break in operations. The water will then be at its stillest. The meter may also be tested at speeds up to 0.1m s⁻¹ in the same run along the tank. After returning the carriage to the starting position, the meter remaining immersed, there is a temptation to start the run at the next higher speed too soon. The surface of the water may be smooth but there may be residual currents at the level of the meter. It is largely by experience with a particular rating tank that the best arrangement of timing of runs can be arranged.

Conditions for satisfactory calibration

To obtain a satisfactory calibration of a current meter several conditions have to be satisfied:

(a) The counting of pulses and the measurements of time and distance are accurate.
(b) The carriage runs smoothly and at constant speed so that oscillatory motion, whether longitudinal or lateral, is not transmitted to the meter. Timing of runs with cable-suspended meters is not started until to-and-fro oscillations initiated during acceleration are damped out.
(c) The method of suspension of the meter is that used during field measurements.
(d) The axis of the meter is parallel to the water surface and to the long dimension of the tank.
(e) Residual motion of the water is negligible.
(f) Measurements are not normally made within a range of speeds where there is an Epper[1] effect. The size of the effect and the range of speeds within which the effect is appreciable vary with the size of the meter and the dimensions of the tank. It is larger with larger meters and may be negligible with miniature meters. For a given meter the effect is larger in a small tank than it is in a large one.

Most of these requirements are obvious. Timing and counting present no problems particularly when automatic pulse counters and timers are used. Vibration of the carriage may occur at particular speeds when rod-suspended meters are calibrated at a particular depth of immersion. The Epper effect occurs for a range of speeds having values near to $\sqrt{(gd)}$. This is the speed of a shallow water wave in water of depth d (g is the acceleration due to gravity). When the carriage has this speed the disturbance caused by the immersed current meter and its suspension equipment moves along the tank with the meter and reduces the rate of revolution of the rotor. At speeds less than the critical velocity the

1   The Epper effect is a phenomenon in which the wave crest produced by a current meter causes an increase in the height of the wetted cross-section and a consequent reduction in the relative velocity.

disturbance may be reflected from the ends of the tank and overtake the meter, the performance of which may be affected over a range of values. The size of the Epper effect may be little more than the uncertainty of a single calibration point. It is a systematic not a random error.

Probably the largest source of error is that due to residual movement of the water. Part of the difficulty may be caused by density currents and part by the distrubance arising from the previous run. Calibration tanks which are underground may be free from density currents which arise from changes in temperature but tanks which are above ground may experience fairly rapid changes of temperature which do not occur simultaneously at all parts of the tank. Residual water movements can be detected by mean of floats but cannot be measured simultaneously with a calibration run. They should not be greater than 1% of the speed of the following calibration run. To conform to this requirement the residual speed of flow of water along the tank at the level of the current meter would need to be less than as follows:

Spead of carriage (m s^{-1})	0.03	0.1	0.25	0.5	1.0	2.0	5.0
Speed of water (mm s^{-1})	0.3	1.0	2.5	5.0	10.0	20.0	50.0

Only when runs are to be made at speeds of more than 3m s^{-1} is it possible to use a current meter to check that the residual movement of water is satisfactorily low. This happens to be the maximum speed at which most current meters are calibrated. Very sophisticated equipment would be required to measure residual motion at still lower speeds. Visual indication of motion can be obtained by observing the surface of the water or by observing a small subsurface float having only slight buoyancy and tethered at the depth of the meter by a fine thread tied to a small sinker resting on the bottom of the tank.

General summary of current meter calibration

(a) Barring accidents the calibration of current meters shows only a small change with time. This is particularly the case for the higher velocities. Changes of calibration at the lower end of the speed range are proportionally higher and depend as much on the cleanliness and lubrication as on any wear and tear of moving parts.

(b) Current meters which embody mechanical arrangements for making and breaking an electric circuit have a higher minimum speed of response and a less consistent performance at low speeds than those having magnets and reed switches; mini-meters have a higher minimum speed of response.

(c) The accuracy of calibration equations is greater than the accuracy of the individual calibration points.

(d) The spread of results when several repeat calibrations are made of one
     meter is much smaller than the spread of results when several current
     meters of one make and type are calibrated.
(e) The uncertainty of group calibration of modern current meters is not large
     however, and such calibrations may be used, and indeed recommended, for
     routine gauging.
(f) The spread of calibration results at relatively high speeds – say over 1 m s^{-1} –
     is less than it is at low speeds. In particular some precision-built meters
     having metal propellers have calibrations such that 19 out of 20 of them
     lie within ±1.5% of the average of the group. With some propellers the
     uncertainty is about ±0.8%. The spread of calibration results for groups
     of meters having plastic propellers of almost neutral buoyancy is between
     ±2 and ±2.5% at the higher speeds. At the lower speeds, where some of
     the main difficulties of flow measurement lie, the spread of results of the
     meters with metal propellers and magnet and reed switch operation is
     also slightly smaller than that of the meters with plastic propellers. The
     performance of the meters with plastic propellers is superior at low speeds
     to that of meters embodying mechanical operation of contacts.
(g) Meters having plastic propellers and water-lubricated bearings require
     little attention and adjustment; they are ideal for measurements of low
     speeds and for use generally at sites which are remote from workshop
     facilities.
(h) Differences in calibration of two well-maintained current meters of the
     same type are systematic. For special studies involving maximum accuracy,
     individual calibrations are preferred.

2.7 Considerations in current meter design

Many current meters still have mechanically operated contacts to indicate
the rate of revolution of the rotor. At low speeds the setting of the contact
mechanism has an influence on the rate of revolution of the rotor. Most meters
now, however, are designed with reed switches or magnetic contacts.

Propellers made of dense metals such as brass need to be more carefully
balanced than those made of light material such as plastics. Some meters may
have propellers made of polystyrene. The Braystoke meter, in particular, was
designed for a propeller having neutral buoyancy which runs on a fixed stainless
steel spindle that rotates in water-lubricated plain bearings, therefore having no
moving oiled parts.

When current meters are used on rod suspension it is necessary to ensure that
the meter is pointing at right angles to the measuring section.

Current meters are normally calibrated with the same suspension equipment
as will be used during the subsequent field measurements. The differences
between calibrations made on rod and cable suspensions depend on the dimensions and shape of section of the rod and on the size and type of sinker weight

used. Some sinker weights are in line with the shaft of the propeller; others are suspended on a hanger bar below the meter. The vertical distance of such a sinker weight below the meter can be of importance.

The sinker weight needs to be sufficient to reduce the vertical angle of trail of the cable to 6°. This not only ensures that the meter will be operating close to the chosen cross-section but also makes certain that the horizontal swivel of both the meter and the weight will not be restricted.

As an example of the effect of suspension on calibration it may be stated that there is about 2% difference between the calibration of an axial flowmeter on a 20 mm rod compared with its calibration on cable suspension using a hanger bar and a 22.7 kg Columbus pattern weight, the points of suspension of the meter and the weight being 0.215 m apart. At a given velocity the propeller of the cable-suspended meter revolves 2% faster than that of the rod-suspended meter.

If a meter, whether suspended by rod or by cable, is free to swivel about a horizontal axis, its freedom to swivel needs to receive attention during mainten-ance and before use the meter requires to be properly balanced.

It is often difficult in the field to see if the current meter propeller is turning evenly at very low speeds and whether the meter is pointing in the correct direction. This is particularly important for cable-supported meters.

If measurements have to be made at a site where the hydraulic conditions are unsuitable, the flow lines may not be at right angels to the cross-section in either the vertical or the horizontal plane and the distribution of velocity in the cross-section will not be symmetrical. Pulsations in the flow may be erratic. All these conditions are different from those under which the current meter was calibrated. If there is pulsating flow then the period of immersion of the meter during each point measurement needs to be longer so as to include an integral number of cycles of pulsation.

If there is any torsion in the suspension cable it is probable that the meter is not heading directly into the stream unless it can swivel freely and inde-pendently of the suspending cable. Few swivels are effective at low velocities and certainly not at the lowest speed that the meter is capable of measuring. This is particularly the case when the weight of the meter and the sinker are both taken on the swivel. Conditions are better when a meter of light weight swivels round a cylindrical bar between the suspension cable and the sinker. In most cases it is better to have no swivel. To ensure the alignment of the meter it should be arranged that the tail unit of the meter and the tail fins of the sinker weight assist each other. A length of chain between the suspension cable and the hanger bar will relieve the torsion in the suspension.

2.8 Care of current meters

To ensure reliable observations of velocity it is necessary to maintain the current meter in good condition. Good maintenance practices may be summarised as follows:

(a) Before and after each discharge measurement examine the meter cups or vanes, pivot, bearing and shaft for damage, wear or faulty alignment.

(b) Clean and oil meters daily when in use (unless they are water lubricated).

(c) Clean the meter immediately after each measurement if the meter is being used in sediment-laden water. For cup-type meters the surfaces to be cleaned and oiled are the pivot bearing, pentagear teeth and shaft, cylindrical shaft bearing and thrust bearing at the cap.

(d) After oiling, spin the rotor to make sure that it operates freely. Ascertain and correct the trouble if the rotor stops abruptly.

(e) Record the duration of spin of a cup-type meter. A significant decrease in the duration of spin indicates that the bearings require replacement more often than other parts.

(f) Keep the pivot and pivot bearing separate when the meter is not in use by means of the raising lever or nut provided.

(g) Replace worn or fractured pivots.

(h) Limit on-site repairs to minor damage only. This is particularly the case with respect to a propeller where small changes in shape can affect the rating. In cup-type meters minor dents in the cups can often be straightened to restore the original shape, but in case of doubt replace the rotor with a new one.

(i) Dispatch to the workshop for repair or reconditioning, badly sprung yokes, bent yoke stems, misaligned bearings, tailpieces and propellers.

(j) With damaged plastic propellers replace with a new propeller.

(k) If a cable suspension is being used check that the conductor cable is adjusted as necessary to prevent interference with meter balance and rotor spin. Check the meter's balance on the hanger bar and the alignment of the rotor when the meter is on the hanger bar.

(l) A meter which is only slightly damaged may still be usable if it is recalibrated. If, however, a group calibration is being used there will be an uncertainty whether a slightly damaged meter will continue to operate within the group tolerance. In such a case it might be better to have the meter check-rated (Section 2.6).

2.9 Procedure for current meter measurement of discharge

Current meter measurements may be classified in terms of the means used to cross the river during the measurements. These are normally by wading, cableway, bridge or boat and the actual method used depends mainly on the depth of flow, the velocity and the length of the measuring section. No matter which method is used or which method is used to suspend the current meter, the principles of measurement described in the previous sections are the same.

Selection of gauging site

For a stage–discharge station, the selection of the site is often dictated by the needs of water management or by the requirements of the hydrometric network. In fulfilling water management needs there is little or no freedom of choice in selecting gauging sites and frequently records need to be obtained under very adverse hydraulic conditions. This is often the case too where spot measurements are required at specified locations.

However, before a permanent stream gauging station is installed, a general reconnaissance is made in order that the most suitable site for the station may be selected. This reconnaissance normally includes an examination of geological and topographical maps of the area. Each possible site is critically examined with reference to the physical characteristics of the channel. It is often useful, if possible, to inspect potential sites during different seasons when flow conditions alter, usually significantly. It is also good practice to install a temporary staff gauge or pressure bulb recorder (Chapter 3) at selected potential sites and to carry out spot current meter gaugings from time to time for interpretation and analysis.

Generally, however, the aim is to select a reach of stream having the following characteristics:

(a) a straight reach with the threads of velocity parallel to each other;
(b) a stable streambed free of large rocks, weeds and protruding obstructions which create turbulence;
(c) a flat streambed profile to eliminate vertical components of velocity.

If all three conditions are not satisfied the best possible reach is selected using these criteria and finally a measuring cross-section is selected.

Current meter measurements by wading

Current meter measurements by wading are preferred if conditions permit. Wading measurements with the current meter supported on a graduated wading rod which rests on the bed of the stream are normally more accurate than those from cableways or bridges, as the operator has more control over the general gauging procedure. This is particularly the case in the selection of cross-section which may not be the usual station measuring section and in the selection of verticals and measurement of depth.

A measuring tape or tag line is stretched across the river at right angles to the direction of flow using 20 to 30 verticals; the spacing of these is determined so that no segment contains more than 10% of the total discharge (Section 2.2). Usually an approximate discharge for this purpose can be obtained from the stage–discharge curve or from previous current meter measurements. The positions of successive verticals used for depth and velocity are located by horizontal measurements from a reference marker (initial point) on the bank, usually defined by a pin or a monument. The gauging starts at the water edge of the near bank, where depth and velocity may or may not be zero. At each chosen vertical the depth is measured and the value used to compute the setting or settings of the current meter depending on the method to be used (usually 0.6 or 0.2 and 0.8 depth). After the meter is in position the rotor is allowed to adjust to the stream velocity before the count of rotor revolution is started – this may take only a few seconds where velocities are over about 0.3 m s^{-1} but a longer period is necessary at lower velocities. A revolution count is then taken at each selected point for a minimum of 60 s but where it is known that the velocity is subject to short-period variations or pulsations it is advisable to continue the observation for at least 3 min, noting the revolution count at the end of each 1 min period. The velocity at the point can then be taken as the average of all the separate readings unless it is apparent that the difference is due to some cause other than pulsation of the flow.

The phenomenon of pulsation has an effect on the measurement of point velocities, and therefore on current meter gauging in general, which is not always fully appreciated. The velocity at any point in the river, even when the discharge is constant and when the surface is apparently smooth and free from surges and eddies, is continuously fluctuating with time. This pulsation is caused by secondary currents developed by hydraulic conditions upstream of the gauging site; for example, by obstructions in the approach channel, by surging produced at riffles or rapids being continued through pooled reaches, or by the acceleration of the water at bends. Generally, the velocity at any point changes in cycles which can range from a few seconds to possibly more than 1 h, and the extent of these pulsations – both short-term and long-term – needs to be known before deciding on the duration of exposure of the meter at each velocity point. For reliable results it is necessary to observe for a time sufficient to average the effects of the pulsation or to determine the velocity limits of the cycle. This is especially important when trying to determine the relationship between mean velocity as determined by different reduced point methods and that obtained by drawing a vertical velocity curve.

To investigate the phenomenon of pulsations, field tests were carried out in the UK in 1978; 40 current meters were run continuously for 24 hours on five verticals on three different rivers (see Further Reading). The following were some of the conclusions:

(a) The effect of pulsations on the measurement of velocity was significant;

(b) For best results, a multimeter or single meter multipoint technique was desirable;

(c) In deep slow flowing water, both the single point and two point methods tended to give biased results which varied with stage;

(d) To reduce scatter about the stage–discharge curve, the time of exposure should be as long as possible and best results were obtained with an exposure time of 3 minutes; and

(e) The discharge computed from simultaneous observations of velocity were generally no better than those computed by consecutive observations.

If the stage is rapidly varying, (and during flood measurements), it is sometimes convenient to reduce the time of exposure of the meter to 30 s. Revolutions may be counted over a fixed time period (30 s, 60 s, 3 min, etc.), or a fixed number of revolutions may be timed, the latter method being normally more accurate because full revolutions are counted whereas in the former method part of a revolution may be lost. This could lead to an error in the observation of low velocities. For wading measurements it is often found useful to have one of the cups of a cup-type current meter painted a distinctive colour on the outside. The revolutions may then be counted manually without the need for an automatic counter. In the case of a propeller meter a spot may be painted on the impeller.

Limitations on wading are imposed by the combination of depth and velocity and by the quality of footing on the bed. The advisability of wading needs to be judged by the operator at each site.

The position of the operator is important to ensure that the operator's body does not affect the flow pattern at or approaching the current meter. The best position is to stand facing one or other of the banks, slightly downstream of the meter and an arm's length from it. The rod is kept vertical throughout the measurement with the meter parallel to the direction of flow (Fig. 2.12).

Wading rods are normally of two types: (1) a rod round in cross-section consisting of several sections depending on the depth of flow and screwed together and (2) a rod known as a top-setting rod consisting of a main rod for measuring depth of flow and a round rod which allows setting the position of the current meter in the flow (Fig. 2.13). Both types of rod have base plates which rest on the streambed during measurements. Wading rods are usually marked in centimetres and measurements made to the nearest 5 mm. The measuring section for a wading measurement can often be modified in order to improve the conditions of flow. Often it is possible to build temporary dykes to cut off dead water and shallow flows in the measuring section, or to improve the measuring section by removing rocks or debris.

Figure 2.12 (a) Current meter gauging by wading; pulses counted using headphones. (b) Performing a gauging by the electromagnetic current meter.

(a)   (b)

Figure 2.13 Current meter wading rods: (a) top-setting, and (b) round rod.

Mean gauge height for current meter measurements

The mean gauge height corresponding to the measured discharge is one of the two coordinates used in plotting the discharge rating curve for gauging stations. An accurate determination of the gauge height is therefore as important as an accurate measurement of the discharge. The correct gauge height for a measurement will be that which is observed at the same time as the gravity centre of the flow is gauged.

When gauging discharge during constant or nearly constant stream stage, there is no difficulty in deciding the gauge height that corresponds with the measured discharge. If the change in gauge height is less than 50 mm during the measurement, the arithmetic mean of the gauge height at the start and end of the measurement can usually be taken as the mean gauge height.

Discharge measurements at time of high water are usually made during a rising or falling stage when a considerable change in the gauge height may occur. The correct gauge height is obtained by computing the weighted mean gauge height, which for a non-recording gauge requires additional observations of stage between the start and end of the measurement. These readings are made at regular intervals, say every 20 or 30 min. The assistance of a gauge reader is usually necessary for obtaining the readings. The mean gauge heights during the set time intervals and the corresponding measured segment discharges are used to compute the mean gauge height of the measurement. The equation used is

$$h = \frac{q_1 h_1 + q_2 h_2 + q_3 h_3 + \dots q_n h_n}{Q} \qquad (2.26)$$

where h = mean gauge height;
     $q_1, q_2, \dots$ = discharge measured in time interval 1, 2, ...;
     $h_1, h_2, \dots$ = mean gauge height in time interval 1, 2, ...;
          Q = total discharge measured.

Current meter measurements from cableways

Cableways are normally used when the depth of flow is too deep for wading, when wading in a swift current is considered dangerous, or when the measuring section is too wide to string a tag line or tape across it.

There are two basic types of cableway:

(a)   Those with an instrument carriage controlled from the bank by means of a winch, either manually or electrically operated (Fig. 2.14).
(b)   Those with a manned personnel carriage, commonly known as a cable car, in which the operator travels across the stream to make the necessary observations; the car may be pulled manually across the river by means of a cable car puller, or it may be electrically operated (Fig. 2.15).

The general gauging procedure is similar except that in the case of the non-manned cableway the instrument carriage, suspended from the track cable, moves the current meter and sinker weight across the stream between the cable supports. The operator remains on the bank and operates the gauging winch which is provided with both distance and depth counters for placing the current meter at the desired position. The electrical impulses from the current meter are returned through the core conductor of the suspension cable and registered on a revolution counter. The manned cableway, on the other hand, is provided with a support for a gauging reel, a guide pulley for the suspension cable, and a protractor for reading the vertical angle of the suspension cable.

The gauging procedure is as follows:

(a)   The water's edge is identified in relation to a permanent initial point on the bank by means of a tag line or by use of the painted marks on the track cable used for spacing the measuring verticals.
(b)   The current meter and the weight are lowered at the first vertical until the bottom of the weight touches the water surface and then the depth counter is set to zero.

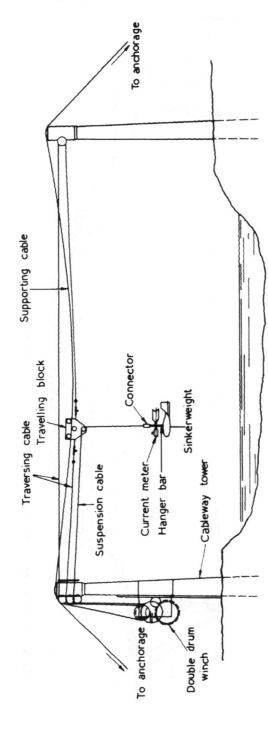

Figure 2.14 Schematic arrangement of unmanned cableway and suspension assembly.

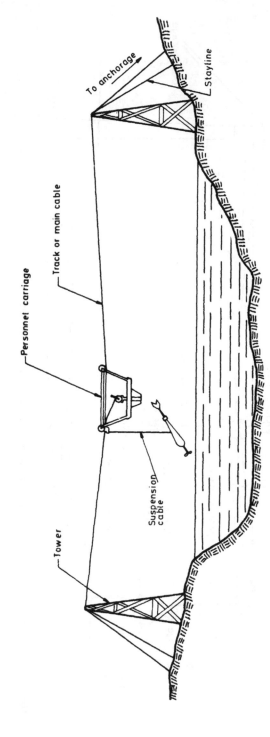

Figure 2.15 Schematic arrangement of manned cableway and suspension assembly.

(c) The current meter assembly is then lowered until the weight touches the streambed, the length of cable played out is read and the sounded depth is recorded.

(d) The velocities are measured at the selected depths in the vertical.

(e) If there is floating drift in the stream channel, the current meter is raised occasionally for control and cleaning. This is always done if there is a sudden drop in the velocity as indicated by the revolution counter. The channel upstream from the gauging section needs to be watched closely for any driftwood or material which could damage the current meter.

(f) If measurements are made where the river is deep and swift, the sinker weight may not be sufficient to maintain the suspension cable in the vertical. This angle is measured by protractor, or other means, in order to correct the soundings to obtain the actual vertical depth (Chapter 5, Section 5.1).

Sinker weights

The size of the sinker weight, attached to the current meter by means of a hanger bar, requires to be sufficient to maintain the current meter suspension cable in an approximately vertical position. It is usually decided on by experience and common weights in operation vary from about 7 to 150 kg depending on the velocity and depth of flow. On the Qintong River in the Zhejiang Province of China, for example, depths are as much as 70 m and velocities are of the order of 7 ms^{-1}, and a sinker weight of 750kg is required. This, however, is an exception but weights of 250kg are common in the Yangtze River catchment. Also in the Yangtze catchment the Bureau of Hydrology in China has introduced a combined sediment sampler and current meter equipped with an ultrasonic depth sounder (Fig. 2.16). The facility of the ultrasonic transducer enables depth to be measured without the necessity of having to make air-line and wet-line corrections although the weight of the sampler requires to be sufficient to place the device at the appropriate location in the vertical to make the velocity measurements.

Sinker weights are designed to a streamline form and furnished with tail fins to orient them parallel with the current so as to cause the minimum interference to the flow. Some designs have the current meter fixed to the nose of the sinker weight (Fig. 2.17) but it is often preferable to have the weight suspended below the meter. In this way it prevents damage to the current meter when the assembly is lowered to the bed to measure depth. The distance between the centre-line of the current meter and the bottom of the weight needs to be taken into account when setting the meter at the chosen depths for the measurement of velocity. Some sinker weights have a ground contact which produces a signal when the weight touches the bed.

(a)

(b)

Figure 2.16 Han River, China. (a) Combined sediment sampler, current meter, 300kg. sinker weight and ultrasonic depth gauge (b) Cableway motor house with 35m tower for a span of 380 m.

Figure 2.17 Current meter and sounding weight with current meter attached to nose of weight. Note ground contact device below weight.

Cableways with instrument carriage

The cableway system with instrument carriage consists of (Fig. 2.14):

(a) supporting towers
(b) track (main) cable
(c) anchorages
(d) staylines (backstay)
(e) towing cable
(f) instrument suspension cable
(g) instrument carriage
(h) double-drum winch or two independent winches.

The supporting towers

The supporting towers are erected one on each bank of the channel. The towers support the main cable at sufficient height as the suspended equipment travels along the main cable between the towers. The tower on the operating bank has pulleys for guiding the suspension cable and the towing cable, and may also have means for securing the winch. The track or main cable passes freely over a saddle on top of the tower on the operating bank with negligible bending movement on the tower. The tower on the opposite bank has a saddle on its top for the main cable and a pulley for the towing cable. The saddles of the two towers require to be at the same level. Instead of a tower, the support on the opposite bank may often consist of a side-hill anchorage where the bank is steep.

Safe and convenient access needs to be available throughout the year so that the hydrologist can inspect the installations on both banks.

The towers are designed to take all loads which are to be supported in addition to their own weight; wind loads need to be included. The pressure on towers due to wind load may vary from 1000 to 2000 kg m^{-2} for towers not exceeding 30 m in height.

The foundations of the tower normally extend from below the frost line to at least 1 m above the general flood level.

The height of the towers should be such as to allow the bottom of the equipment, suspended from the centre of the main cable span, to be not less than about 1 m above the highest flood level to ensure that the cableway does not interfere with navigation along the channel. Aircraft warning signs may be provided according to local regulations.

Various types of construction are used as supporting towers. For long spans, high towers may be used. Steel and timber A-frames are often used as supports when the span is not too large and the height of the support is less than 12 m. An H-beam used as a steel post support has been found to be very satisfactory. This type of support has been used for spans up to 200 m and for heights up to 8 m.

The track or main cable

The track or main cable runs over the saddles on top of the supporting towers and its two ends are fixed to the anchorages. The instrument carriage travels along this cable.

For comparatively short spans, wire rope may be used. For large spans, particularly where a manned carriage is to be supported, special high-strength cables, such as tramway track cables, are necessary.

The horizontal component of the tension in a cable suspended between supports of equal height is given by the formula

$$H = \frac{WS^2}{8D} + \frac{PS}{4D} \qquad (2.27)$$

where H = horizontal component of the tension in the cable (kg);
        W = load per running metre of cable (kg m^{-1});
        S = horizontal span (m);
        D = loaded sag at mid-span (m);
        P = concentrated moving load (kg).

The loaded sag at mid-span of the cable should not exceed 2% of the span.
   The actual tension in the cable is given by the formula

$$T = H \sqrt{\left(1 + \frac{16D^2}{S^2}\right)}.$$

(2.28)

Some typical tensile strength and load values for an instrument carriage cable-way can be given as follows:

(a) Track cable: diameter 15 mm, tensile strength 12 000 kg.
(b) Suspension cable with insulated two-conductor core: diameter 2.5 mm, tensile strength 450 kg.
(c) Load per running metre of cable (weight of cable + wind) for 15 mm, cable: 1.0 kg m^{-1}.
(d) Concentrated moving load (weight of current meter + trolley + pressure head):
      with 25 kg suspended equipment, P = 65 kg;
      with 50 kg suspended equipment, P = 100 kg;
      with 100 kg suspended equipment, P = 170 kg.

A stop is sometimes placed near the far end of the main cable at a known distance to allow for verification of the horizontal measurement given by the distance indicator of the winch.

The anchorages

The anchorages are fixtures to which the track cable and staylines are attached. The anchorages are designed to take up the maximum load for which the cableway is designed and are set in direct line with the track cable and so placed that they can be easily inspected.

   Anchorages are usually constructed of mass concrete whereby the weight of the concrete and the soil resistance to movement are the principal factors in the security of the anchorage. In places where the river banks contain solid rock, anchor-bolts or rods properly set in the rock may be used.

   When a rigid connection is made to an anchorage by means of an anchor-bolt, It is set in a direct line and in the same plane with the connecting stayline so that there will be no bending moment in the anchor-bolt.

The staylines

Staylines (backstays) are cables attached to the top of each supporting post or tower and to the anchorages to counteract the load of the track cable and to ensure stability for the supports. The staylines should be of corrosion-resistant steel and of sufficient strength to maintain the tower in a vertical position under all loading conditions. It is necessary to provide means for adjusting the tension in the staylines.

The towing cable

The towing cable is attached to one of the drums in the double-drum winch, and passes over the sheaves fixed to the towers. The two ends of the towing cable are fixed to the instrument carriage making it an endless circuit to move the carriage across the stream. Alternatively, one end of the towing cable may be attached to the carriage and the other end wound on the drum. The towing cable must have means of adjusting the tension in the cable (turnbuckle) if it makes an endless circuit. In the alternative case, the tension in the cable is given by the weight of the suspended instruments. The towing cable should be corrosion-resistant and as light and as flexible as possible.

The suspension cable

The suspension cable is wound on the second drum in the double winch or on a separate winch, passes over the sheave on the tower on the operating bank, and then passes over the pulley in the instrument carriage. The measuring or sampling instruments are attached to the end of the suspension cable. The suspension cable has an insulated inner core which serves as an electrical conductor for the measuring instruments.

  The suspension cable is made of corrosion-resistant material preformed and reverse laid to prevent spinning and rotation. The cable is made of sufficient strength to suspend the current meter and sounding weight. A breaking strength of five times the maximum load to be used is a sufficient and suitable safety margin to allow for the loading effect of drag and live load during the performance of a measurement. Its elongation when loaded should not exceed 0.5%. The cable should have the minimum diameter consistent with the strength requirement so as to offer minimum resistance to the force of the flow. The cable must be smooth and flexible so that it can take turns without any permanent bends and twists. It is equipped with a suitable connector for attachment of the measuring equipment.

The instrument carriage

Two track pulleys are fixed at the top and one suspension pulley at the bottom of the instrument carriage. The carriage runs on the track cable when pulled from either side. For spans longer than about 100 m, a guide pulley is normally provided to prevent too large a sag in the suspension cable.

The gauging winch

A double-drum winch or two independent winches may be used. In the double-drum winch, the suspension cable is wound on one of the drums and the endless towing cable passes round the other, then over the sheave on the supporting tower on the opposite bank. Alternatively, the towing cable may be wound on the latter drum. Horizontal and vertical travel of the measuring equipment attached to the suspension cable are controlled by a lever which couples either only the suspension cable drum or both drums simultaneously. Each drum has a counter to indicate the released length of cable, one for measuring the horizontal distance travelled by the carriage and the other for indicating the depth of the suspended instrument (Fig. 2.18 (a) (b) and (c). Instead of a double-drum winch, two separate winches may be used for horizontal and vertical movements (Fig. 2.19).

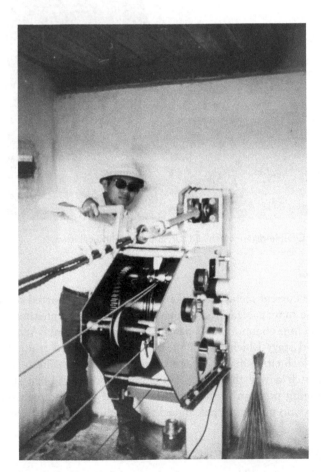

Figure 2.18 (a) Double-drum winch.

Figure 2.18 (b) Double-drum winch and current meter and cableway.

Rod suspension

Instead of the current meter and sinker weight being supported by suspension cable from the main cable, a rod suspension assembly is sometimes used. This cableway system is common in China (Fig. 2.20). The assembly requires a series of pulleys and stays which permit the rod to be operated from the bank by means of a winch either manually or electrically operated. The current meter is fastened in such a way that it can be raised or lowered on a sheave to the required gauging position in the vertical. The rod is maintained vertically by means of a second upstream cable linked to the main cable by stays.

Cableways with personnel carriage

The manned-carriage cableway system consists of (Fig. 2.15):

Figure 2.18 (c) Cableway and tower and meter suspension.

(a) supporting towers
(b) track cable
(c) anchorages
(d) staylines
(e) personnel carriage.

Items (a), (b), (c) and (d) are similar to those described for cableways with an instrument carriage.

The personnel carriage

The carriage, also called a cable car, from which the hydrologist makes the gauging observations, travels on the track cable by means of two track pulleys. The cable car is usually driven manually by cable car pullers. The cable car may be designed for operations from a sitting position or a standing position. One

Figure 2.19 Single-drum winch on truck-mounted crane for gauging from bridge (Mississippi River).

person or two-person cable cars are used. The cable car must be of adequate design and strength to ensure the hydrologist's safety and provide reasonable comfort during the measurement. The cable car is provided with a brake to secure it in all required positions with means of support for the gauging reel. It is equipped with a protractor for measuring the angle of the downstream drift of the measuring equipment.

Some designs of cable car enable the car to be raised or lowered hydraulically or electrically during a gauging. In these cases rod suspension is used instead of cable suspension and the car can be conveniently located just above the river level.

Design and safety considerations for cableways with personnel carriage

Important considerations in the design of cableways are the clear span between the supports, the weight of the track cable and the concentrated load, the loaded and unloaded sag, the effect of changes in temperature and the heights of the supports required for the necessary clearance above extreme high water.

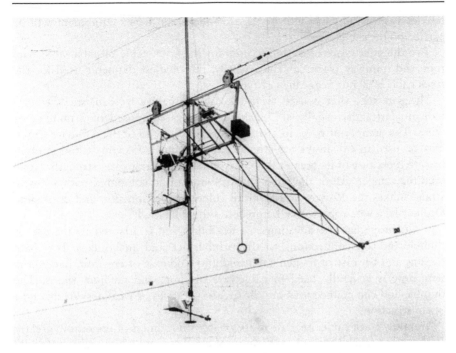

Figure 2.20 Rod suspension from cableway (China).

The design of the track cable consists of the determination of the necessary length, the correlation of sag and allowable stress for any loading that may occur when the cableway is in use, including an allowance for the effect of changes in temperature, and the selection of the size and kind of wire rope or track cable that will meet the requirements most satisfactorily.

The loads to be considered in the design are:

(a) the dead-load weight per metre run of cable which may be the limiting load for long spans
(b) the concentrated load carried by the cable car and
(c) loads caused by wind and ice.

The concentrated load that is carried by the cable car consists of the weight of the car, the equipment and two people, the sum of which is generally taken as 230 kg; also, the additional pull that may be exerted by the suspension cable in case the suspended equipment should become fouled in drifts, etc. must be considered. The suspension cable must break before the track cable, towers or anchorages.

The breaking strength of the suspension cables in general use with the gauging reels may vary from 150 to 450 kg. Thus a concentrated load of 680 kg applied at the point of maximum sag is commonly used in the design of the track cable, except in those instances where it is known beforehand that a heavier suspension cable will be used or additional heavy equipment will be carried on the cable car.

Two different types of cable are used for the track cable of cableways: wire rope and tramway track cable. Generally, the smallest diameter used for the track cable is 20 mm regardless of type.

The wire rope that is used for track cable in cableways consists of several individual strands, usually six, each of which is composed of a number of wires. The number of wires in a strand is generally 7, 19 or 37. A 7-wire strand may be used in cableways for stream gauging, but a 19-wire strand is often preferred because of its greater flexibility and somewhat greater strength. However, the smaller size of the wires in a 19-wire strand compared with a 7-wire strand makes the 19-wire strand more vulnerable to abrasion and corrosion. Ordinarily, a wire rope with a hemp core should be used.

A wire rope has the advantage of flexibility and is adapted to the use of thimble-and-clip connections to the turnbuckles and anchorages. For these reasons and because of its general availability and ease of erection, hemp-core wire rope is generally used in cableways of short and medium spans. The thimble-and-clip connections are the greatest sources of weakness in this type of construction.

Tramway track cable, because of its greater smoothness of operation, greater strength and reliability, higher modulus of elasticity and lower sag, is generally preferred for longer spans. However, the stiffness and lack of flexibility of tramway track cable, compared with wire rope, necessitates the use of socket connections. The ends of the tramway track cable are untwisted and set in the socket by use of molten zinc.

In the design of cableway structures, different safety factors are generally used for the component parts. The parts of the structure for which individual designs are necessary are:

(a) the track cable
(b) the supports
(c) the anchorages
(d) the footings for the supports
(e) the anchorage connections and
(f) the staylines.

With proper design and construction, the uncertainties affecting concrete anchorages and footings can be so reduced that a relatively small factor of safety is adequate. It is therefore customary to design anchorages and footings for twice the working loads that are anticipated. For very favourable conditions where allowable bearing pressures and frictional resistances of the soil are known, the ratio of the design load to the working load may be taken as 1.5:1.

A-frames and towers that are constructed of galvanised lightweight structural members are designed for twice the expected working loads; the allowance for tensile stress used in the design does not normally exceed $1100 \, \text{kg cm}^{-2}$. The L/r ratio for columns and struts should not exceed 120:1 for main compression members and 200 for bracing and other secondary members.

According to the practice of the US Geological Survey, the following maximum allowable tensions are recommended for the main or track cable: one-fifth the breaking strength of galvanised improved plow-steel wire rope and one-fourth the breaking strength of galvanised tramway track cable (see equation (2.28)).

Anchorage connections include the sockets, eyebars, turnbuckles, rods and pins that transmit the tension from the track cable to the fixed anchorage. Experience has shown that these connections are the places of greatest weakness in the structure. Therefore it is recommended that sockets, eyebars and bolts, turnbuckles and anchorage rods are designed for a working load at least 20% greater than the allowable working load of the main cable to which the connection is made. No weldings are permitted on any part of eyebars, turnbuckles or anchorage rods. The specifications should require that each individual part that goes into the finished product, such as an eyebar or a turnbuckle, be forged in one piece. The required minimum breaking strength of the finished eyebar or anchorage rod is specified by the purchaser.

The connections between the wire rope and the turnbuckles and between the turnbuckles and the anchorages are generally made by means of thimbles and clips. The number of clips for each wire rope end should be at least five for the 20 mm diameter rope with a minimum spacing of 120 mm between the clips, and six for the 25 mm diameter rope with a minimum spacing of 160 mm. It is important that the 'live' or long rope rests upon the broad bearing surface of the base of the clip with the U-bolt bearing against the 'dead' or short end of the rope. When unreeling and uncoiling wire rope, it is essential that the reel or coil rotates as the wire rope unwinds. Attempts to unwind wire rope from a reel or coil which is held stationary will result in kinking the rope and ruining it beyond repair. All steel cables must be regularly inspected and lubricated. All other connections and structural components made of steel are protected against corrosion by painting. The sag of the track cable is checked at regular intervals, particularly when great changes in temperature occur, and adjustments made accordingly. Anchorages are regularly inspected and repaired where necessary. Some further examples of cableway installations are shown in Figs. 2.21 and 2.22 (a) and (b).

Current meter measurements from bridges

Although cableways are generally preferred to bridges for current meter measurements, highway or railway bridges are often used to advantage. Bridges do not often offer the right conditions for stream gauging but measurements from them may be necessary where suitable sites for wading or for a cableway are not present. However, contracted sections, piers and other obstructions

Figure 2.21 Electrically operated cable car with current meter and sounding weight.

affect the distribution of velocities and it is therefore necessary to use a larger number of verticals as well as more velocity observations in each vertical, especially close to bridge piers and banks.

Generally there are two types of bridge measurement, namely rod suspension and line suspension.

Figure 2.22 (a) Manually operated cable car with single-drum winch, cup-type Price current meter and Columbus sounding weight.

Rod suspension from bridges

Footbridges may sometimes be used for gauging small streams. Although the procedure for low velocities may be the same as for a wading measurement, at higher velocities it is often advisable to measure the depth in the following manner:

(a) For each selected vertical, a point is established on the bridge.
(b) With this point as an index, the distance to the water surface is measured by lowering the suspension rod until the base plate touches the water.
(c) The rod is then lowered to the bottom of the stream and the rod reading is again noted at the index point. The difference in the readings is the depth of water at the vertical.

Measuring the depth in this manner tends to eliminate errors that may be caused by the piling up of water on the upstream face of the rod.

The natural flow of water is not disturbed when measuring from a footbridge as is often the case when measuring from a boat or by wading.

Line suspension from bridges

From higher bridges and for greater depths, the current meter and weight are suspended on a cable. The cable is controlled by a gauging reel mounted on a bridge crane (Fig. 2.23) or on a bridge board (Fig. 2.24). A handline may be used with the smaller weights. The gauging procedure is essentially the same as that for measurements from a cable car.

No set rule can be given for selecting the upstream or downstream side of a bridge for discharge measurements. The advantages of using the upstream side of the bridge are:

(a)   The hydraulic conditions on the upstream side of the bridge opening are usually more favourable.
(b)   Approaching drifts can be seen and avoided more easily.
(c)   The streambed on the upstream side of the bridge is not likely to be scoured as badly as the downstream side.

Figure 2.22 (b) Cable car operation on the Snake River at Idaho: cup-type current meter and Columbus sounding weight (USGS).

Figure 2.23 Current meter measurement from bridge using bridge crane, Java.

The advantages of using the downstream side of the bridge are:

(a) Vertical angles are easily measured on the downstream side as the sounding line will move away from the bridge.
(b) The streamlines may be straightened out when passing through a bridge opening with piers.

Whether to use the upstream or the downstream side of a bridge for a current meter measurement needs to be decided individually for each bridge after considering the factors mentioned above and the conditions at the bridge, such as location of the walkway and traffic hazards.

Figure 2.24 Bridge board for use in gauging from bridge.

Handline suspension

The handline is a simple device. It is used for making discharge measurements from bridges, using weights up to 20 kg and for velocities up to 2 m s^{-1}. The advantages of the handline are that it is easier to set up, eliminates the use of a gauging reel and the equipment to support the reel, and makes discharge measurements from bridges with vertical and diagonal members quicker and easier. The disadvantages of the handline are that it requires more physical exertion, especially in deep streams, and there is a greater possibility of making errors in determining the depths.

The handline is made up of two separate cables that are electrically connected at a small reel. The upper or hand cable is a heavy two-conductor electric cable, whose thick rubber protective covering makes the cable comfortable to handle. The lower, or sounding, cable is a light reverse-lay steel cable with an insulated core conductor. A connector joins the lower cable with a hanger bar for mounting the current meter and sounding weight. Sounding cable, in excess of the length needed to sound the stream being measured, is stored on the reel. The sounding cable is tagged at convenient intervals with streamers of different coloured binding tapes, each tag being at a known distance above the current meter and sounding weight.

Special equipment for measurements from bridges

Many special arrangements for measuring from bridges have been devised to suit a particular purpose. Truck-mounted cranes are often used for measuring from bridges over larger rivers. Monorail stream gauging cars have been developed for large rivers. The car is suspended from the substructure of bridges by means of I-beam tracks by trolleys and is propelled by a forklift motor having a wheel in contact with the bottom of the beam. The drive mechanism and sounding equipment are powered by a 430 ampere hour, 2000 kg, 12 V battery.

All cranes are designed so that the superstructure can be tilted forward over the bridge rail far enough for the meter and weight to clear most rails. Where bridge members are found along the bridge, the weight and meter can be brought up, and the superstructure can be tilted back to pass by the obstruction.

Cast-iron counterweights weighing 27 kg each are used by the US Geological Survey with four-wheel-base cranes (Fig. 2.25). The number of such weights needed depends upon the size of sounding weight being supported, the depth and velocity of the stream, and the amount of debris being carried by the stream.

A protractor is used on cranes to measure the angle the sounding line makes

Figure 2.25 Bridge crane with four-wheel base showing boom in retracted position. Note protractor at outer end of boom to determine angle of suspension cable.

with the vertical when the weight and meter are dragged downstream by the water. The protractor is a graduated circle clamped to an aluminium plate. A plastic tube partly filled with coloured anti-freeze fitted in a groove between the graduated circle and the plate is the protractor index. A stainless steel rod is attached to the lower end of the plate to ride against the downstream side of the sounding cable. The protractor will measure vertical angles from −25 to +90°. The crane shown in Fig. 2.25 is equipped with protractors at the outer end of the boom.

Bridge boards may be used with a sounding reel and with weights up to 23 kg. A bridge board is usually a plank about 2–3 m long with a sheave at one end over which the meter cable passes and a reel seat near the other end. The board is placed on the bridge rail so that the force exerted by the sounding weight suspended from the reel cable is counterbalanced by the weight of the sounding reel (Fig. 2.24). The bridge board may be hinged near the middle to allow one end to be placed on the sidewalk or roadway.

Current meter measurements from boats

Where the river is too wide for a cableway installation and too deep to wade, discharge measurements are made from boats. One limiting factor in the use of boats is the high velocity of water, especially during floods, as personal safety has to be considered.

Boats used for discharge measurements may vary from a few metres in length to powerful motor survey launches for use on large rivers where the length of measuring section may be several kilometres.

Where the river is sufficiently narrow to use a tag line, this is spanned across the river at the measuring section. The tag line serves the dual purpose of holding the boat in position during the measurement and of measuring the width of river and monitoring the verticals.

The tag line is wound on a reel which is operated from the stern of the boat as it is propelled across the river. On the bank, the slack of the cable is taken up by means of a block and tackle attached to the reel and to an anchored support on the bank.

If there is river traffic, and the river is sufficiently narrow, one member of the gauging team is stationed on the bank ('look-out man') to lower and raise the tag line to permit traffic to pass. The look-out man has a very important safety function to fulfil, especially if the river is one used for rowing boats or regattas, as the tag line can cause serious injury during regatta training periods. Warning signs are advisable in such cases, positioned at appropriate locations upstream of the measuring section. Streamers are also used and fixed on the tag line so that they may be seen by boat pilots.

A permanent supporting cable, where possible, spanned across the river, to which the boat is anchored during discharge measurements, will often prove advantageous. This method is less laborious and safer for the personnel performing the measurement, especially at high flows. Such a cable requires to be erected well above the highest flood stage expected.

Position fixing

If the river is too wide to use a tag line, or if a tag line cannot be used because of river traffic, the boat can be kept on the measuring section by anchoring it with flags or targets positioned on each bank as shown in Fig. 2.26. The position of the boat in the measuring section can be read directly by means of a transit on-line from the bank and a stadia rod held vertically in the boat. The transit may also be placed in a line at right angles to the measuring section, some known distance from it (position E in Fig. 2.26). By measuring the angle a the boat makes with that line (line CE in Fig. 2.26) the position of the boat may be calculated. Alternatively, a sextant may be used from the boat to measure angle θ, as shown in Fig. 2.26.

The position of the boat may also be found by a linear method (Fig. 2.27). Four flags or targets are fixed, two on each bank. One more flag, E in Fig. 2.27, is fixed on one of the banks along a line at right angles to the measuring section and passing through the flag point B, and at a known distance from it. An observer with a flag moves along the bank from C, towards a position N, along a line perpendicular to the measuring section, until the corresponding flag E on the opposite bank, the flag on the boat M and the hand-held flag N are all in a line. The perpendicular distance from the hand-held flag to the measuring section is determined, and the distance, MC, of the boat computed as shown in Fig. 2.27.

The position of the boat can also be fixed linearly from measurements made from one bank only as shown in Fig. 2.28. Two flags C and D on lines

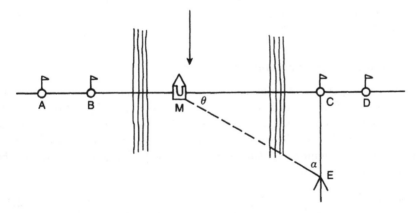

Figure 2.26 Fixing position of boat by transit or sextant: (i) directly by transit from C or D, (ii) by transit from E when MC = CE tan a, or (iii) by sextant from the boat (M) when MC = CE/tan θ.

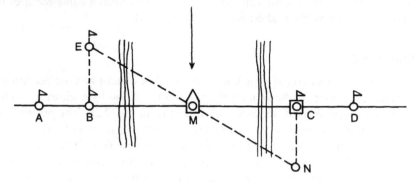

Figure 2.27 Fixing position of boat by a linear method: MC = (CN × BC)/(BE + CN)

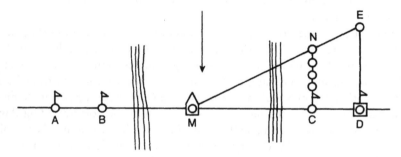

Figure 2.28 Fixing position of boat by a linear method: MD = (DE × CD)/(DE – CN)

perpendicular to the measuring section, and on the same side, are fixed on one bank and the distance, MD, of the boat computed as shown in Fig. 2.28.

At some stations in India, for example, permanent monuments are fixed between D and E and C and N at predetermined locations corresponding to specific boat positions in the measuring section.

Special equipment for measurement from boats

An adequately powered boat is required, and consideration should be given to the possibility of the driver having to return to the required position if he has to move downstream to escape damage from floating debris. The length of the boat requires to be sufficient for it to operate at the maximum speed of flow which may be encountered. A simple guide for the preliminary selection of the length of boat is

$$v = 1.3\sqrt{L} \tag{2.29}$$

where L is the waterline length of the boat (m) and v is the speed of the boat (25% greater than the maximum anticipated velocity of flow). For example, the minimum length of boat required to operate in a river with an anticipated velocity of flow of 2 m s^{-1} would be

$$L = \left(\frac{v}{1.3}\right)^2 \qquad\qquad (2.30)$$

$$= \left(\frac{2.5}{1.3}\right)^2$$

$$= 3.7 \text{ m}, \quad \text{say 4 m.}$$

Catamarans are frequently used because they have the advantage that the winch for supporting the meter can be located along the centreline of the vessel and the observers can stand to both sides of the meter as it is raised and lowered.

Launches used on large rivers are generally equipped for both sediment measurements and discharge measurements and have powerful motors. A typical launch used by the Bureau of Hydrology in China to gauge the rivers in the Yangtze catchment is operated by a 120 hp motor, and is 16 m long, 3.6 m wide and 1.3 m deep with a draught of 1 m and a displacement of 30 t. Four semi-automatic drum winches are available for measurement purposes each capable of measuring depths up to 40 m and velocities of 4 m s^{-1}. The winches are operated by 4.5 and 7.5 kW motors and each is capable of lifting 300 kg sounding weights. The gauging sections vary in length, up to about 2 km.

When a boat powered by a motor is used it is often difficult to maintain it exactly on the transit line throughout the measurement of velocity. The practical solution is often to drive the boat forward very slowly and measurements are commenced as it crosses the transit line. It is then allowed to drift slowly astern and again moved forward to cross the transit line. On the second crossing of the transit line, moving in the same direction, the observer stops recording the velocity. Some slight error is introduced but provided the movement of the vessel is not excessive and is slow the error is negligible.

In a boat measurement, the current meter may be suspended on a rod or on a cable using a bridge board. Specially designed extendable boat booms or boat cranes are available for boat measurements. By means of a boom, the current meter may be placed and operated so as to be reasonably unaffected by any disturbance in velocities that may be caused by the boat itself.

Other current meter methods

There are several other methods which are sometimes used to measure discharge using current meters. These methods are useful when a full current meter gauging is inappropriate because of the time factor involved or when the velocity is too high to take depth soundings or position the meter at the required depths. A loss of accuracy, however, has to be expected.

Two-tenths depth method

In this method the velocity is observed at 0.2 of the depth below the surface and a coefficient applied to the observed velocity to obtain the mean in the vertical. A standard cross-section or a knowledge of the cross-section at the gauging site is used to compute the 0.2 depth when it is impossible to obtain soundings. A significant error in an assumed 0.2 depth position may not be critical because the slope of the vertical velocity curve at this point in deep rivers is usually nearly vertical.

The measurement made by this method is normally computed by using the 0.2 depth velocity observations without coefficients, as though each were a mean in the vertical. The approximate discharge thus obtained divided by the area of the measuring section gives the weighted mean value of the 0.2 depth velocity. Studies of many measurements made by the two-point method show that for a given measuring section the relation between the mean 0.2 depth velocity and the true mean velocity either remains constant or varies uniformly with stage. In either circumstance, this relation may be determined for a particular 0.2 depth measurement by recomputing measurements made at the site by the two-point method using only the 0.2 depth velocity observation as the mean in the vertical. The plotting of the true mean velocity versus the mean 0.2 depth velocity for each measurement will give a velocity relation curve for use in adjusting the mean velocity for measurements made by the 0.2 depth method.

If at a site too few measurements have been made by the two-point method to establish a velocity relation curve, vertical velocity curves are needed to establish a relationship between the mean velocity and the 0.2 depth velocity. The usual coefficient to adjust the 0.2 depth velocity to the mean velocity is about 0.87.

Subsurface velocity method

In the subsurface velocity method, velocity is observed at some arbitrary distance below the water surface. This distance should be at least 0.6 m, and preferably more for deep swift streams to avoid the effect of surface disturbances. The subsurface velocity method is used when it is impossible to obtain soundings and the depths cannot be estimated with enough reliability to even approximate a 0.2 depth setting for a conventional current meter. Coefficients are necessary to convert the velocities observed by the subsurface velocity method to the mean velocity in the vertical. A prerequisite in obtaining these coefficients is to determine the depths during the measurement from soundings made after the stage has receded enough for that purpose. These depths are used with the known setting of the current meter below the water surface to compute the percentage of submergence of the meter during the measurement. The coefficients to be used with the subsurface velocity observations can then be computed by obtaining vertical velocity curves at the reduced stage of the stream.

Integration method

In the integration method the meter is lowered in the vertical to the bed of the stream and then raised to the surface at a uniform rate. During this passage of the meter the total number of revolutions and the total elapsed time are used with the current meter rating table to obtain the mean velocity in the vertical. The integration method is not used with a vertical axis current meter because the vertical movement of the meter affects the motion of the rotor (Section 2.5). The integration method is used to some degree in Russia where good results are reported. The accuracy of the measurement is dependent on the skill of the hydrologist in maintaining a uniform rate of movement of the meter. A disadvantage of the method is the inability of the meter to measure streambed velocities because the meter cannot be placed that low. Coefficients smaller than unity are therefore required to correct the observed integrated velocity.

Measurements have been made in which the entire cross-section has been integrated as a whole. In the course of such measurements the meter is carried across the stream in a zigzag course from top to bottom, the total number of revolutions and total elapsed time being used to compute the mean velocity in the entire cross-section. To obtain the true mean velocity in the cross-section, the current meter must pass through equal elements of the area in equal time intervals. In an irregular cross-section it is therefore necessary that the horizontal movement of the meter be at variable speed to be inversely proportional to depths in the cross-section. Various schemes have been used to compensate for depth irregularity. Because a cable-suspended meter measures the resultant of flow velocity and horizontal velocity of the meter itself, an additional correction is needed for the horizontal velocity of the meter.

The advantage of the integration method is that it can reduce the time of a velocity–area measurement considerably and this can often be important in certain situations. From experience in Russia the recommended moving speed of the current meter is:

(a) for vertical motion, 0.05 m s^{-1};
(b) for horizontal motion, 0.05–0.50 m s^{-1} (but not less than 25% of the average flow velocity).

Figure 2.29 shows diagrammatically various integration methods. In Fig. 2.29(a), the current meter is lowered and raised in each vertical to obtain the mean velocity in each vertical. Figures 2.29(b), and (c) show a typical zigzag motion of the meter to attain the average velocity in the cross-section, found by dividing the total number of current meter revolutions by the corresponding duration. The discharge is then obtained by multiplying this value by the cross-sectional area. In Fig. 2.29(d) the meter is moved horizontally from bank to

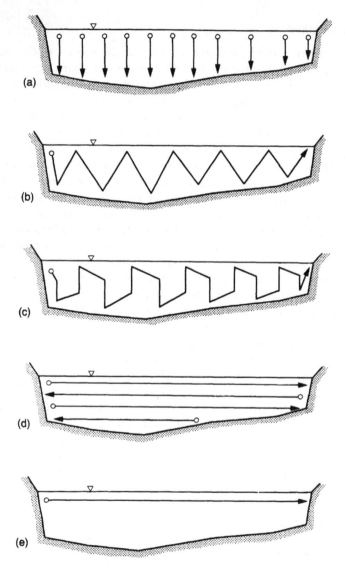

Figure 2.29 Schematic illustration of the integration method of streamflow measurement: (a) current meter moved vertically, (b) and (c) current meter moved across measuring section in zigzag motion, (d) and (e) current meter moved horizontally.

bank in several slices or layers (multilayer method) and in Fig. 2.29(e) only the surface velocity is measured.

For the multilayer method or surface velocity method in Figs. 2.29(d) and (e) the current meter is driven across each layer at a uniform speed dependent on the average velocity in the cross-section. As used in China by the Bureau of Hydrology, this speed is set at about 80% of the average cross-sectional velocity except when this average velocity is less than about 0.5 m s^{-1} when the driving speed is increased to a value not exceeding 1.5 times the average velocity. A sketch of the system as used at Fughou in Fujian Province is given in Fig. 2.30.

The procedure adopted in China is as follows:

(a) The cross-section is divided into four or five layers of equal depth (except for the bottom one), the top layer being taken at about 0.5 m below the surface for small rivers and 1.0 m for large rivers.
(b) The thickness of the layers is between 0.5 and 2 m depending on the depth of flow. The bottom layer is arranged so that it contains less than 5% of the total flow.

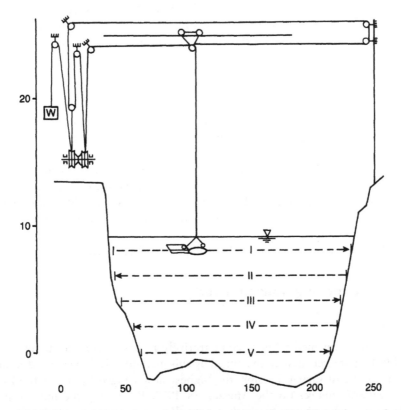

Figure 2.30 Schematic illustration of the integration method adopted for use by the Bureau of Hydrology, Ministry of Water Resources, China.

(c)   The measured widths of the layers are made as near as possible equal to the actual widths so that the blind areas at the banks do not exceed 5% of the total widths.

(d)   The total traversing duration of the current meter normally exceeds 100 s and usually exceeds 200 s. For V-shaped sections, however, the duration is reduced to 50 s for the lower layers. If the bottom layer is very narrow it is preferable to take a point velocity measurement at the centre of the layer.

(e)   During a current meter traverse, the sag in the cableway is compensated for by raising or lowering the suspension cable by means of a micro-processor-controlled regulator.

(f)   The shape of the cross-section is already known or is sounded before measurement begins. The current meter traverses are made from surface to bed.

(g)   To compensate for any oblique flow, forward and backward traverses are made alternately (forward for the first layer, backward for the second, forward for the third, etc., as shown in Fig. 2.30).

(h)   The velocity for each layer is computed by dividing the total number of revolutions for the traverse by the duration of the traverse. This velocity, v_r say, so computed is the vector sum of v, the velocity normal to the measuring section and v_c, the driving speed of the current meter. v is then found by Pythagoras from

$$v = (v_r^2 - v_c^2)^{1/2}. \tag{2.31}$$

(i)   The mean of the width-integrated velocities for two adjacent traverses is taken as the average flow velocity between them. The partial discharge per layer is found from the product of this velocity, the average width of the two adjacent traverses and the difference in height between them.

   The width-integrated velocity of the bottom traverse multiplied by 0.5 is taken as the mean velocity of the lowest traverse. The mean velocity between the highest traverse and the water surface is taken as the average velocity in the top layer.

(j)   The total discharge is the total of the partial discharges.

2.10 The moving boat method

General

The moving boat method is of comparatively recent origin and many of the world's largest rivers have been gauged by this method, which has the advantage of time and cost saving. Possibly the highest flow gauged by the method was some 250,000 cumecs on the Amazon in 1972. Essentially the moving boat method is a velocity–area method. A propeller-type current meter (preferably, but not essentially, a component propeller meter) is suspended from a boat about 1m below the surface and the boat traverses the channel along a pre-selected path normal to the streamflow (Fig. 2.31). During the traverse an echo sounder records the geometry of the cross-section and the continuously operating current meter records the resultant of the stream velocity and the boat velocity. A vertical vane aligns itself in a direction parallel to the water moving past it and an angle indicator attached to the vane assembly indicates the angle between the direction of the vane and the true course of the boat. The velocity, v_b, of the boat is the velocity at which the current meter is being pushed through the water by the boat. The force exerted on the current meter is a combination of two forces acting simultaneously, one force resulting from the movement of the boat through the water and the other a consequence of the streamflow.

The velocity measurements taken at each of the sampling points (verticals) in the cross-section are a vector quantity which represents the relative velocity past the vane and meter. The velocity u_v is the vector sum of v, the required component of stream velocity normal to the cross-section at the sampling point and

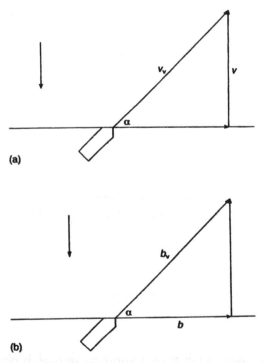

Figure 2.31 General diagrams of velocity vectors. (a) River velocity computation at sampling stations: a, angle of current meter relative to section line; v_v, combined boat and river velocity; v, river velocity; $v = v_v \sin a$. (b) Computation of segment width between sampling stations: a, as above; b_v, relative distance of travel; b, actual distance of travel; $b = b_v \cos a$.

v_b. The principal difference between a conventional current meter measurement as described earlier and the moving boat measurement is in the method of data collection.

A rate indicator unit is used in conjunction with a current meter rating table to obtain v_v, while the angle reading a, representing the angle the vane makes with the cross-section path, defines the direction of the vector.

Theory

The stream velocity, v, perpendicular to the boat at each vertical can be determined from (Fig. 2.31)

$$\bar{v} = \bar{v}_v \sin \bar{a} \tag{2.32}$$

and

$$b = \int \bar{v}_v \cos \bar{a} \, dt \tag{2.33}$$

where b is the distance that the boat has travelled along the true course between two consecutive verticals. It can be assumed that a is approximately uniform over the relatively short distance between verticals and can be treated as a constant. Then equation (2.33) becomes

$$b = \cos \bar{a} \int v_v dt \tag{2.34}$$

and

$$\int \bar{v}_v dt = b_v \tag{2.35}$$

where b_v is the relative distance through the water between two consecutive verticals, as represented by the output from the rate indicator and counter. Then

$$b = b_v \cos \bar{a}$$

and the total width across the channel, B, is

$$B = \Sigma b. \tag{2.36}$$

Finally, d, the stream depth at each sampling vertical, is obtained from the echo-sounder chart, and obtaining, $\bar{v}$, b and d for each vertical across the measuring section, the mid-section or mean-section method is used to compute the discharge.

It is normal practice to make a series of at least six runs, each with 30 to

40 verticals, and average the results, provided the discharge remains constant during the series of measurements.

Alternative methods

The method described above is the one normally used for a moving boat measurement. A second method measures v_b directly. This is done by measuring the distance from the observation points to a fixed point on the bank, from which the width of the traversed segment can be determined, along with the simultaneous measurement of time. From these data, the velocity component of the boat, $\bar{v}_b$, can be computed. By means of the measurement of $\bar{v}_v$, the stream velocity $\bar{v}$ perpendicular to the selected path is determined from

$$\bar{v} = (\bar{v}_v^2 - v_b^2)^{1/2}. \tag{2.37}$$

In this method the angle a is not measured.

In a third method, v_b and the angle a are measured and $\bar{v}$ is found from

$$\bar{v} = v_b \tan a. \tag{2.38}$$

Summarising:

Normal method: measure v_v and the angle a;

Method two: Measure v_b and v_v;

Method three: Measure v_b and the angle a.

There is no record of measurements carried out using method three. A typical boat with angle indicating assembly is shown in Fig. 2.32.

2.11 The electromagnetic method

General

The electromagnetic method of measuring the total flow in an open channel required some 10 years of applied research in the UK. The method requires sophisticated electronics particularly in the field of signal detection and processing but the revolution in microelectronics in the 1970s has made this method an important addition to standard streamflow methods.

A feature of the system is its ability to measure streamflow in weedy rivers and in rivers with silty or moving beds.

The method requires an on-site current meter calibration and a source of electrical power (mains supply at 110 or 240V). The system now being applied, however, has been calibrated in the laboratory of Hydraulics Research Ltd.

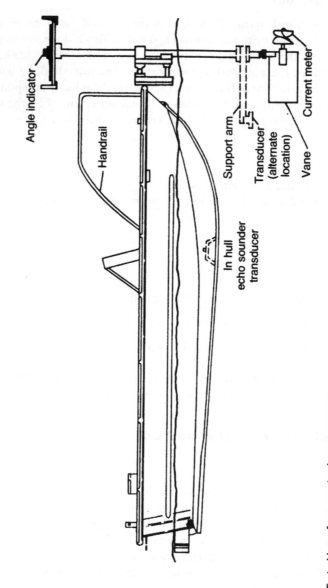

Figure 2.32 Typical boat for a moving boat movement

The electromagnetic streamflow method should not be confused with electro-magnetic point velocity current meters. These have been dealt with earlier in Chapter 2.

Principle

According to Faraday's law of electromagnetic induction, the motion of water flowing in a river cuts the vertical component of the earth's magnetic field and an electromotive force (emf) is induced in the water. This emf can be sensed by electrodes ('probes') on each side of the river and is found to be directly pro-portional to the average velocity of flow in the cross-section. Therefore, unlike the ultrasonic method (Chapter 12) which measures velocity across paths, the electromagnetic method performs an integration over the entire cross-section. However, although measurements have been conducted using the earth's magnetic field, usually in tidal estuaries, these have had a large uncertainty due to the fact that the emf induced by the earth's field is too small to be dis-tinguished from other electrical interference. This interference normally comes from 240V mains, electrical motors and other ambient electrical noise present in some form or another in the ground. Therefore, to induce a measurable potential in the electrodes, a vertical magnetic field is generated by means of a coil buried in the riverbed, or over it, through which an electric current is driven. The potential generated is proportional to the width of the river (m) multiplied by the magnetic field (T) multiplied by the average velocity of flow (m s^{-1}). For a typical coil of 500A turns the induced potential between the opposite banks of the river would be 250 mV for an average velocity of 1 m s^{-1}. Figure 2.33 shows an electromagnetic gauging station installation. It can be seen that there is no construction in the river and the only constructions shown are the access man-holes A, B, C and D used for installing the coil beneath the river and manholes E and F for carrying the cabling from the electrodes and coil to the instrument house (not shown). The staff gauge for stage measurement (to measure depth of flow) is shown on the left bank.

Theory

The electromagnetic technique has been used for many years for gauging full pipes where the area of flow is constant. The basic principle is shown schematically in Figure 2.34. The pipe flow meter is manufactured in units of practically any required diameter sizes, the largest in practice being about 2 m in diameter. The coil is incorporated in the form of a saddle around the pipe section and the electrodes are inserted in the pipe wall flush with the inside wall (see also Chapter 15).

In streamflow measurement, however, the stage varies and therefore the cross-sectional area of flow is not constant. In addition, stream channels are much wider than pipes and hence the electromagnet forms a large part of the

Figure 2.33 Electromagnetic gauging station: coil installed beneath river; bed insulated with
insulation membrane covered with gravel; manholes A, B, C and D are installed
at corners of coil and used for access to ducts; manholes E and F are at
extremities of duct carrying electrode and coil cables to instrument house
(not shown).

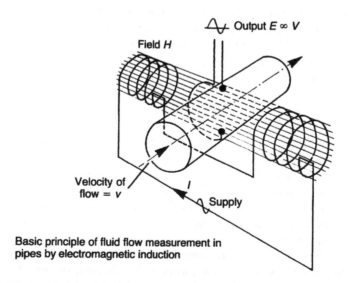

Basic principle of fluid flow measurement in
pipes by electromagnetic induction

Figure 2.34 Basic principle of fluid flow measurement in pipes running full under pressure.

cost; velocities in rivers are normally much slower than in water flowing in full pipes since the former are flowing under gravity whilst full pipes are normally under pressure. In additional, electrical interference is generally higher near rivers than in pipes, particularly at the mains distribution frequency. All of these factors make open channel electromagnetic gauging much more complex than that in full pipes.

Referring to Figure 2.34, Faraday's law of electromagnetic induction relates the length of a conductor moving in a magnetic field to the emf generated, by the equation

$$E = B\bar{v}b \qquad (2.39)$$

where E is the emf generated (V);
   B is the magnetic flux density in teslas (T);
   $\bar{v}$ is the average velocity of the conductor in the cross-section (m s^{-1});
   b is the length of the conductor and is equal to the river width (m).

Now in the case of an operational gauge having an insulated bed and a square coil just wider than the channel, the voltage generated is approximately 0.8 times that given by equation (2.39). This reduction in voltage is caused by the shorting effect of the water upstream and downstream from the magnetic field. Numerically the empirical relation (±3%) is

$$E' \doteqdot \bar{v}bH \qquad (2.40)$$

where E' is the electrode potential (μV) and H is the average magnetic field strength in amperes per metre. (Note that the physical relationship between B and H in free space, air or water is given by

$$B = H \times 4\pi \times 10^{7}$$

where B and H are in different units as above.)

In the ideal case where the magnetic field strength is constant over the entire wetted section, the discharge, Q (m^3 s^{-1}), is given by

$$Q = \bar{v}bd \qquad (2.41)$$

and from equations (2.40) and (2.41) where d is the depth of flow (m)

$$Q = \frac{E'b}{H}. \qquad (2.42)$$

Operationally this equation may take the form

$$Q = K\left(\frac{E'b}{I}\right)^{n} \qquad (2.43)$$

where I is the coil current in amperes and K and n are constants. If the coil is mounted above the channel the water near the bed will move in a less strong magnetic field relative to that near the surface. Normally the relation is expressed in the form

$$Q = (K_1 + h + K_2 h^2)E'/H. \tag{2.44}$$

An on-site current meter calibration may take several forms; it may be a single straight line when transposed logarithmically, or have several straight lines with inflexions or it may be curvilinear.

It can be seen that the discharge from equations (2.42), (2.43) or (2.44) is obtained by the velocity–area principle in which the velocity is inferred from the electrode potential and multiplied by the cross-section area of flow. The area is a function of the recorded water level and the equations contain dimensional constants established empirically. The river width factor does not appear in the equations and is accounted for in the design of the coil, which extends the full width of the channel, and in the empirical calibration.

2.12 A seismic flowmeter for mountain torrents

General

The continuous measurement of discharge in mountain torrents is important not only for information on a precious resource but also for flood warning from upland catchments. Significant physical and practical problems arise in trying to establish a stage – discharge relation and similar problems arise in considering a weir or flume in addition to extremely high Froude[1] numbers (see Further reading). The dilution method, for example, although giving excellent results in many countries and certainly based substantially on turbulence mixing, is used mainly for spot measurements or for checking existing methods. In addition there may be the problem of using chemicals in some rivers. The electromagnetic method is unsuitable because of electromagnetic interference and difficulties of power supply in a hostile environment and entrapped air in the water inhibits the use of the ultrasonic method. Summarising, there has been no satisfactory method of continuously measuring discharge in mountain torrents.

However, research has been proceeding in the UK on a mountain river seismic flowmeter since 2001 (see Further reading). The flowmeter is based on the relation between seismic vibrations and discharge. Over the last few years the research has covered several sites, and continuous measurements are made by spectral analyses on the amplified output from geophones normally installed on

---

1   Froude number. The mean velocity divided by the square root of the product of the mean depth and the acceleration due to gravity. It is dimensionless.

outcrops of bedrock beside the rivers. Where possible, and suitable sections can be found a current meter check is made.

Figure 2.35 shows a typical mountain torrent with geophone installation. In this case, the geophone was simply installed by drilling two 50 mm diameter holes vertically into the top of a large upstanding rock. (In this case two different geophones were installed for comparison purposes).

Seismic recording

The current instrumentation for the seismic flowmeter system utilises Spectral-Lab software running on a laptop PC to perform the Digital Signal Processing (DSP) and data logging functions.

The SpectralLab programme works in conjuncion with the sound card on a laptop computer. The vibration signal from the geophone, suitably amplified by 34dB by a BT26 preamplifier, is plugged into the Line-in or Microphone jack on the back of the sound card. SpectralLAB then uses the sound card to perform an analog-to-digital conversion on the vibration signal. This digitised signal is then passed through a mathematical algorithm known as a Fast Fourier Transform (FFT) which converts the signal from the time domain to the frequency domain. The CPU on the laptop computer is used to perform this transformation. Successive applications of the FFT can be performed on sequential signals to determine the average spectral properties of the vibration signals and these can be logged at set time intervals.

Geophone mounting

Figure 2.35 Showing the turbulent flow adjacent to the geophone location (beneath the pile of stones on the large rock in the foreground).

The initial logging operation consists of averaging 10 spectra and recording the average spectral power values from 60 to 80Hz every 10 seconds (from a sample rate of some 8,000Hz). The data are then filtered to remove spurious noise spikes and averaged over 15-minute intervals.

Figure 2.36 is a typical site (a) showing raw data and (b) processed data for 21 June 2006. Note that undulations in the power data, prior to the rise in flow shown by the stage data, indicate noise from rainfall which caused increased flow.

(a)

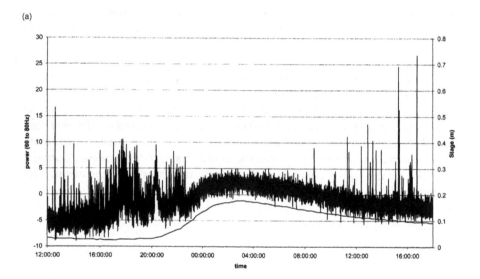

(b)

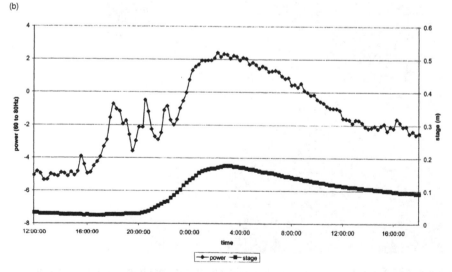

Figure 2.36 (a) Raw and (b) processed vibration data compared to stage for 21 June 2006.

At present the flowmeter is calibrated by current meter but further studies include the effective calibration of the system.

Further reading

Herschy, R.W. (ed.) *Special issue of Flow Measurement and Instrumentation (FMI) on Open Channel Flow Measurement*, 2002 vol. 13 no. 5–6 (December 2002) With the following contributions:

Boiten, W. *Flow measurement structures.*
Dey, S. *Free overfall in circular channels with flat base: a method of open channel flow measurement.*
Godley, A. *Flow measurement in partially filled closed conduits.*
Herschy, R.W. *Editorial.*
Herschy, R.W. *The world's maximum observed floods.*
Herschy, R.W. *The uncertainty in a current meter measurement.*
Holland, P.G. *The importance of glossaries.*
Holland, P.G. *The water framework directive.*
Jones, R.W. *A method for comparing the performance of open channel velocity–area flow meters and critical depth flow meters.*
Lee, M.C., Lai, J.M., Plant, W.J., Keller, W.C. and Hayes, K. *Non-contact flood discharge measurements using an X-band pulse radar (1).*
Lee, M.C., Lai, J.M., Plant, W.J., Keller, W.C. and Hayes, K. *Non-contact flood discharge measurements using an X-band pulse radar (11).*
Marsh, T.J. *Capitalising on river flow data to meet changing national needs – a UK perspective.*
Newman, J.D. and Bennell, J.D. *A mountain stream flowmeter.*
Thomas, F. *Open channel flow measurement using international standards: introducing a standard programme and selecting a standard.*
York, T.H. and Oberg, K.A. Measuring river velocity and discharge with acoustic Doppler profilers.

Archer, D. *Personal communications.* 2007.
Bennell, J.D. *Further work towards the development of a seismic flowmeter for monitoring flow in mountain streams.* Report to the UK Environment Agency 2005.
Bennell, J.D. *Development of a mountain river seismic flowmeter.* Report to the Snowdonia National Park Authority 2006.
Black, A. et. al. Extreme precipitation and runoff. SW Perthshire, Scotland August 2004. *Abstracts volume Royal Geographical Society (with the Institute of British Geographers)* Annual Conference 2005.
Bonacci, O. Several methods of discharge measurement of floods. *Hydrological Sciences Journal*, 28(4), 12 1983.
Chen, R. The multilayer width-integrated velocity measurement method 1982.
Environment Canada. Moving boat method, *Hydrometric field manual, Inland Waters Directorate*, Ottawa 1978.
Gibbard, D. Design and installation of ultrasonic and electromagnetic flow gauging stations. *Flow Measurement and Instrumentation (ed. R.W. Herschy)* 1993.

Green M.J. and Herschy, R.W. New Methods in *Hydrometry: Principles and Practices (ed. R.W. Herschy)*, John Wiley and Sons, Chichester 1978.

Halder, S.K. Developments in hydrometry in India, *IAHS Symposium, Exeter*, Pub. No. 134 1982.

Herschy, R.W. New Methods of river gauging in *Facets of Hydrology* (ed. J. C. Rodda) John Wiley and Sons Chichester 1976.

Herschy, R.W. *Streamflow Measurement, (2nd Ed) The electromagnetic method*, E&FN Spon, London 1995.

Herschy, R.W. *Streamflow Measurement, (2nd Ed) The moving boat method*. E&FN Spon, London 1995.

Herschy, R.W. (ed.) *Hydrometry: Principles and Practices*, Second Edition, John Wiley and Sons Chichester 1998.

Herschy, R.W. and Newman, J.D. The measurement of open channel flow by the electromagnetic gauge. *IAHS Symposium Exeter* Pub. No. 134 pp. 215–217 1982.

Herschy, R.W., Hindley, D.R., Johnson, D. and Tattersall, K.H. The effect of pulsations on the accuracy of river flow measurement. *Department of the Environment Water Data* Unit TM No. 10 1978.

ISO 4369. Moving boat method. *(2004) amendment 1*. Geneva 1979.

ISO 9213. Measurement of discharge by the electromagnetic method, Geneva 2004.

Smoot, G.F. and Novac, C.E. *Measurement of flow by the moving boat method*, US Geological Survey, Washington DC 1969.

Vaiser, V.V., Golubovitch, V.A., Krasyukov, V.A. and Stepanov, B.S. *Measurement of mountain river flow rates by a seismic method*. Meteorologiya Gidrologiya, No. 10, pp. 118–119 1983.

Vinogradov, Y.B., Krasyukov, V.A. and Stepanov, B.S. *A method of measuring the flow rates of water, water containing suspensates and mud flows in natural river beds*: Inventor's certificate no. 539220 (in Russian) Otkr., Izobr., Prom. Obraztsy, Tov, Znaki, 46 1976.

Wang, H. *The design and study of the instrumentation required for the measurement of discharge by the moving boat method*. IAHS Symposium, Exeter, Pub. No. 134 1982.

Weibrecht, V. Kuhn, G. and Jirka, G.H. Large scale PIV measurements at the surface of shallow water flows.

Chapter 3

Measurement of stage

3.1 General

The stage of a stream is the height of the water surface above an established datum plane. The water surface elevation, referred to some arbitrary or predetermined gauge datum, is known as the **gauge height**. Gauge height is often used interchangeably with the more general term 'stage' although gauge height is more appropriate when used to indicate a reading on a gauge. Stage or gauge height is usually expressed in metres and hundredths or thousandths of a metre depending on the accuracy required.

The determination of stage is one of the most important measurements in hydrometry. We shall see in Chapter 13 that the uncertainty in the stage–discharge relation depends largely on the uncertainty with which stage can be measured. In the statistical analysis of the stage–discharge relation, as shown in Chapter 13, stage is taken as the independent variable having negligible uncertainty. Chapter 13 also shows that the uncertainly in the measurement of head over measuring structures can have a significant effect on the discharge. This is due to the effect of the exponent of head in the discharge equations. The discharge equations for measuring structures are therefore more sensitive to the uncertainty in the head measurement than to the uncertainty in the coefficient of discharge.

It can be stated that, in methods of streamflow measurement where a correlation is established between stage, fall or slope and discharge, the uncertainty in the measurement of stage has a significant effect on the overall uncertainty in the record of discharge.

3.2 The reference gauge

The non-recording reference gauge is the basic instrument for the measurement of stage whether at a regular flow measuring station or at a site where only casual observations are made. It can be sited as an outside gauge to allow a direct reading of water level in the stream or as an inside gauge to indicate the level in a stilling well, and is used for setting and checking the water level recorder, for indicating the stage at which discharge measurements are taken, and for emergency readings when the recorder is out of action. There are various forms, the choice being decided by the site conditions and the specific use to which it will be put.

The reference gauge may be read visually one or more times a day if a water level recorder is not part of the installation. The disadvantages of a non-recording reference gauge in this case are the need for an observer and the loss of accuracy of the estimated continuous graph of stage unless observations are taken many times a day. In some large rivers, such as the Yellow River in China, with long sloping shallow banks, however, it is not always feasible to install a water level recorder and resort has to be made to visual observation of the reference gauge.

The reference gauge is normally either a vertical staff gauge or an inclined (ramp) gauge.

Staff gauge

Vertical staff gauges usually consist of standard porcelain enamelled iron plated sections each about 150 mm wide and 1 m long and graduated in units of 10 mm (Fig. 3.1). The sections are screwed to a backing board which is fastened to a suitable support. Slotted holes in the plates are provided for final adjustment in setting the gauge. The standard US Geological Survey vertical staff gauge consists of porcelain enamelled iron sections, each 4 in wide and 3.4 ft long, and graduated every 0.02 ft.

A staff gauge is seldom a stable construction and is often exposed to movement or damage especially during floods. In order to be able to reset the gauge to its correct datum, one or preferably two bench marks are required at a gauging station constructed entirely detached from the staff gauge or its support and secured against movement or destruction.

The gauge is located near the edge of the stream so that a direct reading of the water level may be made (Fig. 3.2). If this is not possible because of excessive turbulence, wind effect or inaccessibility, the observations are made in a suitable permanent stilling bay in which wave action is reduced and the level of the water surface follows that of the stream.

The staff gauge is located at or near to the current meter measuring section without affecting the flow conditions and in a position where there is least possible damage from drift. Bridge abutments or piers are generally unsuitable locations. In order to enable the observer to make readings as near as possible to eye level, convenient access is necessary and in this connection a flight of steps is normally desirable.

A suitable backing for a vertical staff gauge is provided by a board fixed to a wall having a vertical face parallel to the direction of flow. The board is securely attached to the surface of the wall so as to present a vertical face to receive the graduated guage plates. Staff gauges may usually, however, be fixed to piles

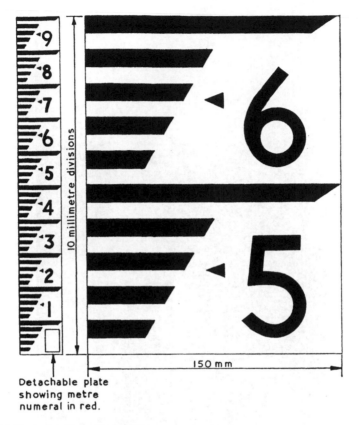

Detachable plate
showing metre
numeral in red.

Figure 3.1 Typical staff gauge markings.

either driven firmly into the riverbed or river banks or set in concrete in order to avoid movement or being washed away during floods. Anchorages are designed to extend below ground level to a level free of any disturbing effects. Provision is made in all cases for easy removal of the gauge plates for maintenance or adjustment.

Where the range of water levels exceeds the capacity of a single vertical gauge section, additional sections are installed on the line of the cross-section normal to the direction of flow (Fig. 3.3).

The inclined gauge

An inclined, or ramp, gauge usually consists of heavy timber securely attached to a permanent foundation. The graduations of an inclined gauge are either marked directly on the surface of the timber or carried on manufactured gauge plates designed to be set to specific slopes. Except where use is made of manufactured gauge plates, an inclined gauge is calibrated *in situ* by accurate levelling from the station bench mark. Usually, various sizes of bronze or brass

Figure 3.2 Vertical staff gauge installation with additional sections, Han River at Xian Tao, China.

strips, pins and numerals are used for the graduations and the assembly is completed in the workshop before finally bolting down the gauge on the prepared foundation (Fig. 3.4).

An inclined gauge is installed to follow the contours of the bank; sometimes a gauge with a single slope is adequate but more usually it is necessary to install an inclined gauge in several sections each with a different slope (Fig. 3.5). As in the case of the staff gauge, it is usually convenient to construct a flight of steps alongside the inclined gauge to facilitate both installation and reading. Reading of the gauge may be improved by the use of a small portable stilling box which helps to dampen wave action (Fig. 3.6). A properly designed inclined gauge is less prone to damage than a staff gauge and normally allows more accurate readings of water level to be made because of its better resolution and its convenience in reading.

Figure 3.3 Staff gauges with additional sections for increased range of stage, Qing River at Gao Ba Zhon, China.

The float–tape gauge

The float–tape gauge (Fig. 3.7) is used mainly as an inside stilling well reference gauge for a water level recorder and consists of a float attached to a counterweight by means of a stainless steel tape. The tape is normally graduated in metres and 10mm divisions and passes over a pulley. The float pulley consists of a wheel about 150 mm in diameter and is grooved on the circumference to accommodate the tape and mounted on a support. An arm extends from the support to a point slightly beyond the tape to carry an adjustable index. The tape is connected to the float by means of a clamp which is also used for making adjustments to the tape reading if the adjustments necessary are too large to be accommodated by the adjustable index. A 250 mm diameter copper float and a 1 kg lead counterweight are normally used.

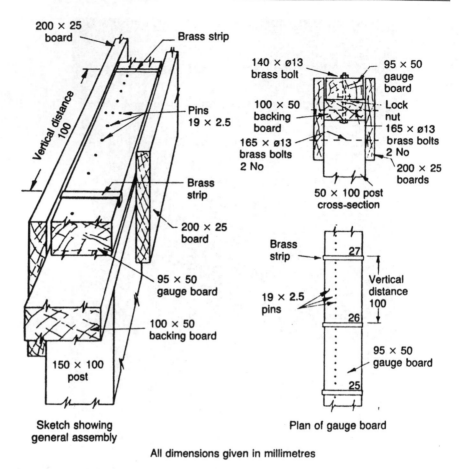

All dimensions given in millimetres

Figure 3.4 Inclined (ramp) gauge: details of typical design.

The electric–tape gauge

The electric–tape gauge (Fig. 3.8), like the float–tape gauge, is used as either an outside or inside reference gauge. It is commonly used at measuring structures as an outside gauge to measure the head over a weir or through a flume where it is operated from a datum plate set into an abutment wall. A staff gauge fixed to the wall of a weir or flume, although useful as a spot check, is difficult to read to the required accuracy because of its awkward location.

A typical hand-held gauge consists of a graduated reel of steel tape and a 9 V (PP3) battery. The gauge is lowered until it makes contact with the water surface which completes an electric circuit and causes a buzzer to sound. The gauge is read to the nearest millimetre against a datum plate.

Figure 3.5 Typical ramp gauge installation.

The wire–weight gauge

The wire–weight gauge, used almost exclusively in the United States, consists of a drum wound with a single layer of cable, a bronze weight attached to the end of the cable, a graduated disc and a counter, all within a cast-aluminium box. In the United States the disc is graduated in tenths and hundredths of a foot; otherwise it is graduated in metres and millimetres. The disc is permanently connected to the counter and to the shaft of the drum. The cable is made

Figure 3.6 Portable stilling box to dampen wave action and facilitate reading of ramp gauge.

of 1 mm diameter stainless steel wire, and is guided to its position on the drum by a threaded sheave. The reel is equipped with a pawl and ratchet for holding the weight at any desired elevation. The diameter of the drum of the reel is such that each complete turn represents 0.3 m movement of the weight. A horizontal checking bar is mounted at the lower edge of the instrument so that when it is moved to the forward position, the bottom of the weight will rest on it. The gauge is set so that when the bottom of the weight is at the water surface the gauge height is indicated by the combined reading of the counter and the graduated disc. The wire–weight gauge is used as an outside reference gauge mounted normally on a bridge parapet, wall or pier.

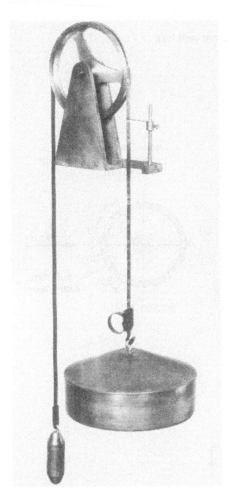

Figure 3.7 Typical float–tape gauge.

Figure 3.8 Typical   electric–tape   gauge (Diptone).

Crest–stage gauge

The crest–stage gauge (Fig. 3.9) is a device for obtaining the elevation of the flood peak of streams; it is simple, economical, reliable and easy to install. The gauge consists of a vertical piece of 50 mm diameter galvanised pipe about 1.5– 2.0 m long containing a wooden or aluminium staff held in a fixed position with relation to a datum reference. The bottom cap has six intake holes located so as to keep the non-hydrostatic draw-down or superelevation inside the pipe to a minimum. The top cap contains one small vent hole and the bottom cap contains granulated cork. As the water rises inside the pipe the cork floats on the water surface. When the water reaches its peak and starts to recede, the cork adheres to the staff inside the pipe, thereby retaining the peak stage of the

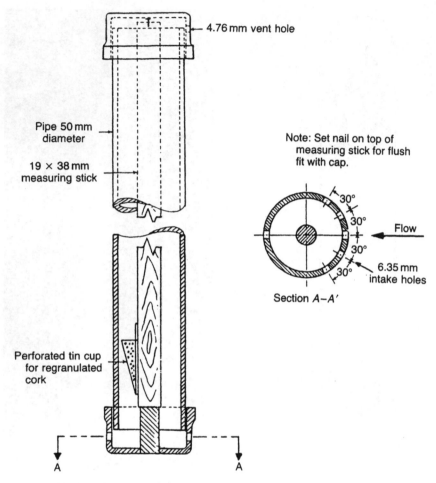

Figure 3.9 Typical crest–stage gauge.

flood. The gauge height of a peak is obtained by measuring the interval on the staff between the reference point and the floodmark. Scaling can be simplified by graduating the staff. The cork is cleaned from the staff before replacing the staff in the pipe in order to prevent confusion with high-water marks that will be left by subsequent peak discharges.

The datum of the gauge is checked by levels run from the bench mark to the top of the staff, the graduated staff being of known length. The gauge is serviced on a regular basis but is not normally removed when the stage is high because when the staff is replaced when water stands high in the pipe the resulting surge of the water displaced by the staff will leave an 'artificial' high-water mark on the staff.

The station bench mark

The station bench mark is set in a position offering maximum security against disturbances and is normally set in a concrete block that extends below the ground surface to a level free from disturbance, or drilled into solid rock. If possible, it is referred to the national Geodetic Survey (Ordnance Survey in the UK) by accurate levelling. To facilitate levelling between the station bench mark and the gauge zero, the bench mark is located in such a position that the transfer of level is carried out by reciprocal levelling or with equally balanced foresights and backsights. The level is transferred from the station bench mark to the gauge zero by a closed levelling circuit, starting and finishing on the bench mark, and with a closing error not exceeding 5 mm. The mean of the two runs is taken as the difference in level between the bench mark and the gauge zero.

Bench marks usually consist of specially made brass bolts or ordinary steel bolts 100–150 mm long and 15 mm in diameter.

If the station is equipped with a water level recorder, an auxiliary bench mark is established on the instrument shelf in order to check the water level recorder for movement and to check the water level in the stilling well.

Gauge datum

The datum of the gauge may be a nationally recognised datum such as mean sea level or an arbitrary datum plane chosen for convenience. An arbitrary datum is selected for the convenience of using relatively low numbers for gauge heights. To eliminate the possibility of minus values of gauge height, it is usual to select the datum, for operating purposes, below the elevation of zero flow on a natural control. Where an artificial control is part of the installation, the gauge datum is usually set at the elevation of zero flow. This also applies to measuring structures.

As a general rule, a permanent datum is maintained so that only one datum for the gauge height record is used for the life of the station. An exception occurs at gauge sites where low flow stages have a negative gauge height. In this situation a change in gauge datum to eliminate the negative numbers is normally carried out. This avoids possible confusion involving the algebraic sign of the gauge heights.

3.3 Water level recorders

The record of stage is normally produced by a water level recorder ('logger') actuated by a float and counterweight or tensator spring system working within a stilling well, the movement of the float being used to operate a recording mechanism such as a pen or head which can produce either an analogue record on a chart or a digital record. Essentially the recorder consists of a time element and a water height element which operate together and produce on the chart or tape a record of the rise and fall of the water surface with respect to time. The time element consists of a clock actuated by a spring, weight or electrical mechanism, driving in the case of an autographic recorder either a chart drum or a recorder pen, or in the case of a digital recorder, rotating a cam which initiates the recorder pen. The water height element consists of a float, combined with a counterweight or tensator device, and some form of mechanical linkage to connect either directly or through reduction gearing to the recording device.

An encoding or digitising device is used to convert the analogue measurement of water level to a digital output. The recording device can be a pen or pencil, in the case of an autographic record.

In solid-state loggers, water level data is transmitted from the sensor (float and tape) via a shaft encoder to a solid-state medium in the form of a cartridge, cachette or similar module.

Autographic recorder

The autographic recorder is the recorder used worldwide and supplies a continuous trace of water stage with respect to time on a chart. Usually the gauge height element moves the pen or pencil stylus and the time element moves the chart, but in some recorders this is reversed. The range of gauge height scales is from 1:1 to 1:50, common recording scales being 1:1, 1:2.5, 1:5, 1:10 and 1:20. Timescales vary according to the chart design but are normally not less than 48 mm for 24 h.

Most autographic recorders can record an unlimited range in stage by a stylus-reversing device or by unlimited rotation of the drum.

Most strip chart recorders will operate for several months without servicing. Drum recorders require attention at weekly or monthly intervals.

It is important that the chart be cut by the manufacturer in such a manner that the direction in which the height is recorded is in accord with the machine direction of the paper or material of which the chart is made. In the case of strip chart recorders however, the timescale is always in the machine direction since it is impossible for it to be otherwise.

Typical autographic water level recorders are shown in Figs 3.10 and 3.11.

Operation of float-operated water level recorders

In any recording system using a float to sense water level there are possible sources of error which can result in a difference between actual and recorded water levels. Some of these errors can be minimised by careful design of the instrumentation and by proper operational procedures, but it is unlikely that all can be eliminated and precautions are necessary to reduce such errors to

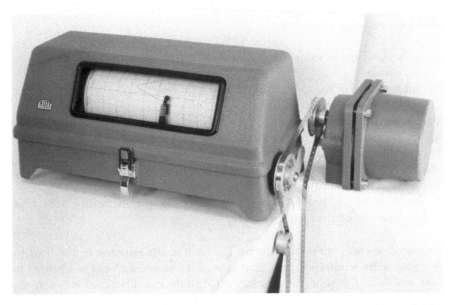

Figure 3.10 Horizontal drum autographic recorder, Ott type X, with shaft encoder for digital remote transmission.

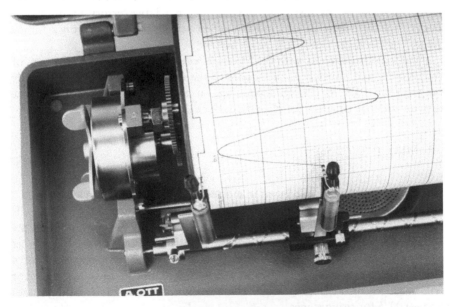

Figure 3.11 Horizontal drum autographic recorder, Ott type X, with reversing tracing mechanism. Recording of water levels not limited by length of drum. Line traced at a distance of 5 mm from edge of chart indicates rising stage; line traced at a distance of 8 mm from edge of chart indicates falling stage.

acceptable limits. Errors caused by the float system can result from various combinations of possible faults:

(a) a change, from its initial setting, in the depth of submergence of the float in the water;
(b) submergence of the counterweight and float line;
(c) backlash in the gearing;
(d) friction in the mechanism;
(e) inadequate diameter of the float or badly matched float and counterweight;
(f) overriding or displacement of wires on the float or counterweight pulleys;
(g) overriding or displacement of the wires on the pen carriage movement;
(h) kinks in the float suspension cables;
(i) build-up of silt on the float pulley affecting the fit of the float tape perforations on the sprockets.

A number of these faults can cause a lag both in the response of the float to changing water levels and in the response of the recording head to changes in float position. The result of this lag is to produce a difference between the indicated height on rising and falling stages, the sign of the error being dependent on the conditions of stage at the time of the original setting of the recorder (see also Chapter 5). The error can be systematic and is of particular importance during a period of recession to low summer levels when comparatively small errors in water level can produce large relative errors in discharge. Because of this, there is some advantage in setting the recorder on a falling stage.

Float lag varies directly with the force (F) required to move the mechanism of the recorder and inversely as the square of the float diameter (D). F commonly ranges from 0.03 to 0.15kg depending on the type and condition of the instrument. The equation for maximum float lag error (MFLE) is

$$\text{MFLE} = (0.002\ 56)\,\frac{F}{D^2}$$

where MFLE and D are expressed in metres, and F is expressed in kilograms.

If we assume a value of 0.08 kg for F and a value of 0.2m for D, MFLE equals 0.005m. If the recorder is set to the true water level while the float is rising, the record on a falling stage will be in error by +0.005m. If, however, the index is set at the true level at a stationary stage – that is, at the peak or trough of a changing stage, or when the valves in a stilling well are closed – then the error will be halved on a changing stage. The error will be +0.0025 for falling stages and −0.0025 for rising stages.

Line shift

With every change of stage a part of the float tape passes from one side of the float pulley to the other, and the change in weight changes the depth of flotation of the float. The correction for this error depends on the magnitude of change in stage (H) since the last correct setting of the recorder, the unit weight (u) of the tape, and the float diameter (D). The error will be positive (+) for a rising stage and negative (−) for a falling stage. The equation for line shift error (LSE) is

$$\text{LSE} = (0.002\ 56)\frac{u}{D^2}\Delta H$$

where LSE, D and H are expressed in metres, and u is expressed in kilograms per metre. If we assume a value of 0.013kg m^{-1} for u, a value of 15m for ΔH and a value of 0.2m for D, LSE equals 0.0124m.

Submergence of the counterweight

When the counterweight and any part of the float line become submerged as the stage rises, the pull on the float is reduced and its depth of flotation is increased. The converse is true when the submerged counterweight emerges from the water on a falling stage. Thus the error caused by submergence or emergence is opposite to that of the line shift error and tends to compensate for the line shift error. The submergence error is dependent on the weight of the counterweight (c) and the float diameter (D). The equation for submergence error (SE) is

$$\text{SE} = (0.000\ 118)\frac{c}{D^2}$$

where SE and D are expressed in metres, and c is expressed in kilograms. If we assume a value of 0.6 kg for c and a value of 0.2 m for D, SE equals 0.0018 m.

Although not related to errors inherent in a float-operated stage recorder, it might be mentioned here that error in recorded stage may be caused by expansion or contraction of the stilling well of a tall gauge structure that is exposed to large temperature changes. For example, a steel well 25m high, exposed to an increase in temperature of 40°C, will have its instrument shelf raised 0.012 m, assuming, of course, that the instrument shelter is attached to the well.

Backlash is caused by badly cut or badly paired gears and produces an error on either the rising or the falling limb of the hydrograph. Friction is a function of the design of the mechanism and, more particularly, the quality of the maintenance given, and has to be overcome by the driving force developed by the float. The driving force depends on the buoyancy of the float and, on a float of regular diameter, varies directly with the depth of flotation and with the square of the diameter at the water surface. With changing water levels, the depth of flotation is affected by the change of weight induced by the shift of the suspension line from one side of the float pulley to the other or by submergence of part of the line and its counterweight. To some extent the friction factor increases with the weight of the pulley shaft which varies directly with the mass of the counterweight. For the best results, therefore, the instrument needs to be maintained to a high standard, the float needs to be as large as practicable and the counterweight needs to be as light as possible, consistent with being able to fully tension the suspension cables or tapes, which require to be light and flexible.

There are many other sources of error which require to be recognised and taken account of. On analogue recorders, for example, the main sources of error are:

(a)   improper setting of the chart on the recorder drum;
(b)   improper joining of the chart edges;
(c)   distortion and/or movement of the chart paper;
(d)   improper setting of the pen on the chart;
(e)   distortion or misalignment of the chart drum;
(f)   faulty operation of the pen system;
(g)   clock inaccuracy.

From the point of view of accuracy and consistency most analogue recorders, if selected after careful consideration of the conditions in which they have to operate, can be expected to sense and record a changing water level with sufficient accuracy for most streamflow purposes. It must be realised, however, that this will be so only with a high standard of maintenance and adequate checking. The performance of all instruments deteriorates with age and water level recorders are no exception.

On each visit, the field operator checks that the recorder is functioning properly, the clock is at the correct time and the record shown on the chart is legible and accurate. The reference gauge(s) is/are read, the reading compared with that shown by the recorder and the information entered on a check sheet with the relevant times. The use of a portable stilling box will increase the accuracy of reading if surface waves are present.

On a chart recorder the gauge readings can be related exactly to the chart trace by raising and lowering the float wire to mark the chart with a vertical line. If the readings do not agree with an acceptable tolerance, further investigation is carried out until the source of the error has been found. This is done systematically by reference to the errors listed above, but particular attention should be paid to the intake pipe and stilling well to ensure that they are not obstructed and to the float to check that it is not stuck, damaged or has debris lodged on it.

Procedure for routine chart changing

After installation of the recorder a routine is introduced for changing charts or tapes, taking check readings of stage from the reference gauge and providing regular maintenance. The procedure adopted depends on the ability and experience of the personnel employed.

To ensure efficient and continuous operation of the recorder it is advisable to furnish each observer with printed instructions detailing a specific routine which has to be followed. The exact nature of the routine depends on the type of instrumentation but consists essentially of checking the condition of the recorder and its installation, or recording on the check sheet details of reference gauge readings, and of chart or tape changing. It is imperative that this is done on a regular basis so that a breakdown does not last long enough to cause a significant gap in the water level record. Where possible, additional visits are made between chart-changing periods to verify and record that the recorder is operating satisfactorily. The following routines are designed for specific types of analogue recorders but the same principles apply generally.

Analogue recorders

(a) Read the outside reference gauge, cleaning if necessary.
(b) Read the inside reference gauge if one is installed.
(c) Read the indicated level on the chart and compare with the above. If the readings do not agree find the cause and remedy it.
(d) Check that the clock is running and read the time indicated by the pen on the chart.
(e) Enter all readings of water level, recorder time and clock time on the check form and/or the chart. For this purpose the operator is provided with a reliable watch checked to local time.
(f) Mark the chart with a short vertical line by raising the float wire.
(g) Remove the stylus from the chart.
(h) Remove the chart drum from the recorder.
(i) Remove the chart from the drum by cutting cleanly with a sharp blade. Do not cut at the joint as it is essential to be able to examine the joint to determine any error.
(j) Place the new chart on the drum, making sure that it fits properly on the rim and that it matches at the joining edges.
(k) Rewind the clock.
(l) Check the stylus assembly to ensure that it is working properly and recharge with ink if necessary.
(m) Replace the chart drum on the recorder.
(n) Check the float and counterweight assembly and clean the float if necessary.
(o) Clean and oil the recorder mechanism according to manufacturer's instructions.

(p) Rotate the drum anticlockwise to eliminate backlash.

(q) Reset the stylus on the chart at the correct time and level.

(r) Enter the readings of water level and time on the new check form and/or on the new chart.

(s) Before leaving the station check that the instrument is working properly.

Comparison of inside and outside water levels

In most countries the outside reference gauge is considered as the base gauge and the water level recorder set and checked against this gauge. However, in some countries the practice is to use an inside gauge in the stilling well such as the float gauge, for checking and resetting the recorder. In this case the inside gauge is the reference gauge. A reference mark or reference point of known elevation with respect to gauge datum is often used as an inside gauge and the stage determined by measuring from the reference mark down to the water surface in the stilling well.

The reason for the practice of using the inside reference gauge rather than the outside reference gauge as the base gauge for the recorder is that the water level recorder gauge heights are used to determine discharge. If differences exist between inside and outside stages, these differences will be known only for those times when both gauges are read. If the outside gauge is used as the base gauge, corrections, known or assumed, require to be applied as necessary to recorded gauge heights to convert them to outside stages. If there is a significant difference between inside and outside gauge readings, the stage–discharge relation is first developed on the basis of outside gauge readings observed when discharge is measured. The relation is then adjusted to correspond with inside gauge readings observed at the times of discharge measurement.

Clearly the problem does not arise if both outside and inside gauge readings always correspond, which is the aim in good design practice of stilling wells and intake pipes.

Servomanometer or servobeam balance (bubble gauges)

The gas purge (bubbler) gauge offers advantages over the more commonly used float-operated recorders in certain circumstances. The instrument does not require a stilling well and can be located at some distance from the point of measurement. All pressure-actuated water level recorders operate on the principle that the water level is directly proportional to the pressure at a fixed point below the water surface. A gas purge system is used to transmit the pressure at a dip tube or orifice from beneath the surface to a point on the bank. The principle is shown diagrammatically in Fig. 3.12 where p_a is the atmospheric pressure, H is the difference in elevation between the instrument and the orifice

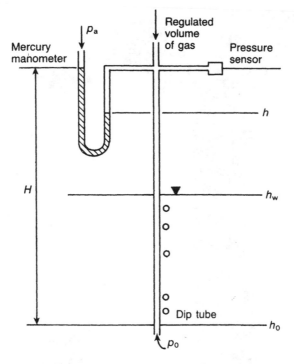

Figure 3.12 A simple gas purge system connected to a differential mercury manometer and to a pressure sensor. H, difference in elevation between instrument and orifice; h, height of pressure side of instrument; h_w, height of water column above orifice; h_o, elevation of orifice; p_o, pressure at orifice; p_a, atmospheric pressure at instrument.

and h is the height of the pressure side of the instrument. Now if ρ is the density of mercury and p_r is the recorded pressure, it can be shown that

$$p_r = \rho g(H - h) \tag{3.1}$$

where g is the acceleration due to gravity. Major units of the bubble gauge are shown in Fig. 3.13.

In the servomanometer-type gauge, reaction pressure is sensed by a manometer. Mercury is normally used as the manometer liquid in order to limit the physical size of the equipment. The manometer has a pressure sensitivity of 1/13.6 times that of water and can be used to sense ranges in gauge height in excess of 35 m. Figure 3.14 shows diagrammatically one form of manometer gauge in which changes of reaction pressure cause changes in the level of the mercury in the reservoir and hence in the monitoring column. The level of the mercury in the monitoring column is 1/13.6 times that of the change in the natural water level. The mercury surface in the column is followed by a needle/

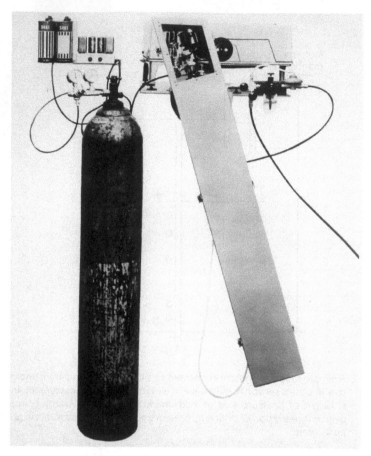

Figure 3.13 (a) Major units of bubble gauge interfaced to Leupold and Stevens Type A recorder.

point gauge, driven by a servo which in turn is coupled to a digitiser or chart recorder.

In another form, the needle/point gauge contact is fixed, and the mercury reservoir is positioned by a servosystem to maintain a null position at the mercury surface contact point. The servomotor is again coupled to a digitising unit or to the drive shaft of an analogue recorder or shaft encoder.

Several servomanometer-type bubbler gauge sensors, differing in detail from those described above, are commercially available.

Sensing orifice

The sensing orifice or dip tube is located in such a position that the head of water above it represents the stage to be measured. It is positioned where it is

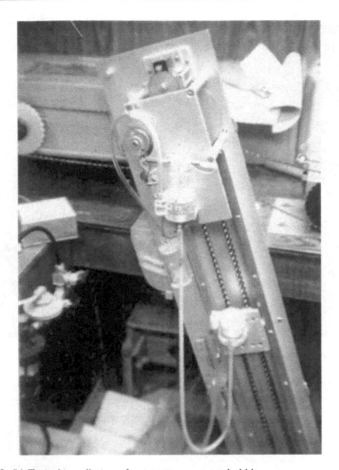

Figure 3.13 (b) Typical installation of a servomanometer bubble gauge.

unlikely to be silted over or where there is unlikely to be undue turbulence or strong current action. It is recommended that the orifice be orientated at right angles to the direction of flow. In unstable streambeds it may be advantageous to place the orifice in a vented well point driven into the bed.

The orifice is fixed in position but normally incorporates a facility for simple recovery for inspection and testing. When replaced it is seated carefully in its former position. The orifice is designed to minimise the effects of surface tension on the formation of bubbles at widely differing gas flow rates and under widely differing pressures.

From the source of the gas supply, the tube is given a continuous downslope to the measurement point. When the indicator or recorder is situated some distance from the point of measurement, the gas supply enters the system near the orifice. When two tubes are used, one for the gas supply, the other linked to the reaction pressure sensor, they are joined by a T-connector, from whose

Figure 3.13 (c) Ott Nimbus bubbler system with recorder chart.

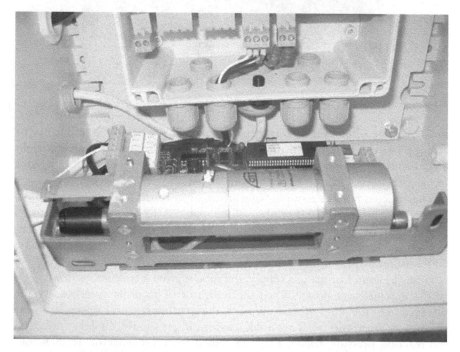

Figure 3.13 (d) Bubble gauge with Hydro-Logic Isodag Hawk XT logger.

Figure 3.13 (e) Bubble gauge bubble pipe of bubbler installed at a dam.

vertical leg a single tube extends to the orifice. The T-connector is located above the normal high-water elevation to prevent the entry of water into both legs of the system if pressure loss occurs. Additional valves are provided so that the static pressure tube can be separately purged, if necessary.

Gas supply

A supply of inert gas, usually nitrogen, or air is fed to the orifice at a pressure slightly greater than the maximum head to be measured. A continuous flow of gas to the orifice is necessary to prevent the liquid entering when the level is rising. At a constant rate of gas flow, the pressure in the system will rise at the same rate of increase as the head of water above the orifice.

The flow of gas from the orifice is adjusted so that on average one bubble per second is emitted. The rate of emission, however, is never less than one bubble

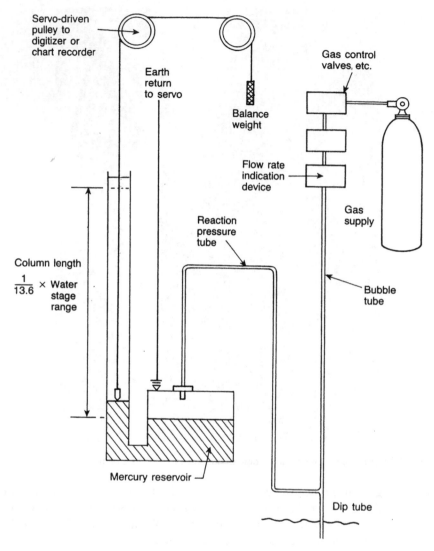

Figure 3.14 Diagrammatic illustration of the gas purge (bubbler) servomanometer water level gauge.

per 3 s. A faster rate may be needed when the level variation is rapid. When the gas flow is adjusted correctly, the recording gauge will fluctuate slightly each time a bubble leaves the orifice. This fluctuation is an indication that the instrument is operating satisfactorily.

Where air is used as the operating gas, it may be supplied using an air compressor. This arrangement is only practical if a mains source of electricity supply is readily available. Normally a cylinder of compressed nitrogen gas is used for supply purposes.

A log of the gas feed rate, gas consumption and gas cylinder replacement is maintained in the instrument house to ensure a continuous supply of gas and to help in checking for leakage in the system. There can be no serious leak in the gas purge system if the manometer operates to indicate stage correctly, and if the gas consumption, based on the average bubble rate over a period of time, corresponds with the gas consumption computed from the decrease in cylinder pressure. If a gas leak is evident, its location can be determined by isolating various parts of the gas purge system by the sequential closing of valves.

Effects of temperature changes

There are several ways in which temperature changes can affect the accuracy of data produced by pressure-actuated stage devices. Changes in the temperature of the water flow produce changes in water density. The temperature effect is small even for large changes (less than 0.5% for 30°C change) and as water temperature generally does not change quickly, the effect is virtually eliminated when the instrument is inspected and the correct stage noted.

Another temperature effect is that caused by temperature changes of the gas in the gas purge system. An instrument will tend to under-register more as gas temperature increases. This effect cannot be evaluated rigorously, as the gas temperature will change as the gas moves through the bubble tubing to the orifice.

The servomanometer is subject to a temperature effect caused by density changes in the mercury used in the system. This effect will cause the instrument reading to be incorrect and the servomanometer will over-register for an increase in temperature and under-register for a decrease. Generally this error is eliminated by observing the correct level at the time of visits to site.

Since the density of the water which the sensor is to measure will vary with temperature and also with chemical and silt contents, either an automatic or a manual means of compensating for these changes is provided.

Atmospheric pressure effects

The gas purge system is subject to the effects of changes in atmospheric pressure throughout the period of operation of the instrument. These changes are less than ±1%; the purge system is not sensitive to changes of this magnitude.

Variations in gas friction

Friction created by the flow of gas through the bubble tubing results in the pressure at the manometer being slightly higher than that at the orifice. The effect is aggravated by using long lengths of narrow-bore tubing. Inaccuracies due to variation in gas friction can be eliminated by using two gas tubes – one to feed gas to the bubble orifice, the other to act as a static pressure tube which transmits pressure from a point near the orifice back to the manometer.

As a conservative criterion for determining when the use of dual tubing is desirable, it is suggested that if an error no greater than 3 mm results from a 100% increase in bubble rate a single bubble tube will be satisfactory. During rapid rises in stage the instrument will lag if the bubble feed rate is too low. Manufacturers normally state the bubble rates required for a range of rates of increase of stage. Generally the rate of increase of stage that a given bubble rate will support increases directly with stage.

Effects of flow past the orifice

In general when the orifice points downstream the readings may display a small negative error of up to about 3% under a wide range of velocities in smooth and turbulent flows. When the orifice points upstream the readings may display a small positive error of up to about 4% in velocities of up to 3 m s^{-1} but above this velocity errors increase rapidly. The proper placement of the orifice is therefore essential for an accurate record of stage and it is preferable to set the orifice at right angles to the flow and better still to mount it flush with the wall of the mounting structure.

Effects of rapid changes in stage

The response of the gas purge system is primarily related to the depth of water over the orifice. As the depth increases, the effects of wave action tend to be increasingly damped. Furthermore, some instruments such as the servomanometer and some servobeam balance units have an electronic delay circuit that tends to eliminate the effects of wave action.

When the pressure at the orifice drops, the gas purge system responds almost instantly by releasing a burst of bubbles at the orifice. The water level recorded by the instrument will then fall to the correct level within the response time of the system. In the case of increasing pressure, water is drawn into the purge tubing and is expelled gradually as the gas pressure builds up in the system. The response of the instrument to a rise in stage may therefore be slower than its response to a fall in stage. Increasing the bubble rate will shorten response time.

The rate of response of the servomotor also characterises the response time of the instrument as a whole. For example, the servomanometer may move the mechanism quickly enough to follow a change in stage of about 20 mm s^{-1}. For some very flashy streams, however, it may be necessary to use a higher than normal power supply with a special servocontrol to ensure that the servomotor can keep pace with the change in stage. A typical bubble gauge linked to a strip chart autographic recorder is shown in Fig. 3.15 and an installation sketch is shown in Fig. 3.16.

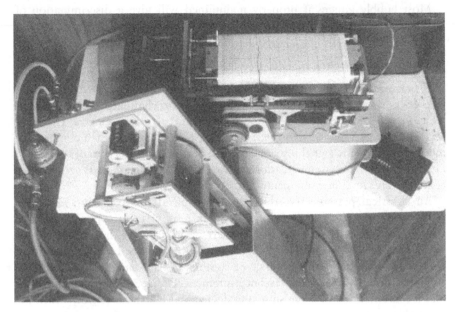

Figure 3.15 Overhead view of bubble gauge installation linked to strip chart recorder.

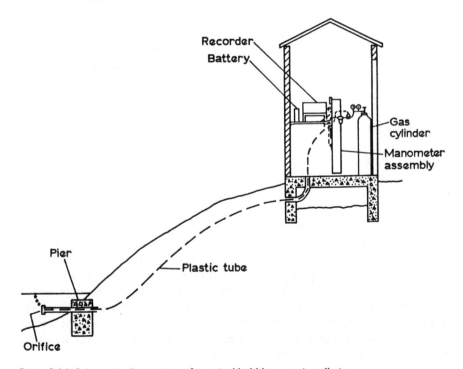

Figure 3.16 Schematic illustration of a typical bubble gauge installation.

Most bubble gauges if properly maintained will give a discrimination of ±10 mm or better.

Pressure transducers

The pressure transducer is a general term applied to devices which convert changes in water pressure and hence water level into changing electrical signals which can be recorded remotely from the point of measurement. There are several different types of transducer distinguished in the way each converts the mechanical pressure signal into an electrical output. Several types of transducer are described and recommendations made on the most suitable types for open channel water level measurement.

An electrical pressure transducer may be considered as having two main components: the force summing device which responds to the phenomenon to be measured, and the sensor which converts the output of the force summing device into the electrical signal. There are several designs for each component part which may be used in a range of combinations. Some systems are more suitable than others for water level measurement. Figure 3.17 shows some of the many combinations of force summing device and electrical sensor, which are preferable for water level sensing. Many of the transducers incorporate compensation for temperature-induced errors. Not all designs of pressure transducer are suitable for water pressure measurement.

The transducer may contain signal processing electronics which change the low-level sensor output into a form suitable for transmission over long distances. For short-distance transmission or high-output transducers, the sensor output is connected directly to the recorder. All current transducer design represents a compromise between transducer stability and electronic stability.

In all transducers, water pressure and hence level is measured with respect to one of three references; this reference gives the transducer classification:

(a) vented gauge – reference to atmospheric pressure
(b) sealed gauge – a fixed known pressure
(c) absolute – vacuum.

Since all open channel flows are subject to atmospheric pressure, the vented gauge type is the most suitable for application to water level measurement. If sealed gauge or absolute gauge types are used, atmospheric pressure compensation arrangements are required. The vented gauge type incorporates a vent pipe (generally integral with the electrical signal output cable) whose open end is secured in a position above the highest water level and capped to prevent the entry of any particulate matter or insects and other organisms. Precautions need to be taken to eliminate the occurrence of condensation in the vent tube and the reference chamber of the transducer.

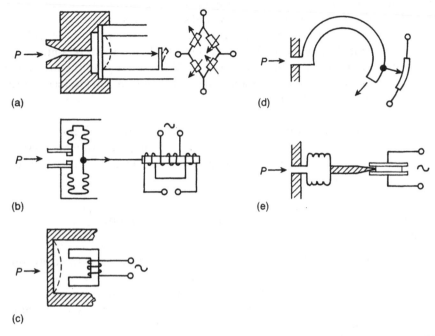

Figure 3.17 Possible electric pressure transducers for water level measurement. (a) Welded diaphragm operating: strain-gauged beam (shown); discrete gauged diaphragm; diffused semiconductor diaphragm; wire-wound element. (b) Capsule operating: variable differential transformer (shown); variable capacitor. (c) Integral diaphragm and: variable inductance sensor (shown); oscillating wire. (d) Bowdon tube and variable resistor. (e) Bellows and vibrating quartz crystal.

Transducer selection for water level measurement

The following are the criteria necessary in the choice of transducer:

(a) *Pressure range*. A transducer is selected whose permitted maximum over-pressure rating is in excess of the greatest head likely to be encountered at the measurement site.

(b) *Non-linearity*. The deviation of a calibration curve from a straight line; independent based or zero based.

(c) *Hysteresis*. The deviation between an ascending and a descending calibration curve.

(d) *Span error*. The deviation from a predetermined output change measured between zero and full-scale pressure.

(e) *Repeatability*. The difference between the outputs at identical pressures during successive pressure cycles.

(f) *Thermal zero and span shift*. The effects of zero shift and span shift due to temperature changes within the operating range.

(g) *Long-term stability*. The zero and span errors occurring over 1 year.

(h) *Thermal stability*. The errors of zero and span as a result of thermal cycling.

(i) *Suitability for measurement in water*. Care should be taken in transducer selection; its housing assembly and cable sheathing should be suitable for immersion in water to pressures at least 50% greater than the maximum pressures being measured. Suitable anchorage points require to be provided for cables liable to flexure.

The final choice of transducer for water level measurement will result from a best compromise of the criteria outlined above. Amongst the suitable pressure sensors in current manufacture, the highest performance is achieved by the resonating quartz transducer.

Sources of error in pressure transducer measurement of water level

The error response results from the effect of flow past the transducer pressure port and may be reduced to a minimum by aligning the transducer in such a way that the flow takes up a direction parallel to that of the pressure sensing surface. Alternatively, the transducer may be mounted in static conditions, for example in a stilling well.

The accurate functioning of the pressure transducer is subject to the effects of waves and other disturbances. These should be minimised using hydraulic or electronic techniques. Variations of water density can cause errors of depth measurement, for example in tidal estuaries or in heavily sediment-laden flows. Sea water of 35 parts per thousand salinity has a density 2.5% greater than fresh water. Water with a sediment content of the order of 1000 parts per million exhibits a density of 0.1% above that of freshwater. Parallel measurements of these variables are necessary to effect appropriate correction to the observed pressure values.

Ultrasonic water level gauge

In the ultrasonic water level instrument (Fig. 3.18), the sensor is located at either a fixed point above the water surface or a fixed point below the surface depending on design. In operation the instrument generates an acoustic pulse which is directed by the sensor at the water surface. The pulse is reflected from the surface back to the sensor. The time of travel from transmission to reception is electronically measured by the instrument. From a knowledge of the velocity of sound in air or water, as the case may be, the distance between sensor and reflecting water surface can be computed. The time measurement is converted electronically into a distance signal, which can be adapted to drive an analogue recorder, or coded into digital form for recording on magnetic tape or other form of digital 'memory' device.

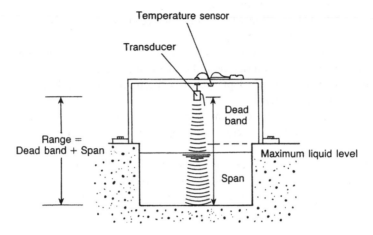

Figure 3.18 Schematic illustration of a typical ultrasonic water level gauge.

In order to obtain data of acceptable accuracy certain corrections and adjustments are necessary. The velocity of sound in air or water changes with density. Hence temperature, humidity and the presence of other gases all affect the accuracy of the method. Corrections can be readily made for temperature changes. The velocity of sound varies by approximately 1% for each 5°C change in temperature. Compensation for other variables is not easily effected.

In order to ensure effective reflecting characteristics, the beam width of the ultrasonic pulse is narrow, thus avoiding extraneous reflections from surrounding objects. Maximum echo strength is obtained when the reflecting water surface is flat and at right angles to the direction of travel of the pulse; if waves occur on the surface of the water during measurement it is likely that no echo pulse will be received by the sensor due to scattering. Similarly, if transmission takes place through air moving across the direction of travel of the pulse, it is possible that little or no echo signal will be received due to deflection of the beam. For a given wavelength of emission, the strength of the echo signal is not only dependent on the above conditions, but also on the reflective properties of the surface to be measured and the distance between sensor and surface.

The acoustic transmitter and receiver are normally encapsulated in a single unit. The initial and echo pulses are both routed through the one sensor which is linked by coaxial cable to the pulse generator and control system. Temperature compensation is also normally incorporated into the sensor but, because of the high thermal capacity of the sensor block, the instrument will not readily respond to rapid temperature changes.

The sensor range and span vary according to instrument but individual sensors may be tailored to specific ranges and spans.

The electronic control circuitry performs a number of functions:

(a) generation of the acoustic pulse;
(b) electronic measurement of the time between acoustic pulse transmission and reception of the echo signal;
(c) adjustment of the time value for changes in temperature as sensed by a temperature correction device in the ultrasonic sensor;
(d) adjustment system to adapt the instrument to site conditions of range, span and echo strength;
(e) conversion of the adjusted timing signal (measure of the distance from the sensor to the water surface) to the appropriate signal to drive a recording device;
(f) arrangement to switch the equipment to the quiescent state of operation between sampling, thus saving battery power;
(g) memory system to prevent erroneous recording in the event of the occurrence of a 'lost echo'.

The data recording device may be:

(a) a servo-operated chart recorder
(b) a magnetic tape recorder
(c) a solid-state memory recording device (Section 3.5).

A number of operational features are considered important:

(a) ability to calibrate the instrument manually under carefully arranged conditions, with all appropriate controls;
(b) ability to check the echo strength;
(c) ability to check the battery voltage under load conditions;
(d) ability to adjust the frequency of measurement.

The electronic control and recording systems may be located at some distance from the sensor. This may result in sensor and recording devices being operated under differing environmental conditions. With the control equipment and recorder connected by coaxial cable only, it is possible to effect a high level of sealing in the equipment thus obviating problems resulting from humidity and dust.

3.4 Optical shaft encoders

The linear movements of a conventional float or point gauge surface level sensor, which respond to variations of the water surface level, are converted into angular movements of a shaft and are monitored by a 'shaft angle transducer'.

The term 'shaft angle transducer' is a generic one; there are a variety of devices which give a variable electrical signal output in response to the rotation of a shaft. For reliability and high resolution the most suitable form of transducer is the optical shaft encoder. Optical shaft encoders are classified as 'incremental' or 'absolute' and their output can be displayed and recorded on electromechanical or electronic solid-state recorders.

The principal features of an incremental shaft encoder are shown in Fig. 3.19. The device comprises a glass disc mounted on a shaft, the rotation of which is to be monitored. A number of radiating lines etched on to the glass disc divide them into alternating opaque and transparent sectors. Divided in this way, they are referred to as 'gratings'. Light from a 'point' source illuminates an area of the grating and is received by an array of light-sensitive cells placed on the opposite side of the gratings. On rotation of the disc the cells are alternately exposed to the light and shielded from it. The output from the light-sensitive cells is a series of pulses whose resolution depends upon the number of lines comprising the gratings. Since incremental encoders produce only one pulse per unit of angular rotation, i.e. one grating interval, signal integration is necessary to provide a continuous measure of shaft position. Since a rise or fall in water surface level will result in forward or reverse rotation of the shaft, it is essential

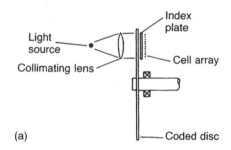

(a)

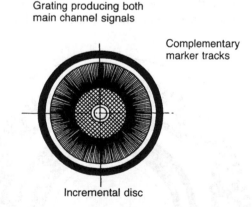

(b)

Figure 3.19 Schematic illustration of principal features of an incremental optical shaft encoder.

that the encoder has the capability of subtracting as well as adding incremental pulses. The disposition of the detector array and the design of the associated electronics can achieve this function. To present the output from an incremental encoder as a digital display or to record it in a form suitable for subsequent retrieval and examination, it is necessary to convert the encoder output into a recognisable code by a further stage of electronic translation.

The resolution of the incremental shaft encoder, or the smallest detectable rotation of the shaft, will depend upon the number of lines comprising the grating; this is commonly as high as 10 000 and may even exceed 30 000. Further enhancement of resolution can be achieved by electronic multiplication.

The absolute optical shaft encoder shares a similar principle to the incremental encoder. In this instrument the grating is replaced by a glass disc carrying a series of circular tracks which are divided into a regular pattern of opaque and transparent areas, as shown in Fig. 3.20. A small area of each track is illuminated. On the opposite side of the disc is a light-sensitive cell which monitors each track. As the coded disc rotates, the light-sensitive cells corresponding to the transparent areas of the track will transmit a signal, whilst the light-sensitive cells corresponding to the opaque areas will transmit no signal. The pattern of 'signal' and 'no signal' or 'on' and 'off', as defined by the disposition of opaque and transparent areas of the tracks, is a binary device which can be readily displayed or recorded.

To drive the shaft encoder a conventional float or other sensor may be employed. The sensitivity of the shaft encoder system is limited by the sensitivity of the primary sensor and incremental movements of ±1 mm in the tape or wire-drive can be detected and transmitted.

Data retrieval

Extracting data from recorder records is sometimes difficult and tedious. It is not easy to achieve the necessary accuracy. Automatic coordinate plotting equipment (digitisers) may sometimes be used to advantage for taking data from charts.

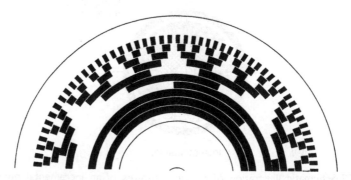

Figure 3.20 Schematic illustration of an absolute optical shaft encoder (half disc shown).

At gauging stations, solid-state loggers are now used in which data are recorded, stored, and subsequently replayed on command and the data transmitted over a telemetry link.

3.5 Solid-state recorders (loggers)

A new generation of recording devices is available capable of accepting and storing data within an entirely solid-state system. Such devices have no moving parts, and, as a result, are not subject to the shortcomings of mechanical or electromechanical devices. The advantages of solid-state storage include no wearing parts, unlimited life, physical robustness and portability, low power consumption and easy data retrieval. They represent significant advances in data collection and storage technology and contribute significantly to general scientific advancement.

In simplest form, the function of the solid-state recorder comprises:

(a) the receipt of an input signal produced by a suitable sensor;
(b) the conditioning of that signal to compatibility with the recording format;
(c) recording of time-control/parity information;
(d) recording of data in a solid-state memory;
(e) site identification data recording in the data memory; a recorder may be purpose-made for a single dedicated application.

In more sophisticated form the mode of operation of the recorder may be controlled by an integral central processing unit which executes the process entirely electronically. The mode of operation of the processor may be changed by suitable alteration of the instructions within the recorder circuit systems. Elements of the system contain:

(a) a unique timing control system, programmed to correct for month length and leap years;
(b) input signal processing which controls the input signal conditioning (signal identification, signal damping, signal values relating to stored calibrations);
(c) sampling strategy (basic control of the sampling frequency with the option of more frequent sampling at critical values);
(d) transfer of data to memory.

Input signals

Being a digital device, the solid-state recorder accepts suitable signals of a digital form. Analogue signals require appropriate conditioning through an analogue-to-digital converter with suitable encoding.

The 'logger' may need to store primary values of input signals but record only a single time-integrated value within the memory. Similarly it is necessary to damp out any rapid oscillations in input values if these are not representative of the true variations in water levels.

Solid-state recorders have a facility for inputting suitable test data in order to check the correct functioning of the equipment. Such test data are designed to produce a complete operational check of all systems of the recorder. It is also possible to check the state of internal power supplies and to input station identification data in order to label the data memory.

Data memory

The data memory may be integral with the instrument or may be detachable for ease of return to the processing centre. The memory store may be continuously powered or powered only during the actual write/store cycle. Special safeguards are available to protect data in a memory requiring continuous power support.

Some memory devices are completely non-volatile in nature and these will retain data for an indefinite period. In others data can only be retained for a limited period before degradation sets in. Increased data storage accuracy is obtained as the number of bits in a word is increased. The stored data word usually includes a sign bit and a parity bit for subsequent checking. The data memory is capable of accepting an unlimited number of read/erase cycles.

There are several alternative methods for returning stored data to the processing centre. The choice will usually be determined by economic considerations and site accessibility. A need for continuous uninterrupted operation will also influence the technique used. The following alternatives are available:

(a) the complete logger is returned to the processing centre for data extraction;
(b) a separate data storage unit is taken to the 'logger' and data transferred to a secondary store for subsequent return;
(c) the storage device (cartridge) itself is extracted from the 'logger' and returned, while it is replaced by an unused memory. (If the memory requires to be continuously powered, the storage device contains its own internal power supply.)

The potential flexibility of solid-state recorder systems enables a greater breadth of data to be gathered than previously. With appropriate system design the following information can be gathered and processed for storage:

(a) site location and serial number;
(b) actual data and time of record commencement (date corrected for month length and leap years; time corrected for specific time zones or energy-saving arrangements);
(c) data inputs received from the water level sensor to be processed and

recorded as representative water levels related to the datum of the recorder;

(d) date and time of each water level recording check signal at beginning and end of each record batch, which may include the total number of level recordings within the memory;

(e) in order to simplify a complete understanding of the collected data it may be advantageous to record site observations of air and water temperatures.

Typical solid-state recording systems for stage measurement at a streamflow station are shown in Fig. 3.21. In Fig. 3.22 solid-state recorders (loggers) are shown which are normally used at sewage treatment and water works. These are usually programmed to give digital displays of head, flow and total flow. Streamflow recorders are normally battery operated whereas in sewage treatment and water works 'loggers' are mains operated.

Figure 3.21 (a) Yangtze River at Hueng Ling Miao, China showing logger.

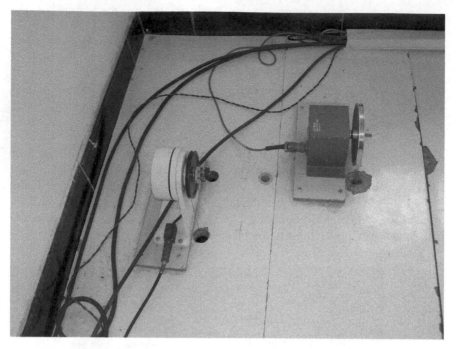

Figure 3.21 (b) Yangtze River at Hueng Ling Miao, China showing shaft encoders.

3.6 Stilling wells and intakes

The function of a stilling well is to accommodate the water level recorder and protect the float system, to provide within the well an accurate representation of the mean water level in the river and to damp out natural oscillations of the water surface. The function of the intakes is to allow water to enter or leave the stilling well so that the water in the well is maintained at the same elevation as that in the stream under all conditions of flow, and to permit some form of control with which to limit lag and oscillating effects within the well.

The well itself can be constructed from any suitable building material such as concrete blocks, bricks or stone, or fabricated from sections of concrete, steel, fibreglass, galvanised iron or similar type of pipe. It can be set into the bank of the stream or directly in the stream when attached to a rigid support such as a bridge pier or abutment. When placed in the bank it should have a sealed bottom to prevent seepage into or leakage out of the chamber and should be connected to the stream by one or more intake pipes; when placed directly in the stream the intakes may take the form of holes or slots cut in the well itself. It is essential that the well remains stable at all times and it must therefore be firmly founded when placed in the bank and firmly anchored when standing in the stream (see Fig. 3.23). Its dimensions should allow unrestricted operation of

Figure 3.21 (c) Han River at Xian Tao, Wuhan, China showing logger.

all equipment installed in it and it is recommended that where a single float is used within the well the clearance between walls and float should be at least 75 mm, and where two or more floats are used clearance between them should be at least 150 mm. In silt-laden rivers it is an advantage to have the well large enough to be entered and cleaned.

The well and all construction joints of well and intake pipes are made water-tight so that water can enter or leave only by the intake itself. The well should be vertical within acceptable limits and have sufficient height and depth to allow the float to travel freely the full range of water levels. The bottom of the well should be at least 300 mm below the invert of the lowest intake pipe to provide space for sediment storage and to avoid the danger of the float grounding at times of low flow. The well itself should not interfere with the flow pattern in the approach channel and if set behind a control should be located far enough upstream to be outside the area of draw-down to the control.

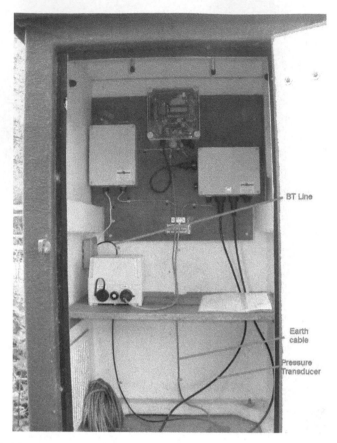

Figure 3.21 (d) Typical logger (Environment Agency, Thames UK).

To ensure continued operation of the system if the lowest intake pipe becomes blocked it is often advisable to install two or more intakes, one vertically above the other. The lowest intake is located at least 150 mm below the lowest anticipated stage. In cold climates this intake should be below the frost line.

Stations installed on streams carrying a significant amount of fine sediment (silt and clay) should be provided with a means of cleaning the stilling well and intake pipes. The following methods are in general use:

(a) Flushing devices whereby water under considerable head can be applied to the well end of the intake pipes. Ordinarily, the water is raised from the well to an elevated tank by use of a hand-pump. The water is then released through the intakes by the operation of valves. Water may also be raised from the stream by use of a length of hose connected to the pump or carried in buckets if a pump is not installed.

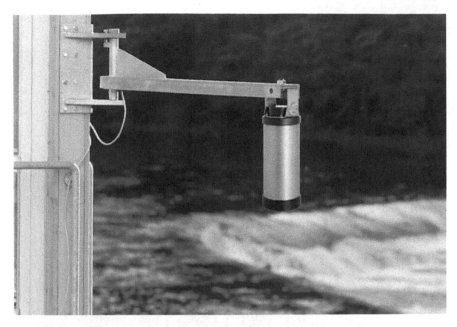

Figure 3.21 (e) Ott Kalesto radar water level gauge.

(b) Pumping water through the intake pipes.
(c) Building up a head of water and stirring up deposited silt in the well with a stationary hand-pump or a small engine-driven portable pump to force an obstruction out of the intakes.
(d) Hand cleaning of well and intake pipes by the use of shovel and bucket, and flexible steel rods.

Figure 3.24 shows a recommended design for a stilling well and Figs 3.25–3.31 show gauging stations with different types of stilling well. Stilling wells will often fill with sediment, especially those located in arid or semi-arid regions. A well placed on a stream carrying heavy loads of sediment must be cleaned out often. In such cases, a sediment trap can greatly reduce the work of removing the sediment. A sediment trap consists of a large box-like structure located between the intake and the stilling well. Inside it, baffles are fitted to promote settling before the sediment reaches the stilling well. The trap is made to open for easy access and removal of the deposited material.

To reduce and eliminate the silting problem of stilling wells, the vertically hung pipe well has proved practical (Fig 3.28 and 3.29). This is a 40 cm diameter steel pipe provided with a so-called hopper bottom that will, in general, keep itself clear of silt. Figure 3.32 shows a hopper bottom of a 40 cm diameter steel pipe well with a 25–50 mm inlet hole at the point of the hopper. If such a well is placed in flowing water, the increased velocity under the point of the hopper

Figure 3.22 Typical solid-state loggers used at sewage treatment works and water works to record head and flow.

Figure 3.23 Stilling well and instrument house overturned by flood (Queensland, Australia).

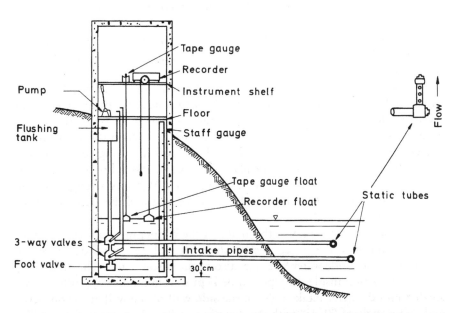

Figure 3.24 Typical design of stilling well installation in river bank with two intakes and flushing valves.

Figure 3.25 Streamflow station (UK) showing instrument house over stilling well, ramp
gauge with access steps and cableway. Note upper intake pipe installed as a
precaution should lower intake become choked with sediment or debris.

prevents the inlet from being choked from the outside, and the steep slope of the
bottom combined with small oscillations and fluctuations of the water surface
will not permit the deposition of silt inside.

Sometimes not even a hopper bottom can keep the well clear of silt. In these
cases, a self-seating cone with an inlet hole at the apex may be provided where
the cone closes a larger hole in the hopper bottom. In order to remove the
accumulated silt, the cone is pulled up by means of a chain (brass or bronze)
and churned up and down until the silt has been worked through the large hole.
When clear, the cone is reset in its place (Fig. 3.33). Instead of a cone, a light
weight may be provided as illustrated in Fig. 3.34. With this arrangement, the
chain is worked up and down through the intake hole and the accumulated silt
cleared away.

The velocity of the stream at the intake of the hopper must not exceed about
1 m s^{-1}, otherwise the water level in the well will be drawn down. Superelevation of the inside water level may also occur at some sites. The draw-down
or super-elevation inside the pipe can be kept to a minimum by drilling a few
small 8 mm diameter intake holes in the side wall of the well at various levels
and at an angle of 30–45° with the direction of flow. In general, for a stilling
well mounted on a pier or wall, the angle should be 38–40°. However, the

Figure 3.26 Streamflow station (Indonesia) showing instrument house of plastered brick, stilling well of heavy concrete pipe, bank protection of masonry and ramp gauge with access steps.

self-cleaning effect of the hopper bottom will be reduced when the side wall of the well is punctured and it is no longer watertight.

Intake pipes should be laid at 90° to the direction of flow and preferably at a slight slope to prevent air pockets from forming inside the pipe. A valve may be fitted at the well end to control surge in the well.

If the velocity past the end of the intake is high, draw-down of the water level in the stilling well may occur. In order to reduce this draw-down, static tubes are often placed on the stream end of the intake pipes. The static tube consists of a short length of pipe, 0.5 m long, attached to a 90° elbow on the end of the intake pipe and extending downstream in the same horizontal plane as the intake. The end of the tube is capped and water enters or leaves through holes drilled in its side-wall (Fig. 3.24). Draw-down is dealt with in Chapter 5.

Figure 3.27 Streamflow station Han River Xian Tao, Wuhan, China, instrument house and hydrometric conference centre with satellite and radio communication.

A well gauge is normally installed inside the stilling well to provide a check if there is a free flow of water between the stream and the well. The datum of this inside gauge is the same as for the outside reference gauge in the stream.

In cold climates the well is protected from the formation of ice which could restrict or prevent the free operation of the float system. This is usually done by the use of well covers, subfloors to act as a frost barrier, heaters or oil on the water surface. When oil is used the oil surface will stand higher than the water level in the stream and a correction is applied when setting the recorder. When discharge in the river increases, the water level in the channel rises and water flows from the channel into the stilling well. If this flow is appreciable, head losses in the connecting pipe including any control fittings used (valves, outlet tees, static tubes, etc.) can cause a difference between water level in the river and that in the well. This is due to stilling well lag and for any given combination of

Figure 3.28 Streamflow station (USA) with hung stilling well of corrugated steel pipe fixed to bridge pier with hoop brackets.

well size, intake size and control fittings the magnitude will be a function of the rate of change of stage. For a given rate of change of stage the amount of lag can be determined by this relationship. Lag is dealt with in Chapter 5.

To eliminate surging effectively it may be necessary to restrict the cross-sectional area of the intakes to 0.1% of the cross-sectional area of the well, whereas to reduce lag effect to acceptable limits the ratio may have to be at least

Figure 3.29 Streamflow station (Indonesia) with hung stilling well of steel pipe from bridge with 'look-in' instrument shelter and access ladders to vertical staff gauge.

1%. This will depend on site conditions, the type and length of intakes, and the surface area of the well. Because of this, no firm rule can be laid down for determining the best size of intake but it is advisable to make the connection too large rather than too small, as a restriction may be added if found necessary. As a general guide, the total cross-sectional area of the intakes should not be less than 1% of the cross-sectional area of the well.

3.7 Instrument house

Some suitable form of housing is essential to protect the recording and measuring equipment from the elements and from unauthorised attention or vandalism, and also to give shelter to the servicing and gauging staff who operate the station. The size and quality of the housing adopted will depend on conditions at the site and on the type of equipment to be installed in it, but in general the minimum required is a well-ventilated, weatherproof and lockfast hut set on a stable foundation and of such dimensions as will permit normal servicing of the

Figure 3.30 Streamflow station Yangtze River at Hueng Ling Miao, China showing instrument house and satellite and radio antennas.

recorders and, where relevant, the operation of the cableway traversing gear. Generally dimensions of 2 m × 2 m × 2 m are a minimum requirement.

There are many locations, however, where timber huts over a concrete well are adequate and may indeed have certain advantages – for example, in times of hard frost – but various forms of construction such as brick, stone or reinforced concrete are used to suit local conditions. There should be provision for heating the hut in winter conditions, for excluding vermin and insects, and for restricting the entry of water vapour from the well. Some typical instrument houses are shown in Figs 3.26–3.31.

The recorder should normally be mounted on a rigid table firmly fixed to the foundations of the hut but independent of the framework of the building. It should be placed so that the float and counterweight have ample clearance from the sides of the stilling well walls throughout their full range of travel, and in a

Figure 3.31 Streamflow station Qing River at Gao Ba Zhon, China showing instrument
house, satellite and radio communication antennas.

position which allows convenient access for the service staff for chart or tape
changing and general maintenance. Where a cover is placed over the well, the
holes or slots for the various cables and tapes are located accurately to eliminate
any risk of rubbing or fraying.

In humid climates it is advisable to have instrument houses well ventilated
and to have a close-fitting floor to prevent the entry of water vapour from the
well. Screening is used over vent holes and other openings in the instrument
house to prevent the entry of insects, birds, rodents and reptiles.

3.8 Telemetering systems

Telemetering systems are used when current information on stage (or discharge)
is needed at frequent intervals or it is impractical to visit the gauging station
each time the data are needed. Current or real-time data information is usually
necessary for reservoir operation, flood forecasting, prediction of flows, and for
current data reporting. However, it is becoming apparent in some countries that
telemetering the recorded data direct to the central office computer is a cost-
effective alternative to manually collecting the chart or solid-state module on a
regular basis (Fig 3.21).

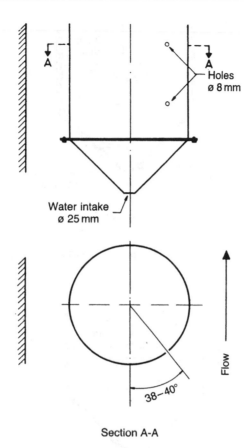

Section A-A

Figure 3.32 Hopper bottom for pipe stilling well.

It is not intended to provide in this section a detailed description of the various telemetry systems available but a short introduction is appropriate.

A schematic of a telemetry system for streamflow showing the main divisions into sensors, logger, modem, communication link and a PC at the base station is shown in Fig. 3.35.

Systems of data transmission

System 1

An observer at the gauging station mails the data or initiates a radio or telephone call to the central office based on pre-arranged criteria and supplies instantaneous readings only.

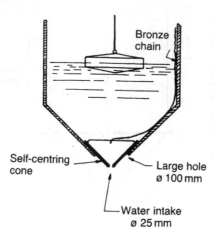

Figure 3.33 Section of hopper bottom for pipe stilling well with cone and chain for cleaning out sediment deposits.

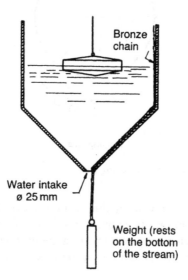

Figure 3.34 Section of hopper bottom for pipe stilling well with weight and chain for cleaning out sediment deposits.

System 2

Central office interrogates the remote automatic station by telephone, radio or radio telephone and receives single discrete values as often as interrogated. It is possible to have an automatic dialling device in the central office which interrogates and records responses at regular intervals.

System 3

Automatic equipment installed at the gauging station is programmed to initiate a telephone, satellite or radio call to supply selected instantaneous observations.

System 4

An impulse is transmitted automatically by telephone or radio for a specified unit of change of a parameter (for example each 10 mm change in stage).

System 5

An impulse is transmitted automatically by telephone or radio at predetermined intervals of time. (Systems 4 and 5 can be combined and unit changes in the measured parameter stored and transmitted at predetermined intervals of time.)

System 6

Data are transmitted and recorded digitally on a continuous basis.

System 7

Data are transmitted on a continuous basis over radio or telephone by equipment that converts observations to a continuous tone or frequency. This information is reconverted at the central office (frequency modulation).

System 8

Remote field sites equipped with satellite sensors, encoder and radio transceiver transmit coded information to the central office where the information is decoded and fed to the computer that initiated the sequence calls.

Table 3.1 outlines the advantages and disadvantages of each system. Systems 1–7 refer specifically to terrestrial methods (radio and land line) in common use. Included in System 8, however, is the facility of data transmission by satellite and today there are many hundreds of gauging stations using this method of telemetry.

Selection of system

Considerations to be given in selecting an automatic data transmission system are as follows:

(a) Where a station is located in an area with difficult access it is preferable to have automatic water-level recording and telemetering of the data.
(b) In many cases because of rigorous local climatic conditions the operation of on-site mechanical recorders may be difficult and it is more reliable to transmit the data electronically to a central office.
(c) Consideration is given to the time between observations and receipt of the

Table 3.1 Summary of telemetering systems

System[a]	Advantages	Disadvantages	Remarks
1	Simple sensors used and malfunction known immediately. Inexpensive as communications paid for only when used. If necessary can be interrogated from central office.	Continuous record of events not available. Central office telephone or radio link may be overloaded at certain critical periods.	Telephone calls can be received and recorded on answering service equipment and transcribed later.
2	Instantaneous information can be obtained as required even in isolated areas.	No continuous record. Difficult to determine malfunction of water level recorder. In addition to radio or telephone, equipment at outstation requires a device to answer calls for information automatically.	Method of answering may be voice recording, short tones for hundredths, tenths, units, etc. In some instruments a memory can be included so that upon interrogation continuous or extreme data can be received for a predetermined period prior to the call. When using FM radio the cost of the radio and antenna is normally the main portion of total cost.
3	Immediate alert of extraordinary hydrometric conditions.	No continuous record. Difficult to determine malfunction of water level recorder. Special device necessary to initiate calls based on predetermined criteria.	As this is an alert-oriented unit, it is usually on a private line to a receiver who is always on-call.
4	Complete record of change of events as they occur.	Malfunction of water level recorder can be detected only after a certain interval of time. For example, due to malfunction no change may be detected in parameter and therefore no signal transmitted. Reliable power supply important.	Information received usually on an autographic recorder.
5	A total record is obtained.	Reliable power supply very important.	
6	A total record is obtained immediately.	Distance from water level recorder to user is limited to a few thousand metres.	

7	A total record is obtained immediately.		Systems using telephone line can cover any distance using signal repeaters (amplifiers) along the way. FM radio usually applies to line-of-sight distance. AM radio can be used over greater distances but is more subject to atmospheric conditions.
8	System is completely automatic.	Expensive and complex in service	FM radio links or satellites can be used for transmission or retransmission. Systems usually include microprocessor which is programmed for data processing.

[a] See text for details.

data for management purposes; time required to process and analyse the data; speed with which changes in water level or flow take place; accrued benefits of forecasts from telemetered data; cost due to lack of or delay in receiving information.

(d) When designing a system the three main components to develop are:

    (i) measuring and encoding equipment
    (ii) the transmission link
    (iii) receiving, decoding and analysing.

(e) The type of transmission link used is determined by the frequency band requirements and the economics. Availability locally of any one of the alternative choices is a constraint. Possible choices for transmission links include dedicated land lines, commercial telephone or direct radio links and satellite links.

Dedicated land lines

These are perhaps the simplest to install when relatively short distances are involved and no commercial lines already exist. Land lines are typically able to transmit frequencies of up to 300 Hz without special techniques. Time division and frequency multiplexing can be used to provide more economic use of the transmission line. (Multiplexing is the process of using one transmission line to transmit several measurements.)

Commercial telephone lines

When the distances involved are long and existing telephone systems are available, use can be made of these. Equipment is available to enable the instrument to simulate the behaviour of a relatively normal subscriber to the telephone service. Measurements and commands can be transmitted to and from the remote site.

Commercial teletype lines

In this system, data are transmitted via teletype lines to the receiving equipment in the central office.

Radio links

Radio links are used when the number of calls required exceeds those economically practical over land lines or when distances or terrain prevent the economic installation of wires. Distances of up to hundreds of kilometres may be spanned by radio transmitters, depending on the techniques used. At the higher frequencies, the transmitter and receiver require a clear line-of-sight transmission path. This limits the range unless repeater stations are installed.

Satellite links

The weather satellites GOES (East and West), GMS, METEOSAT and GOMS form a system of geostationary satellites contributing to the World Weather Watch and Global Atmospheric Research Programme. These satellites afford a virtually complete imagery coverage of the earth's surface. They are placed in an orbit coincident with the earth's equatorial plane at a height of about 36 000 km and since they revolve around the earth at the same speed as the earth rotates, they appear to be stationary. In the imagery mode they have a constant view of almost a full hemisphere. They also have a communications facility which may be used for telemetering purposes. The data are transmitted to the satellite by a small radio or data collection platform (DCP) (Fig. 3.37).

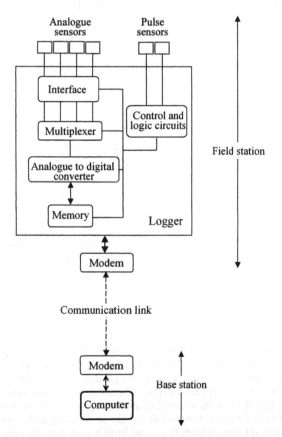

Figure 3.35 Schematic of a telemetry system for streamflow showing the main divisions into sensors, logger, modem, communication link and PC at the base station.

A DCP is an electronic device, containing a small UHF transmitter, for collecting, storing and transmitting observations and measurements of the physical, chemical and biological properties of the oceans, rivers, lakes, solid earth and atmosphere (Figs. 3.36 and 3.37). Data from the sensor are encoded through an interface and transmitted in the 402 MHz band to the satellite. In the European METEOSAT system these data are relayed from the satellite to the European Space Agency Operations Centre in Germany and retransmitted to the user's receiving station. A similar facility is available through the (two) GOES satellites, but not through the Japanese GMS. The data can also be received by telex through the global telecommunications system. The receiving station consists of a low-cost receiver unit with a 1.5 m antenna (Fig. 3.38). An alert facility also available on the DCP is triggered at any predetermined threshold level.

The microprocessor in the DCP offers the facility of preprocessing before transmission. For example, in some DCPs, stage is converted to discharge by means of the stage–discharge relation programmed into the processor.

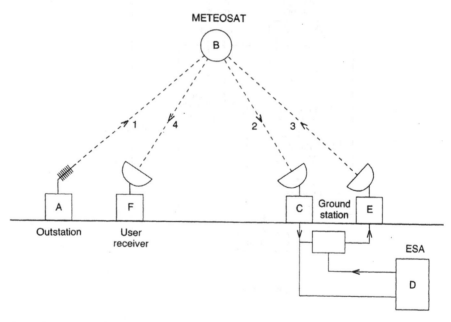

Figure 3.36 The METEOSAT DCP telemetry system. The outstation (A) transmits its measurement data to METEOSAT (B) along path 1 at set time intervals. The satellite immediately retransmits these data to the ESA ground station in Germany (C) along path 2. From here the data are sent by land line to Darmstadt (D), 40 km away, where they are quality controlled, archived and returned to the satellite from a second dish antenna (E) along path 3 for retransmission to users (F) along path 4. The signal level is such that the user can receive the data by a 1.5 m dish antenna. Other satellite systems are used for telemetry communication including those in USA, China, India and Japan.

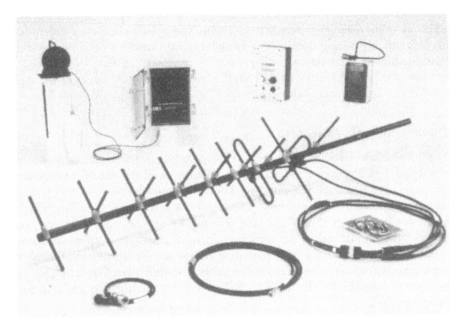

Figure 3.37 Basic parts of a DCP system for stage measurement showing pulley, float and counterweight, optical shaft encoder and DCP. The antenna and fixings are shown in the foreground.

Figure 3.38 Satellite receiving dish antenna (1.5 m diameter) for METEOSAT. The antenna receives streamflow data from the satellite within 4 min. of transmission and is on-line to a central computer.

Satellite systems have many advantages over terrestrial systems, not least of which is their reliability. Terrestrial systems often fail when they are most needed, notably during floods. The availability of a satellite on a permanent basis is, however, crucial to an operational telemetry system. Nevertheless transmission of hydrometric data by satellite now offers a cost-effective alternative to existing systems of telemetry.

Quality control of site data

The installation of a telemetry system does not reduce the need for strict on-site quality control at the gauging station nor does it avoid the need for site visits although it may reduce the number. The early fault-finding capability of a telemetry system has obvious advantages. A fault at a gauging station, which might not be discovered until the next scheduled visit, can, with telemetry, be attended to immediately. The introduction of a telemetry system does not necessarily mean that the on-site recording system may be withdrawn. Indeed it is an advantage to maintain a record at a gauging station using a data 'logger' as back-up in case of telemetry failures. Monthly visits may then be adequate for most purposes.

Further reading

Herschy, R. W. Hydrological data collection by satellite. *Proc. Instn Civ. Engrs Part 1*, 68, 759–71 1980.

Herschy, R. W. Towards a satellite-based hydrometric data collection system. *IAHS Symposium, Exeter* Publication No. 134 1982.

Herschy, R. W. (ed.) *New Technology in Hydrometry*. Adam Hilger, Bristol 1986.

Herschy, R.W. *Hydrometry: Principles and Practices, (ed.) Second Edition*, John Wiley and Sons, Chichester 1998.

ISO 4366 *Liquid Flow Measurement in Open Channels: Echo Sounders for Water Depth Measurements*. ISO, Geneva, Switzerland 1979.

ISO 4373 *Liquid Flow Measurement in Open Channels: Water Level Measuring Devices*. ISO, Geneva, Switzerland 1995.

ISO 6419/1 *Liquid Flow Measurement in Open Channels: Hydrometric Data Transmission Systems – General* ISO, Geneva, Switzerland 1992.

Lambie, J. C. Measurement of flow–velocity–area methods. In *Hydrometry: Principles and Practices* (ed. R. W. Herschy). John Wiley and Sons, Chichester 1978.

Shope, W. G. and Paulson, R. W. *Real time data collection via satellite for water management*. ASCE Convention, Exposition, Florida, October 1980.

Strangeways, I. C. *The Telemetry of Hydrological Data by Satellite. Institute of Hydrology, UK*, Report No. 112 1990.

Strangeways, I. C. Satellite transmission of water resources data. *Proceedings of the WMO Regional Workshop on Advances in Water Quality Monitoring*. Vienna 1994.

Strangeways, I.C. Transmission of Hydrometric Data by Satellite. In Hydrometry: Principles and Practices Second Edition (ed. R. W. Herschy). John Wiley and Sons, Chichester 1999.

Strangeways, I.C. *Measuring the Natural Environment, Second Edition*, Cambridge University Press 2003.
World Meteorological Organization *Manual on Stream Gauging*. WMO Report No. 13, Pub. No. 519 1980.

Chapter 4

The stage–discharge relation

4.1 General

Both the stage and the discharge of a stream vary most of the time. In order to obtain a continuous record of discharge the stage is recorded and the discharge computed from a correlation of stage and discharge. This correlation, or calibration, is known as the stage–discharge relation.

The operations necessary to develop the stage–discharge relation at a gauging station include making a sufficient number of discharge measurements and developing a rating curve by plotting the measured discharges against the corresponding stages and drawing a smooth curve of the relation between the two quantities. Discharge measurements are carried out over the range of stage variation in order to establish the rating curve as quickly as possible. Normally the lower and medium stages present little difficulty but discharges at the higher stages may take some time and resort may require to be made to careful extrapolation until such time as the higher discharges are available to be measured.

If the channel is stable, comparatively few measurements may be required although very few rivers have completely stable characteristics. The calibration therefore cannot be carried out once and for all, but has to be repeated as frequently as required by the rate of change in the stage–discharge relation.

It is therefore the stability of the stage–discharge relation that governs the number of discharge measurements that are necessary to define the relation at any time. In order to define the stage–discharge relation in sand-bed channels, for example, several discharge measurements a month may be required because of random shifts in the stream geometry.

Sound hydrometric practice requires that the discharge curve is determined as rapidly as possible after the establishment of a new station. Unless this is done, and the curve maintained, the record of stage cannot be converted into a reliable record of discharge.

4.2 The station control

An analysis of the stage–discharge relation and the construction of the rating curve require an appreciation of the functioning of the channel control.

In order to have a permanent and stable stage–discharge relation the stream channel at the gauging station must be capable of stabilising and regulating the flow past the station so that for a given stage the discharge through the measuring section will always be the same. The shape, reliability and stability of the stage–discharge relation are normally controlled by a section or reach of channel at, or downstream from, the gauging station and known as the **station control**. The geometry of the station control eliminates the effects of all other downstream features on the discharge at the measuring section. The channel characteristics forming the control include the cross-sectional area and shape of the stream channel, the channel sinuosity (meanders and loops), the expansions and restrictions of the channel, the stability and roughness of the streambed and banks, and the vegetation cover, all of which collectively constitute the factors determining the channel conveyance (see Chapter 8).

In terms of open channel hydraulics, a station control is a critical depth control, generally termed a **section control** (Fig. 4.1) if a critical flow section exists a short distance downstream from the gauging station, or a **channel control** (Fig. 4.2) if the stage–discharge relation depends mainly on channel irregularities and channel friction over a reach downstream from the station. A control is **permanent** if the stage–discharge relation it defines does not change with time, otherwise it is impermanent and generally called a **shifting control**.

Figure 4.1 Section control (Norway). Low-water part of control is sensitive while high-water part is non-sensitive.

Figure 4.2 Channel control (Indonesia), moderately stable, probably shifting during and
after major floods.

From the standpoint or origin, a control is either artificial or natural, depending
on whether it is man-made or not.

Natural controls vary widely in geometry and stability. Some consist of a
single topographical feature such as a rock ledge across the channel at the crest
of a rapid or waterfall so forming a **complete control** (Fig. 4.3) independent of
all downstream conditions at all stages of flow. Some controls are formed by a
combination of two or more features, such as a rock ledge combined with a
channel constriction. Some are V-shaped and thus sensitive to changes in discharge; some are U-shaped and less sensitive to changes in discharge. Some
consist of two or more interacting controls each effective over a particular range
of stage and termed **compound controls**. A common situation is that a section
control is effective at low flow only and is submerged by a channel control at the
higher discharges. Some controls consist of a long reach of stable bed extending
downstream as the stage increases. In general, the distance covered by such a
control varies inversely with the slope of the stream and increases as the stage
rises. The tendency for a control to extend downstream as the stage rises has a
significant effect on the stage–discharge relation. As the stage increases, low
water and medium water controlling elements are drowned out and new downstream elements are successively introduced into the station control causing a
straightening out of the typical parabolic curvature of the rating curve and
at times even causing a reversal of this curvature. For rivers with very flat
slopes the station control may extend so far downstream that backwater

Figure 4.3 Complete section control (Indonesia) with upstream pool stable and sensitive.

complications, which do not exist at lower stages, are introduced at higher stages. The simplest and most satisfactory control is the complete control mentioned above. First, it ensures permanence; second, it creates a pool or forebay in which a gauging station is often easily installed; third, favourable conditions for carrying out current meter measurements may frequently be found within the reach of such a pool; and, fourth, the point of zero flow is conveniently located and surveyed in this situation. Whenever practical, this type of control is utilised for a stream gauging station. Most natural controls, however, are liable to shift, albeit even slightly. A shifting control exists where the stage–discharge relation changes frequently, either gradually or abruptly, because of changes in the physical features that form the station control. The controlling features may be modified by a number of factors, of which the principal ones are:

(a) scour and fill in an unstable channel;
(b) growth and decay of aquatic (weed) growth;
(c) formation of ice cover;
(d) variable backwater in a uniform channel;
(e) variable backwater submerging a control section;
(f) rapidly changing discharge;
(g) overbank flow and ponding in areas adjoining the stream channel.

The corresponding stage–discharge curves for the above hydraulic conditions are given in Fig. 4.4 and a brief discussion of each follows (see also Chapter 5).

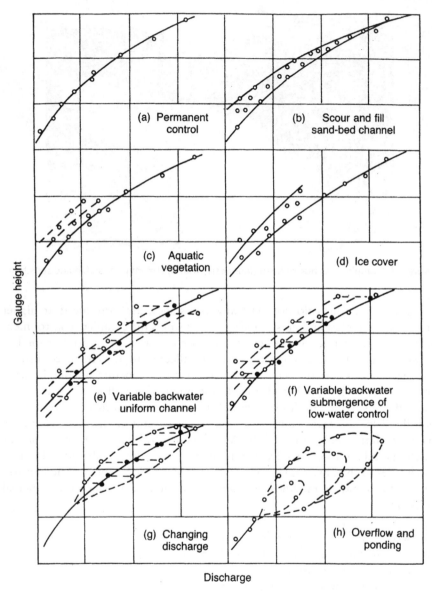

Figure 4.4 Rating curves for different hydraulic conditions (see also Fig. 4.5).

Permanent control (Fig. 4.4(a))

If the control is a permanent one, occasional discharge measurements need to be made for verification. The stage–discharge relation for a permanent control can be expressed as a simple exponential function.

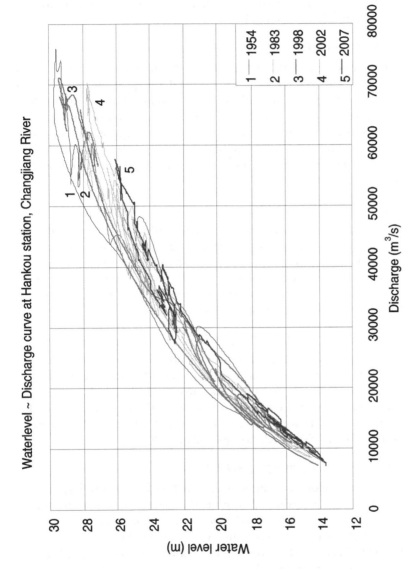

Figure 4.5 Complex stage–discharge relation: Yangtze River at Wuhan. A selection of complex stage–discharge curves for the years 1954, 1983, 1998, 2002 and 2007; 1954 was the largest Yangtze flood (75,000 cubic metres per second), the second largest was in 1998 (71,100 cubic metres per second), the 2002 curve was immediately before the affect of the upstream Three Gorges Dam (70,300 cubic metres per second) and the 2007 curve was the normal curve after the completion of the Three Gorges Dam (57,700 cubic metres per second).

Sand-bed channel (Fig. 4.4(b))

The movement of fluvial sediment, particularly in channels in alluvium, affects the conveyance, the hydraulic roughness, the channel sinuosity and the energy slope. This makes the determination of a stage–discharge relation difficult. In addition, since the movement is erratic, determination of the temporal variation of the stage–discharge relation is complex.

Aquatic vegetation (weeds) (Fig. 4.4(c))

The growth of weeds decreases the conveyance of the channel and changes the roughness with the result that the stage for a given discharge is increased. The converse is true when the weeds die and the stage–discharge relation gradually returns to its original condition. The change in weed growth is closely observed over the growing season and determined by a series of discharge measurements. Normally the development of a family of stage–discharge curves for different conditions of weed growth presents the best means of gauging rivers with prolific weed growth.

Ice cover (Fig. 4.4(d))

Ice in the measuring section increases the hydraulic radius and the roughness and decreases the cross-sectional area. As with weed growth, the stage for a given discharge is increased. The effect of ice formation and thawing is complex and the temporal stage–discharge relation can only be determined by a series of discharge measurements, using stage, temperature and precipitation records as a guide for interpolation between measurements.

Variable backwater – uniform channel (Fig. 4.4(e))

If the control reach for a gauging station has within it a weir or a dam, a diversion or a confluent tributary which can increase or decrease the energy gradient for a given discharge, variable backwater, is produced. That is, the slope in a reach is increased or decreased from the normal. In this case a second gauge is installed below the control section in order to measure the fall for developing a stage–fall–discharge relation (see Chapter 9).

Variable backwater – submergence (Fig. 4.4(f))

Some channel reaches below gauging stations contain local control sections such as falls, rapids or a dam which determine the stage–discharge relation at low flows, but which may be submerged at times by inflow from a tributary downstream or by the operation of a dam. As in the case of rating a station with uniform channel and variable backwater, a second gauge is installed below the control in order to measure the fall.

Rapidly changing discharge (Fig. 4.4(g))

At some stations, generally those of low energy slope, the stage–discharge relation is affected by the rate of change of discharge. If the discharge is increasing rapidly, it will be greater than that for zero rate of change and, conversely, if it is rapidly decreasing, it will be less.

Out-of-bank flow (overbank) and ponding (Fig. 4.4(h))

At many gauging stations there may be significant out-of-bank flow and ponding areas on the flood plain adjacent to the stream channel. During increasing discharge, a part of the flow goes into these areas, increasing the slope and discharge relative to stage. Conversely, when the discharge decreases, water returning to the channel from the flooded areas causes backwater and the discharge for a given stage is significantly decreased. Each flood produces its own loop rating. No satisfactory method is available to develop a single rating under these conditions. A loop rating is required for each flood and is determined by a series of discharge measurements for the channel flow and for the out-of-bank flow, the sum of the two giving the total flow.

4.3 The simple stage–discharge curve

The general procedure in establishing the stage–discharge curve is as follows. The discharge measurements are plotted on arithmetic graph paper with discharge on the horizontal scale (abscissa) and the corresponding gauge height on the vertical scale (ordinate). If a measurement of discharge was not made at steady stage, the mean gauge height during the measurement is used. The plotted observations are labelled in chronological order, and rising and falling stage during the measurement are indicated by distinguishing symbols if necessary (Fig. 4.6).

The relation should be defined by a sufficient number of measurements suitably distributed throughout the range in stage, taking into account the shape of the stage–discharge relation. Ideally, the number and spacing of the observations are made to conform to the relative frequency of flow at the various stages. That is, the number of observations at various subranges is in proportion to the probable occurrence of discharge at these same ranges, covering the whole range of discharge for which the relation is plotted.

Nevertheless, in practice, it is desirable to have as many observations as possible at the extreme ranges, both at the low flow and at the high flood stages.

The curve of relation, the rating curve, is drawn evenly and smoothly through the scatter of plotted data points.

Although all current meter discharge measurements are checked and considered correct before plotting, observations which plot more than, say, 5% in discharge off the curve should again be checked for possible uncertainties. Particular attention is paid to the need to adjust or weight the gauge height, to the correct current meter rating, and to errors in the computation. With respect

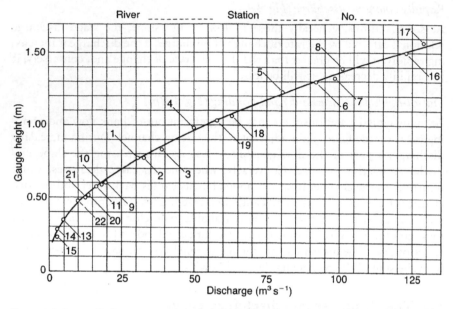

Figure 4.6 Stage–discharge curve, first plotting.

to the latter, it is useful to make a plot of the cross-sectional area of flow and the mean velocity against gauge height for each discharge measurement. Such plots reveal the presence of an error and where it is located in the computation, either in the velocity or in the cross-sectional area. If no apparent error is found to be caused by the above, the condition of the control is investigated before the measurement is discarded or a shift correction applied if applicable. From the above it is evident that a copy of the stage–discharge curve is conveniently available in the instrument house.

Fitting the stage–discharge curve – visual estimation

There are several methods of fitting a median curve to observed or measured data points. This can be performed, however, quite satisfactorily by visual estimation of the plot with the aid of drafting curves which are usually designed to conform to parabolic equations. Very often the trend of discharge measurements plotted on graph paper follows a particular drafting curve due to the fact that stream discharge tends to vary as some power of the depth of flow.

The criterion used when fitting a median curve to observations by visual estimation is that there are about the same number of plus and minus deviations. A deviation is considered negative for a measurement lying above the curve and positive when lying below the curve.

4.4 The logarithmic method

In many cases the stage–discharge curve may be established by plotting the logarithms of stage against the logarithms of discharge. The use of logarithmic graph paper obviates the necessity of computing the logarithms and the plotting of the observations is performed in the same manner as before. There are certain important advantages in using the logarithmic method:

(a) The logarithmic form of the rating curve can be developed into a straight line, or straight-line segments, by adding or subtracting a constant value (datum correction) to the gauge height logarithmic scale.

(b) The straight-line graph can be described by a simple mathematical equation that is easily handled by pocket calculator or computer.

(c) The straight-line graph may be conveniently analysed for uncertainties.

(d) A percentage distance off the curve is always the same regardless of where it is located. Thus a measurement that is 10% off the curve at high stage will be the same distance away from the curve as a measurement that is 10% off at low stage.

(e) It is easier to identify the range in stage for which different controls are effective.

(f) The gauge height scale may be conveniently altered by halving, doubling or adding a percentage to the scale (Fig. 4.7). The curve will merely shift position but retain the same shape.

(g) The curve can be easily extrapolated, if necessary, but caution is required in extrapolation at either the top or the bottom end of the curve. If the curve is a single segment and the control is stable, then extrapolation may be performed with more confidence than if the curve is made up of several segments.

Theory of the logarithmic curve

The stage–discharge relation may be expressed by an equation of the form

$$Q = C(h + a)^n \tag{4.1}$$

which is the equation of a parabola where Q is the discharge, h is the gauge height, C and n are constants, and a is the stage at zero flow (datum correction). This equation may be transformed by logarithms to

$$\log Q = \log C + n \log (h + a) \tag{4.2}$$

which is in the form of the equation of a straight line

$$y = nx + C \tag{4.3}$$

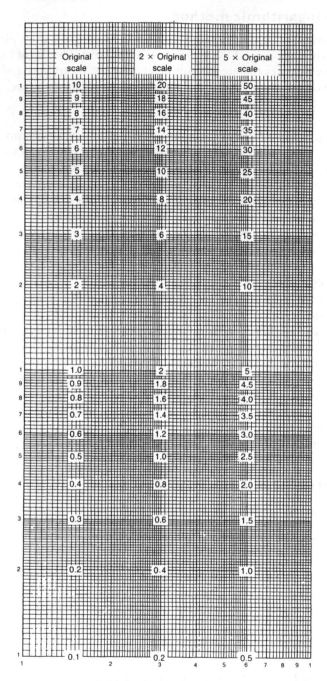

Figure 4.7 Illustration of how scales of logarithmic graph paper can be transposed.

where n is the gradient and C is the intersection of the line on the y-axis. By plotting Q against $(h + a)$, therefore, on double logarithmic graph paper a straight line is obtained.

Often two or more straight lines may be required to fit the data and it is usually possible, initially, to decide on the approximate location of the break points of each range by a careful investigation of the controls. The actual break points may be determined by solving the two equations concerned for Q and h or by purely graphical means. For very irregular channels, or for non-uniform flow, equation (4.1) cannot be expected to apply throughout the whole range of stage. Sometimes the curve changes from a parabolic to a complex curve or vice versa and sometimes the constants and exponents vary throughout the range.

The logarithmic rating equation therefore is seldom a single straight line or a gentle curve throughout the entire range of stage at a gauging station. Even if the same channel cross-section is the control for all stages, a sharp break in the contour of the cross-section causes a break in the slope of the rating curve. Also the other constants C and a in equation (4.1) are related to the physical characteristics of the stage–discharge control.

If the control section changes at various stages, it may be necessary to fit two or more equations, each corresponding to the portion of the range over which the control is applicable. If, however, too many changes in the parameters are necessary in order to define the relation, it is possible that the logarithmic method may not be suitable and a curve fitted by visual estimation can be employed as previously described.

The first derivative of equation (4.1) is a measure of the change in discharge per unit change in stage, that is the first derivative gives the first-order differences of the discharge series. The first derivative is

$$\frac{dQ}{dh} = Cn(h + a)^{n-1}. \tag{4.4}$$

Second-order differences are obtained by differentiating again. The second derivative is

$$\frac{d^2Q}{dh^2} = Cn(n - 1)(h + a)^{n-2}. \tag{4.5}$$

An examination of the second derivative shows that the second-order differences increase with stage when n is greater than 2 (section control), and decrease with stage when n is less than 2 (channel control).

Normally in graphical analysis the dependent variable, Q in equation (4.1), would be plotted on the vertical axis and the independent variable, h, plotted on the abscissa. It has been a tradition in stream gauging, however, that this procedure is reversed while still retaining Q as the dependent variable and taking n, the slope, as the cotangent instead of the tangent.

The geometry or shape of the channel section is reflected in the slope, n, of the stage–discharge equation (4.1). This property is a useful indicator when carrying out a preliminary survey at a new site. The following are approximate relations between n and channel sections:

(a) for a rectangular channel section, $n = 3/2$
(b) for a concave section of parabolic shape, $n = 2$
(c) for a triangular or semicircular section, $n = 5/2$.

Changes in channel resistance and slope with stage, however, will affect the exponent n. The net result of these factors is that the exponent in equation (4.1) for relatively wide rivers with channel control will generally vary from about 1.3 to 1.8. For relatively deep narrow rivers with section control, the exponent n will almost always be greater than two and may often exceed a value of three.

The stage of zero flow

The datum correction a is the value of the stage at zero flow and corresponds to the lowest point on the low water control. It is defined as the gauge height at which flow over the control ceases. Usually this stage does not coincide with the zero of the gauge unless this is specifically set to the lowest level of an artificial control or the crest of a measuring structure. The point of zero flow is therefore easily determined for artificial controls and in those cases where the control is well defined by a rock ledge.

The stage of zero flow is determined by subtracting the depth of water over the lowest point on the control from the stage indicated by the gauge reading. If the gauge is at some distance from the control, an adjustment is made for the slope. The difficulty in determining the point of zero flow is in establishing the lowest point on the control, as not all controls are easily identified. Generally a cross-section is surveyed across the stream at the first complete break in the slope of the water surface below the measuring section. This is usually the location of the upstream lip of the low water control. For a channel-controlled gauging station, the maximum depth directly opposite the gauge will give a reasonable approximation of the depth to be subtracted from the gauge reading in order to obtain the stage of zero flow.

The position of the point of zero flow is best determined at time of low water when rivers can often be waded. In those cases, however, where the control section is difficult to identity, it may be located by surveying a close grid of spot levels or by running a sufficient number of cross-sections over the area of the assumed control section or reach.

It should be noted that when a quantity has to be added to the gauge heights in order to obtain a straight line, a is taken as positive, that is the zero of the gauge is in this case positioned at a level above the point of zero flow. Conversely when a quantity has to be subtracted from the gauge heights, a is taken as negative and in this case the zero of the gauge is positioned at a level below the point of zero flow. When the zero of the gauge coincides with the level of the point of zero flow then *a* is zero. The three cases are given in Fig. 4.8

There are four methods of estimating the datum correction *a* apart from making a field survey. However, if at all possible, the estimates are verified by field investigation.

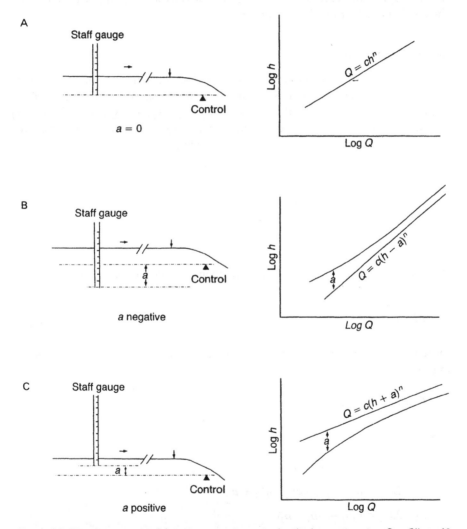

Figure 4.8 The three cases of datum correction *a* in the discharge equation $Q = C(h + a)^n$. (A) Zero of staff gauge at level of lowest point of control $a = 0$; (B) zero of staff gauge below lowest point of control *a* is negative; (C) zero of staff gauge above lowest point of control *a* is positive.

Trial and error procedure

All discharge measurements are plotted on double logarithmic graph paper ('log–log paper') and a median line is drawn through the scatter of observations. Usually this line will be a curved line. Various trial values, one value for each trial, are added or subtracted to the gauge heights of the measurements until the plot obtained forms a straight line. The trial value forming the straight line is the required value of a (Fig. 4.9). All the plotted observations may be used in the trial operation. However, it is better to use only a few points selected from the median line first fitted to the points.

Arithmetic procedure

All discharge measurements are plotted on log–log paper. If the mean line drawn through the observations results in a curve then a datum correction is required (Fig. 4.10). Three values of discharge, Q_1, Q_2 and Q_3, are selected in geometric progression, that is two values Q_1 and Q_3 are chosen from the curve and the third value Q_2 is then computed according to

$$Q_1/Q_2 = Q_2/Q_3 \qquad\qquad (4.6)$$

then

$$Q_2^2 = Q_1 Q_3. \qquad\qquad (4.7)$$

If the corresponding gauge heights read from the plot are h_1, h_2 and h_3, then from

$$Q = C(h - a)^n \qquad\qquad (4.1)$$

(note that a is negative since the curve in Fig. 4.10 is 'concave up') and substituting for Q in equation (4.7)

$$C^2(h_2 - a)^{2n} = C(h_1 - a)^n C(h_3 - a)^n.$$

Therefore

$$(h_2 - a)^2 = (h_1 - a)(h_3 - a)$$
$$h_2^2 - 2ah_2 + a^2 = h_1 h_3 - ah_3 - ah_1 + a^2$$
$$h_2^2 - 2ah_2 = h_1 h_3 - a(h_1 + h_3)$$
$$a(2h_2 - h_1 - h_3) = h_2^2 - h_1 h_3.$$

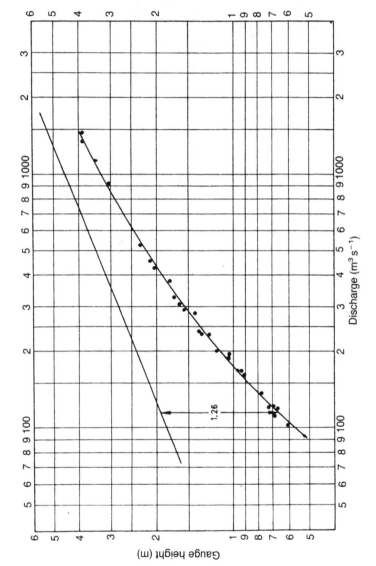

Figure 4.9 Trial and error method of determining *a*.

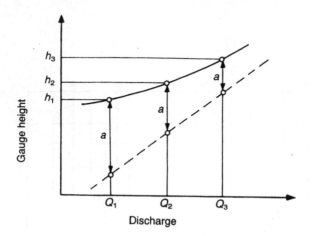

Figure 4.10 Transforming a curved line into a straight line on logarithmic graph paper: $a = \dfrac{h_1 h_3 - h_2^2}{h_1 + h_3 - 2h_2}$ and $Q_2^2 = Q_1 Q_3$.

Hence

$$a = \frac{h_1 h_3 - h_2^2}{h_1 + h_3 - 2h_2}. \tag{4.8}$$

If the curve is 'concave down' (*a* positive) then proceeding as above since

$$Q = C(h + a)^n,$$

$$a = \frac{h_2^2 - h_1 h_3}{h_1 + h_3 - 2h_2}. \tag{4.9}$$

Example

In Fig. 4.12, if Q_1 is taken as 200 m³ s⁻¹, and Q_3 as 1000 m³ s⁻¹

$$Q_2 = (200 \times 1000)^{1/2} \text{ m}^3 \text{ s}^{-1}$$
$$= 447 \text{ m}^3 \text{ s}^{-1}$$

and the corresponding *h* values are

$$h_1 = 1.150 \text{ m}$$
$$h_2 = 2.041 \text{ m}$$
$$h_3 = 3.260 \text{ m}.$$

The curve in Fig. 4.12 is 'concave down', therefore

$$a = \frac{h_2^2 - h_1 h_3}{h_1 + h_3 - 2h_2}$$

$$= \frac{4.164 - 3.749}{4.410 - 4.082}$$

$$= \frac{0.415}{0.328}$$

$$= 1.26 \text{ m.}$$

Graphical procedure

As above, three values of discharge in geometric progression are selected but this time from a plot on arithmetic graph paper. Let the points be A, B and C as illustrated in Fig. 4.11. Vertical lines are drawn through A and B and horizontal lines drawn through B and C intersecting the verticals at D and E, respectively. Let DE and AB meet at F. Then the ordinate of F is the value of a which takes the sign as before.

The last two methods are based on the assumption that the lower part of the stage–discharge relation, including the selected points, is part of a parabola. In most cases this assumption holds and the methods will give acceptable results on condition that there are enough discharge measurements available to satisfactorily define the curvature of the lower part of the rating curve.

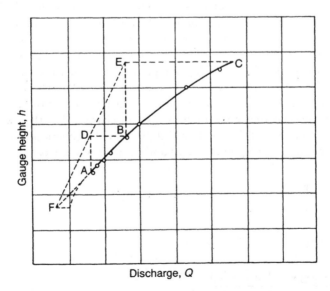

Figure 4.11 Graphical determination of a.

Computer programme procedure

If automatic data processing by computer is available, the datum correction can be conveniently found by a computer programme which is designed essentially to carry out an iteration procedure until the best straight line is obtained. In the iteration programme, different values of *a* are tested against the correlation coefficient for the line of best fit. The value of *a* giving the maximum correlation coefficient is the value selected. Usually about 20 or more iterations are necessary. This is a purely mathematical procedure and probably gives the best results but every endeavour is made to ensure that the result can be confirmed by a site investigation.

Estimating the constants C and n

After a straight-line plot of the discharge measurements on log–log graph paper has been obtained, the constants C and n of the rating equation (4.1) can be computed by any of three methods, namely arithmetically by a least squares procedure, or graphically.

The stage–discharge relation is first analysed from a plot on log–log graph paper in order to establish whether the rating curve is composed of one or more straight-line segments, each having its own constants C and n. The constants for each separate segment are calculated separately.

Arithmetic procedure

A series of discharge measurements at a gauging station is given in Table 4.1. The datum correction has been computed as + 1.26 m. It is required to determine the constants C and n and therefore the rating equation.

To the gauge heights of the measurements given in Table 4.1, 1.26 m is added and the observations plotted on log–log graph paper, the gauge heights on the vertical scale and the discharges on the horizontal scale. The plot defines a straight line (Fig. 4.12). Select two points on this line as far from each other as possible but within the range of measured discharges. Let the two selected points (Q, h) be (97, 1.80) and (1300, 5.00). Now the equation of the straight line passing through two points (x_1, y_1) and (x_2, y_2) is written as

$$\frac{y - y_1}{x - x_1} = \frac{y_2 - y_1}{x_2 - x_1}.$$

(4.10)

It follows that

$$\frac{\log y - \log y_1}{\log x - \log x_1} = \frac{\log y_2 - \log y_1}{\log x_2 - \log x_1}.$$

(4.11)

Table 4.1 Stage and discharge data for calculating the stage–discharge equation

Observation no.	h (m)	Q (m³ s⁻¹)	Observation no.	h (m)	Q (m³ s⁻¹)
1	1.55	300	15	3.87	1374
2	1.44	287	16	2.33	540
3	1.26	235	17	3.49	1152
4	1.05	193	18	3.93	1452
5	0.73	125	19	2.03	440
6	0.69	113	20	1.61	306
7	0.70	124	21	2.13	469
8	1.70	340	22	1.37	246
9	0.96	169	23	1.05	189
10	0.94	168	24	0.91	163
11	1.35	240	25	0.79	139
12	1.17	202	26	0.68	120
13	1.79	387	27	0.61	104
14	3.09	930	28	0.53	94.6

In the present case, after changing the notation and inverting both sides of equation (4.11), it follows that

$$\frac{\log Q - \log Q_1}{\log (h + a) - \log h_1} = \frac{\log Q_2 - \log Q_1}{\log h_2 - \log h_1}. \tag{4.12}$$

Substituting the given values in equation (4.12) gives

$$\frac{\log Q - \log 97}{\log (h + 1.26) - \log 1.80} = \frac{\log 1300 - \log 97}{\log 5.00 - \log 1.80}. \tag{4.13}$$

Then

$$\frac{\log Q - 1.9868}{\log (h + 1.26) - 0.2553} = \frac{3.1139 - 1.9868}{0.6990 - 0.2553} = 2.54$$

and

$$\log Q = 2.54 \log (h + 1.26) + 1.3383.$$

Therefore

$$Q = 21.79(h + 1.26)^{2.54}. \tag{4.14}$$

Equation (4.14) is therefore the rating equation for the stage–discharge curve shown in Fig. 4.12.

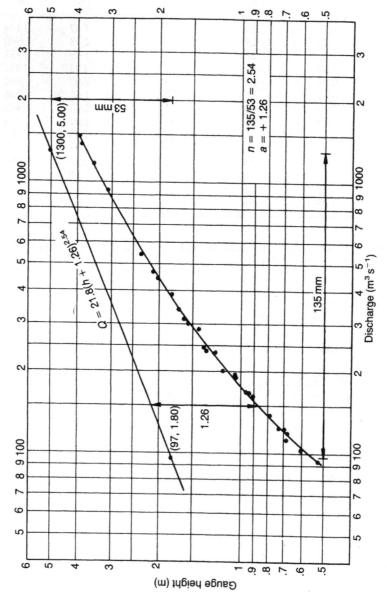

Figure 4.12 Stage–discharge curve established by the logarithmic method.

Least squares procedure

The values of C and n may be computed by the method of least squares; that is, the sum of the squares of the deviations between the logarithms of the measured discharge and the estimated discharge from the curve should be a minimum.

The values of C and n are obtained from the following equations

$$\Sigma Y - N(\log C) - n\Sigma X = 0 \tag{4.15}$$

$$\Sigma(XY) - \Sigma(X)(\log C) - n\Sigma(X^2) = 0 \tag{4.16}$$

where ΣY is the sum of all the values of $\log Q$
$\quad\Sigma X$ is the sum of all the values of $\log(h + a)$
$\quad\Sigma(X^2)$ is the sum of all the values of the square of X
$\quad\Sigma(XY)$ is the sum of all the values of the product of X and Y
$\quad N$ is the number of current meter observations.

In order to illustrate the method, the data in Table 4.1 are prepared as shown in Table 4.2. Substituting the calculated values from Table 4.2 into equations (4.15) and (4.16) gives

$$68.0506 - 28 \log C - n\, 12.0182 = 0 \tag{4.17}$$

$$30.4351 - 12.0182 \log C - n\, 5.6430 = 0. \tag{4.18}$$

Solving the above equations for n and C gives $n = 2.53$ and $C = 22.10$ which is in close agreement with equation (4.14) developed by the arithmetic method.

The least squares method is the method conveniently employed in computer processing but the stage–discharge curve is normally first plotted on log–log graph paper for examination. As already discussed, the observations may not plot as one straight line but as two or more straight-line segments differing in slope with each segment having its own equation. If break points are apparent they are best fed into the computer programme from the log–log plot rather than developing a programme to perform this exercise.

Graphical procedure

The graphical method of determining C and n is simple and normally as effective in giving results as good as the two methods described above.

The value of n is found by scaling the horizontal and vertical projections of the line and calculating this ratio. In Fig. 4.12 the horizontal projection has been scaled as 135 mm and the vertical projection as 53 mm. The value of n is therefore calculated as

Table 4.2 Tabulation of data for determination of the constants C and n[a]

No.	h	Q	$h + a$	$\log Q = Y$	$\log (h + a) = X$	(XY)	(X^2)
1	1.55	300	2.81	2.4771	0.4487	1.1115	0.2013
2	1.44	287	2.70	2.4579	0.4314	1.0603	0.1861
3	1.26	235	2.52	2.3711	0.4014	0.9518	0.1611
4	1.05	193	2.31	2.2856	0.3636	0.8310	0.1322
5	0.73	125	1.99	2.0969	0.2989	0.6268	0.0893
6	0.69	113	1.95	2.0531	0.2900	0.5954	0.0841
7	0.70	124	1.96	2.0934	0.2923	0.6119	0.0854
8	1.70	340	2.96	2.5315	0.4713	1.1931	0.2221
9	0.96	169	2.22	2.2279	0.3464	0.7717	0.1200
10	0.94	168	2.20	2.2253	0.3424	0.7619	0.1172
11	1.35	240	2.61	2.3802	0.4166	0.9916	0.1736
12	1.17	202	2.43	2.3054	0.3856	0.8890	0.1487
13	1.79	387	3.05	2.5877	0.4843	1.2532	0.2345
14	3.09	930	4.35	2.9685	0.6385	1.8954	0.4077
15	3.87	1374	5.13	3.1380	0.7101	2.2283	0.5042
16	2.33	540	3.59	2.7324	0.5551	1.5168	0.3081
17	3.49	1152	4.75	3.0615	0.6767	2.0717	0.4579
18	3.93	1452	5.19	3.1620	0.7152	2.2615	0.5115
19	2.03	440	3.29	2.6435	0.5172	1.3672	0.2675
20	1.61	306	2.87	2.4857	0.4579	1.1382	0.2097
21	2.13	469	3.39	2.6712	0.5302	1.4163	0.2811
22	1.37	246	2.63	2.3909	0.4200	1.0039	0.1764
23	1.05	189	2.31	2.2765	0.3636	0.8277	0.1322
24	0.91	163	2.17	2.2122	0.3365	0.7444	0.1132
25	0.79	139	2.05	2.1430	0.3118	0.6682	0.0972
26	0.68	120	1.94	2.0792	0.2878	0.5984	0.0828
27	0.61	104	1.87	2.0170	0.2718	0.5482	0.0739
28	0.53	94.6	1.79	1.9759	0.2529	0.4997	0.0640
Σ				68.0506	12.0182	30.4351	5.6430

[a] $N = 28; a = +1.26.$

$$n = \frac{135}{53} = 2.54$$

which is the same as the value of n calculated by the arithmetic method.
    The value of C is given by the numerical value of the discharge when

$(h + a) = 1$ since by equation (4.2)

$\log Q = \log C + n \log (h + a)$

and when $(h + a) = 1$, $n \log (h + a) = 0$ and therefore $Q = C$.
    Referring to Fig. 4.12, it will be observed that the logarithmic stage–discharge curve does not go through the line $(h + a) = 1$. If the line is projected,

however (not shown in Fig. 4.12), it is found that when $(h + a) = 1$, $Q = 21.9$ and therefore $C = 21.9$. Otherwise, knowing n, C may be found by solving the rating equation (4.1) for C as follows:

From

$$Q = C(h + a)^{2.54}$$

when

$$Q = 355$$

$$(h + a) = 3.00$$

and

$$C = \frac{355}{3^{2.54}}$$

$$= 21.79.$$

Therefore

$$Q = 21.79(h + 1.26)^{2.54}. \tag{4.19}$$

4.5 Examples of logarithmic stage–discharge curves

Table 4.3 gives details of a stage–discharge curve where the current meter gaugings and corresponding stage values are shown in columns 2 and 3, respectively. These values are plotted on log–log graph paper in Fig. 4.13 and a curve drawn through the points. From an examination of this curve and an inspection in the field it was clear that the high water control became operative at a gauge height of 0.306 m. The curve therefore had two segments. The computer programme procedure gave a values for the lower range of 0.015 m and for the upper range 0.063 m (the trial and error procedure would give similar results). These values of a are scaled off graphically in Fig. 4.13 and two straight lines of relation established as shown. The ordinate scale for these lines now becomes $(h - a)$ (both a values are negative). The values of the slope, n, of each line are scaled off as shown and the C values obtained on the abscissa at $(h - a) = 1$. It will be observed that in order to obtain C for the lower line, the line requires to be projected to $(h - a) = 1$.

The equations of the lines of relation are therefore

$$Q = 47.3(h - 0.015)^{2.22} \tag{4.20}$$

Table 4.3 Example of logarithmic two-range stage–discharge curve with the equations of the curve computed by the least squares method

(1) Observation no.	(2) Q, gauging (m³ s⁻¹)	(3) Stage h (m)	(4) (h + a) (m)	(5) log Q, (Y)	(6) log (h + a) (X)	(7) XY	(8) X²
Range I (a = −0.015)							
1	0.121	0.085	0.070	−0.9172	−1.1549	1.0593	1.3338
2	0.136	0.086	0.071	−0.8665	−1.1487	0.9953	1.3195
3	0.166	0.093	0.078	−0.7799	−1.1079	0.8641	1.2274
4	0.187	0.098	0.083	−0.7282	−1.0809	0.7871	1.1683
5	0.268	0.111	0.096	−0.5719	−1.0177	0.5820	1.0357
6	0.385	0.130	0.115	−0.4145	−0.9393	0.3893	0.8823
7	0.449	0.139	0.124	−0.3478	−0.9066	0.3153	0.8219
8	0.490	0.146	0.131	−0.3098	−0.8827	0.2735	0.7792
9	0.783	0.171	0.156	−0.1062	−0.8069	0.0857	0.6511
10	0.764	0.172	0.157	−0.1169	−0.8041	0.0940	0.6466
11	1.148	0.202	0.187	0.0599	−0.7282	−0.0436	0.5303
12	1.289	0.213	0.198	0.1103	−0.7033	−0.0776	0.4946
13	1.932	0.249	0.234	0.2860	−0.6308	−0.1804	0.3979
14	2.065	0.262	0.247	0.3149	−0.6073	−0.1912	0.3688
15	2.628	0.287	0.272	0.4196	−0.5654	−0.2372	0.3197
16	2.738	0.294	0.279	0.4374	−0.5544	−0.2425	0.3074
Σ				−3.5308	−13.6391	4.4731	12.2845

Range 2 ($a = -0.063$)

1	2.965	0.306	0.243	0.4720	-0.6144	-0.2900	0.3775
2	3.073	0.318	0.255	0.4876	-0.5934	-0.2893	0.3521
3	3.437	0.327	0.264	0.5362	-0.5784	-0.3101	0.3345
4	3.727	0.340	0.277	0.5714	-0.5575	-0.3186	0.3108
5	4.049	0.356	0.293	0.6073	-0.5331	-0.3238	0.2842
6	4.356	0.380	0.317	0.6391	-0.4989	-0.3188	0.2489
7	5.580	0.427	0.364	0.7466	-0.4389	-0.3277	0.1926
8	5.779	0.445	0.382	0.7619	-0.4179	-0.3184	0.1746
9	6.046	0.462	0.399	0.7815	-0.3990	-0.3118	0.1592
10	6.219	0.463	0.400	0.7937	-0.3979	-0.3158	0.1583
11	7.249	0.502	0.439	0.8603	-0.3575	-0.3076	0.1278
12	7.651	0.526	0.463	0.8837	-0.3344	-0.2955	0.1118
13	10.740	0.646	0.583	1.0310	-0.2343	-0.2416	0.0549
14	13.508	0.725	0.662	1.1306	-0.1791	-0.2025	0.0321
15	13.943	0.744	0.681	1.444	-0.1669	-0.1910	0.0279
16	13.742	0.754	0.691	1.1380	-0.1605	-0.1826	0.0258
17	17.714	0.867	0.804	1.2483	-0.0947	-0.1182	0.0090
18	17.855	0.886	0.823	1.2518	-0.0846	-0.1059	0.0072
19	18.532	0.897	0.834	1.2679	-0.0788	-0.0999	0.0062
20	20.320	0.958	0.895	1.3079	-0.0482	-0.0630	0.0023
21	21.950	1.009	0.946	1.3414	-0.0241	-0.0323	0.0006
22	26.052	1.101	1.038	1.4158	0.0162	0.0229	0.0003
Σ				20.4184	-6.7763	-4.9415	2.9986

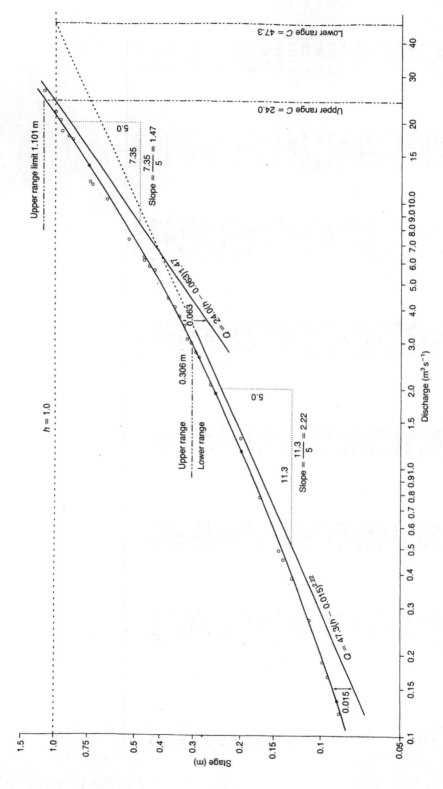

Figure 4.13 Example of logarithmic stage–discharge curve having lower and upper segments (see Table 4.3).

for values of h up to 0.306 m, and

$$Q = 24.0(h - 0.063)^{1.47}. \tag{4.21}$$

The equations may also be found by the least squares method from equations (4.15) and (4.16) as follows. The values of X, Y, XY and X^2 are shown in Table 4.3, columns 5–8, and from equations (4.15) and (4.16)

$$\Sigma Y - N(\log C) - n\Sigma X = 0$$

$$\Sigma(XY) - \Sigma(X)(\log C) - n\Sigma X^2 = 0.$$

For the lower range

$$-3.5308 - 16 \log C + n\, 13.6391 = 0 \tag{4.22}$$

$$4.4731 + 13.6391 \log C - n\, 12.2845 = 0, \tag{4.23}$$

equation (4.22) × 13.6391/16 gives

$$-3.0098 - 13.6391 \log C + 11.6266n = 0, \tag{4.24}$$

equation (4.23) + equation (4.24) gives

$$0.6579n = 1.4633 \tag{4.25}$$

$$n = 2.224. \tag{4.26}$$

Substituting $n = 2.224$ in equation (4.22) gives

$$-3.5308 - 16 \log C + 2.224 \times 13.6391 = 0 \tag{4.27}$$

$$16 \log C = 26.8025 \tag{4.28}$$

$$C = 47.326. \tag{4.29}$$

The rating equation is therefore

$$Q = 47.326(h - 0.015)^{2.224}. \tag{4.30}$$

For the upper range

$$20.4184 - 22 \log C + n\, 6.7763 = 0 \tag{4.31}$$

$$-4.9415 + 6.7763 \log C - n\, 2.9986 = 0, \tag{4.32}$$

equation (4.31) × 6.7763/22 gives

$$6.2891 - 6.7763 \log C + 2.0872n = 0, \tag{4.33}$$

equation (4.32) + equation (4.33) gives

$$0.9114n = 1.3476 \tag{4.34}$$

$$n = 1.4786. \tag{4.35}$$

Substituting $n = 1.4786$ in equation (4.31) gives

$$20.4184 - 22 \log C + 6.7763 \times 1.4786 = 0 \tag{4.36}$$

$$- 22 \log C = - 30.4378 \tag{4.37}$$

$$C = 24.185. \tag{4.38}$$

The rating equation is therefore

$$Q = 24.185(n - 0.063)^{1.4786}. \tag{4.39}$$

If the rating equations are established entirely by graphical means, it is often convenient to obtain a by the graphical procedure already described. If the stage–discharge relation consists of one segment only, the a value so obtained will normally be acceptable as producing the required straight line relation. If more than one segment is involved the datum corrections for the upper segments, found by the graphical procedure, may be considered as initial trial values and may require modification to produce a straight line of relation. This is demonstrated in Fig. 4.14 for the above example.

For the lower segment (Fig. 4.14), Q_1 and Q_3 have been taken as 0.50 m³ s⁻¹ and 2.50 m³ s⁻¹, respectively, then (equation (4.7))

$$Q_2^2 = Q_1 Q_3$$

$$= 0.5 \times 2.50$$

$$Q_2 = 1.12 \text{ m}^3 \text{ s}^{-1}.$$

The datum correction of $- 0.015$ m so obtained is the same as that obtained by computer.

For the upper segment Q_1 and Q_3 have been taken as 5 m³ s⁻¹ and 20 m³ s⁻¹, then (equation (4.7))

$$Q_2^2 = Q_1 Q_3$$

$$= 5 \times 20$$

$$Q_2 = 10 \text{ m}^3 \text{ s}^{-1}.$$

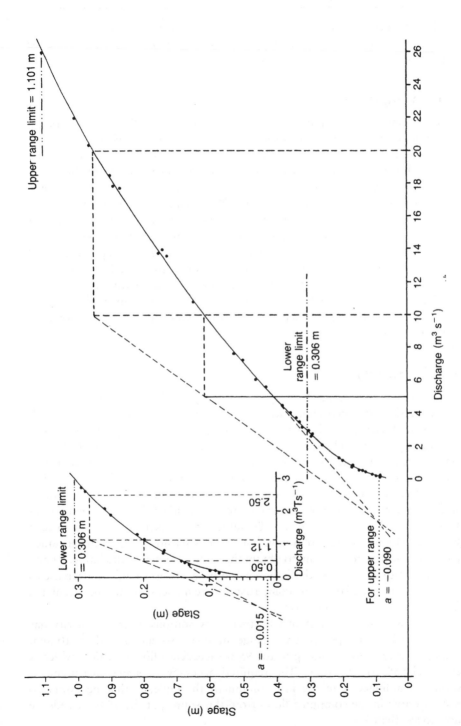

Figure 4.14 Example of determining *a* by the graphical method (see also Fig. 4.11).

The datum correction of –0.090 m so obtained, however, does not agree with that obtained by computer (mainly because of scaling) and requires to be adjusted to –0.063 to obtain the best straight line of relation.

4.6 Rating table

When a computer programme is used to convert stage to discharge, the rating equation is normally built into the programme. However, when the conversion is performed manually a rating table is constructed. A typical rating table is shown in Table 4.4. This table has been constructed from the rating equations in the previous example and the break in the stage– discharge curve can be seen in the $(h - a)$ column (column 2) at gauge height 0.300 m.

The rating table shown gives discharge values (Q_c) for every 10 mm increment in gauge height (h) and intermediate values are obtained by interpolation.

For a stage–discharge curve plotted on arithmetic graph paper column 2 would not be required, otherwise the construction of the table is the same, but in this case Q_c values are taken directly from the stage–discharge curve.

When the rating table has been constructed as shown in Table 4.4 it is often more convenient to prepare a working table in the form shown in Table 4.5.

4.7 Sensitivity

The sensitivity of a stage–discharge relation is an indication of the extent of the response to an increase in discharge by an increase in stage. Where a small increase in discharge produces a relatively large increase in stage the relation is said to be sensitive; where a relatively large increase in discharge produces a small increase in stage, the relation is said to be non-sensitive. The degree of sensitivity affects the record of the station at all stages of the flow and is reflected in the accuracy of the flow data. It follows that a sensitive record of stage can be converted more accurately into a record of discharge than a non-sensitive one. For example, it may be sufficient at a sensitive station to read water level to an accuracy of 5 mm, but at a less sensitive station the accuracy required may have to be 1 mm to obtain the same accuracy of discharge. At any station sensitivity varies with discharge and in most cases tends to decrease as discharge decreases. For a true comparison to be made, it should be calculated for each station at the same level.

Table 4.6 shows a sample of UK streamflow stations where sensitivity has been related to the percentage change in flow associated with a 10 mm change of stage corresponding to the 95% exceedence flow that flow which is exceeded 95% of the time. The required accuracy of stage measurement, shown in millimetres in the last column of the table, is that necessary to restrict errors in the computed flows from this source, at the 95% exceedence flow to less than 5%.

Table 4.4 Typical rating table (part)

(1) Stage h (m)	(2) (h − a) (m)	(3) Q_c $(m^3 s^{-1})$	(4) ΔQ_c $(m^3 s^{-1})$	(5) Δh (m)	(6) Increment for 1 mm stage $(m^3 s^{-1})$
0.080	0.065	0.108			
0.090	0.075	0.149	0.041	0.010	0.004
0.100	0.085	0.197	0.048	0.010	0.005
0.110	0.095	0.252	0.055	0.010	0.005
0.120	0.105	0.315	0.063	0.010	0.006
0.130	0.115	0.386	0.071	0.010	0.007
0.140	0.125	0.464	0.078	0.010	0.008
0.150	0.135	0.551	0.087	0.010	0.009
0.160	0.145	0.646	0.095	0.010	0.010
0.170	0.155	0.749	0.103	0.010	0.010
0.180	0.165	0.861	0.112	0.010	0.011
0.190	0.175	0.981	0.120	0.010	0.012
0.200	0.185	1.110	0.129	0.010	0.013
0.210	0.195	1.248	0.138	0.010	0.014
0.220	0.205	1.395	0.147	0.010	0.015
0.230	0.215	1.551	0.156	0.010	0.016
0.240	0.225	1.716	0.165	0.010	0.017
0.250	0.235	1.890	0.174	0.010	0.017
0.260	0.245	2.073	0.183	0.010	0.018
0.270	0.255	2.266	0.193	0.010	0.019
0.280	0.265	2.469	0.203	0.010	0.020
0.290	0.275	2.681	0.212	0.010	0.021
0.300	0.285	2.902	0.197	0.010	0.020
0.310	0.247	3.057	0.181	0.010	0.018
0.320	0.257	3.242	0.185	0.010	0.019
0.330	0.267	3.432	0.188	0.010	0.019
0.340	0.277	3.624	0.192	0.010	0.019
0.350	0.287	3.819	0.195	0.010	0.020
0.360	0.297	4.018	0.199	0.010	0.020
0.370	0.307	4.219	0.201	0.010	0.020
0.380	0.317	4.424	0.205	0.010	0.021
0.390	0.327	4.632	0.208	0.010	0.021
0.400	0.337	4.843	0.211	0.010	0.021
0.410	0.347	5.057	0.214	0.010	0.021
0.420	0.357	5.274	0.217	0.010	0.022
0.430	0.367	5.494	0.220	0.010	0.022
0.440	0.377	5.717	0.223	0.010	0.022
0.450	0.387	5.942	0.225	0.010	0.023
0.460	0.397	6.171	0.229	0.010	0.023
0.470	0.407	6.402	0.231	0.010	0.023
0.480	0.417	6.636	0.234	0.010	0.023
0.490	0.427	6.873	0.237	0.010	0.024
0.500	0.437	7.112	0.239	0.010	0.024

Table 4.5 Alternative rating table (part) in millimetre steps

Stage h (m)	Discharge ($m^3\ s^{-1}$)									
	0.000	0.001	0.002	0.003	0.004	0.005	0.006	0.007	0.008	0.009
0.080	0.108	0.112	0.116	0.120	0.124	0.128	0.132	0.136	0.140	0.144
0.090	0.149	0.153	0.157	0.162	0.167	0.172	0.177	0.182	0.187	0.192
0.100	0.197	0.202	0.207	0.212	0.217	0.222	0.228	0.234	0.240	0.246
0.110	0.252									
0.120										
—										
—										
—										
1.101										

Table 4.6 Table showing the results of a sensitivity analysis of a sample of UK streamflow stations

Gauging station number	River	Station type	Catchment area (km²)	Mean flow (m³s⁻¹)	95% ile flow (m³s⁻¹)	Sensitivity error (%)	Required accuracy of stage measurement (mm)
004001	Conon	VA	961.8	45.62	8.59	5.5	9
007001	Findhorn	VA	415.6	13.22	2.05	13.9	4
012001	Dee	VA	1370.0	36.40	8.40	5.2	10
015006	Tay	VA	4587.1	158.10	42.84	1.9	26
021009	Tweed	VA	4390.0	76.71	14.02	4.5	11
023001	Tyne	VA	2175.6	43.87	5.44	6.5	8
024005	Browney	CB	178.5	1.73	0.34	13.7	4
024009	Wear	FV	1008.3	14.78	3.29	7.8	6
025019	Leven	FV	14.8	0.20	0.06	25.0	2
027035	Aire	VA	282.3	6.04	0.52	15.9	3
027041	Derwent	C	1586.0	17.53	4.92	5.5	9
027051	Crimple	FV	8.1	0.11	0.01	54.0	1
027055	Rye	C	131.7	2.36	0.55	22.1	2
028003	Tame	VA	408.0	5.84	2.70	3.3	15
028012	Trent	VA	1129.0	12.52	5.04	3.6	14
028025	Sence	C	169.4	1.51	0.25	22.4	2
028044	Poulter	C	65.0	0.33	0.17	21.2	2
031006	Gwash	C	150.0	0.86	0.29	23.3	2
033012	Kym	CB	137.5	0.63	0.02	65.0	<1
036006	Stour	FL	578.0	2.83	0.50	7.9	6
038007	Canons Brook	FL	21.4	0.20	0.05	32.0	2
039016	Kennet	C	1033.4	9.65	3.98	6.4	8
039019	Lambourn	C	234.1	1.72	0.79	13.3	4

Table 4.6 Continued

Gauging station number	River	Station type	Catchment area (km²)	Mean flow (m³s⁻¹)	95% ile flow (m³s⁻¹)	Sensitivity error (%)	Required accuracy of stage measurement (mm)
039020	Coln	C	106.7	1.34	0.38	21.3	2
043005	Avon	C	323.7	3.43	1.15	8.9	6
048005	Kenwyn	CC	19.1	0.38	0.05	15.6	3
049004	Gannel	C	41.0	0.69	0.10	38.2	1
052010	Brue	C	135.2	1.89	0.26	21.5	2
053017	Boyd	FV	48.0	0.57	0.05	27.5	2
054004	Sowe	C	262.0	2.94	1.03	8.6	6
054012	Tern	FV	852.0	7.09	2.41	4.2	12
054019	Avon	C	347.0	2.50	0.48	15.0	3
056001	Usk	VA	911.7	27.67	4.34	5.1	10
065005	Erch	C	18.1	0.60	0.09	45.0	1
090003	Nevis	VA	76.8	6.28	0.57	8.8	6

Station types
VA   Velocity–area   C   Crump weir
CB   Compound broad-crested weir   CC   Compound crump weir
FV   Flat vee weir   FL   Flume

Note: The 'sensitivity error' referred to in the table relates to the percentage change in flow associated with a 10 mm change of water level at a stage corresponding to the 95% exceedence flow. The 'required accuracy of stage measurement' is that necessary to restrict errors in computed flows (from this source), at the 95% exceedence flow, to less than 5%.

4.8 Semi-logarithmic graph paper

On this type of graph paper values of h are plotted linearly on the ordinate and discharge values plotted logarithmically on the abscissa. Referring to Fig. 4.15

$$\frac{h - h_1}{h_2 - h_1} = \frac{\log Q - \log Q_1}{\log Q_2 - \log Q_1} \tag{4.40}$$

$$= \frac{\log Q/Q_1}{\log Q_2/Q_1}. \tag{4.41}$$

Therefore

$$\frac{\log Q}{Q_1} = \frac{\log Q_2/Q_1}{h_2 - h_1}(h - h_1) \tag{4.42}$$

and letting

$$\frac{\log Q_2/Q_1}{h_2 - h_1} = A \tag{4.43}$$

then

$$\frac{\log Q}{Q_1} = A(h - h_1). \tag{4.44}$$

Therefore

$$\frac{Q}{Q_1} = 10^{A(h - h_1)} \tag{4.45}$$

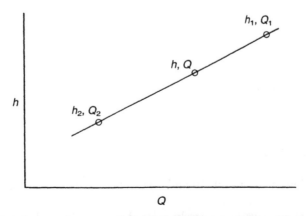

Figure 4.15 Determining the stage–discharge equation on semi-logarithmic graph paper. ('h' plotted linearly and 'Q' plotted logarithmically).

and

$$Q = \frac{Q_1}{10^{Ah_1}} \times 10^{Ah} \tag{4.46}$$

and letting

$$\frac{Q_1}{10^{Ah_1}} = B \tag{4.47}$$

then

$$Q = B \cdot 10^{Ah}. \tag{4.48}$$

An example of a stage–discharge curve at the Ganges at Faraka, plotted on semi-logarithmic graph paper is shown in Fig. 4.16. It can be seen that the curve has three break points dividing the relation into four straight-line segments having a gauge zero at 14.5m. The rating equations are derived as follows:

(1) The following values are obtained from the curve and equations (4.43), (4.47) and (4.48)

$$Q_1 = 1500 \quad h_1 = 0.50$$
$$Q_2 = 2500 \quad h_2 = 1.60$$

$$A = \frac{\log Q_2/Q_1}{h_2 - h_1} = \frac{\log \dfrac{2500}{1500}}{1.60 - 0.50} = 0.202$$

$$B = \frac{Q_1}{10^{Ah_1}} = \frac{1500}{10^{0.202 \times 0.50}} = 1190$$

then

$$Q = B \cdot 10^{Ah}$$
$$= 1190 \cdot 10^{0.202h}$$

and

$$Q = 1190 \cdot 1.59^{h}. \tag{4.49}$$

Similarly rating equations (2), (3) and (4) can be calculated and these are shown in Fig. 4.16.

Stage–discharge curves are rarely plotted on semi-logarithmic graph paper but may have some advantages in exceptional cases.

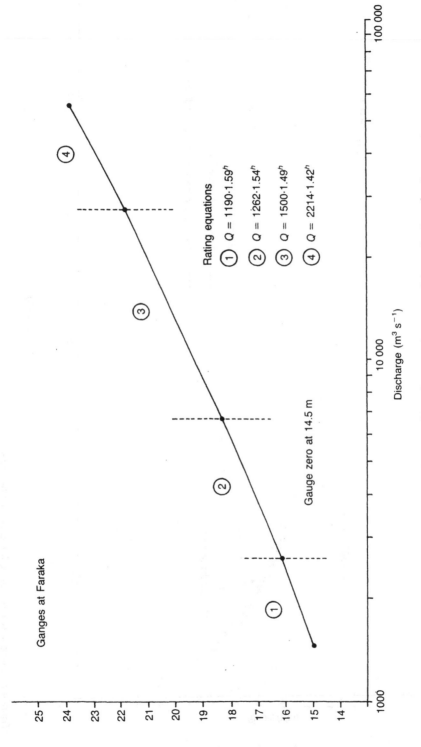

Figure 4.16 Example of stage–discharge curve plotted on semi-logarithmic graph paper and determination of rating equations.

Ganges at Faraka

Gauge zero at 14.5 m

Discharge ($m^3 s^{-1}$)

Rating equations

1   $Q = 1190 \cdot 1.59^h$

2   $Q = 1262 \cdot 1.54^h$

3   $Q = 1500 \cdot 1.49^h$

4   $Q = 2214 \cdot 1.42^h$

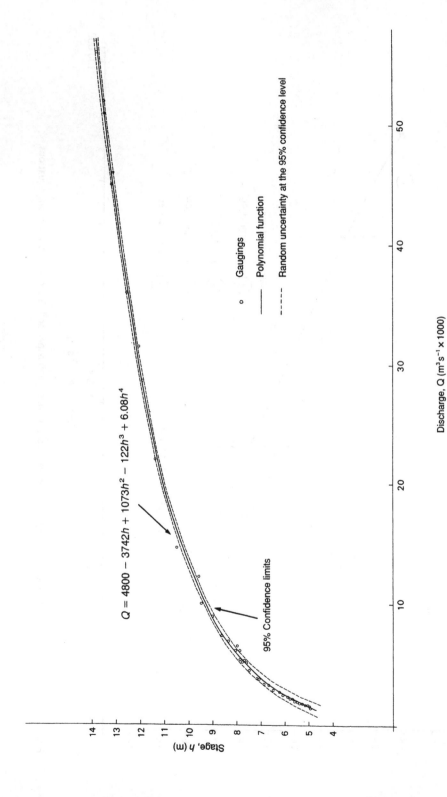

$Q = 4800 - 3742h + 1073h^2 - 122h^3 + 6.08h^4$

95% Confidence limits

Gaugings
Polynomial function
Random uncertainty at the 95% confidence level

Stage, h (m)

Discharge, Q (m³ s⁻¹ × 1000)

Figure 4.17 Stage–discharge curve fitted by fourth-degree polynomial.

4.9 Polynomial curve fitting

Values of h are plotted on the ordinate and corresponding discharges on the abscissa on arithmetic graph paper. The general polynomial expression is

$$Q = b_0 + b_1 h + b_2 h^2 + \ldots + b_m h^m \text{ m}^3 \text{ s}^{-1}. \tag{4.50}$$

Procedures for fitting a quadratic, cubic or higher-degree polynomial to a set of stage–discharge data using the least squares criterion, and of assessing the uncertainty associated with the resulting stage–discharge curve are described in ISO 7066/2. Because it is generally not practicable to carry out this type of curve fitting and assessment of uncertainty without using a computer, ISO 7066/2 provides a FORTRAN programme for this purpose. An example of a fourth-degree polynomial fitting for the curve in Fig. 4.16 is presented in Fig. 4.17. To compare these figures graphically it should be noted that the ordinate scale in Fig. 4.17 requires to be adjusted by adding the value of 10 to each stage value. Using the equations for comparative purposes we have, for the polynomial

$$Q = 4800 - 3742h + 1073h^2 - 122h^3 + 6.0811 \text{ m}^3 \text{ s}^{-1} \tag{4.51}$$

and at a stage value of 10 m, say,

$$Q = 4800 - 3742 \times 10 + 1073 \times 10^2 - 122 \times 10^3 + 6.0811 \times 10^4$$
$$= 13\,490 \text{ m}^3 \text{ s}^{-1}. \tag{4.52}$$

From Fig. 4.16 the corresponding stage is $(20 - 14.5) = 5.5$ m and from rating equation (3) (see Fig. 4.16)

$$Q = 1500 \times 1.49^{5.5}$$
$$= 13\,450 \text{ m}^3 \text{ s}^{-1}. \tag{4.53}$$

As these results are within about 0.3% of each other, it would appear that the polynomial procedure may have some potential in fitting stage–discharge curves having break points or inflexions which cannot be treated satisfactorily by other means. However, some user experience is still required with this method before it is accepted as an alternative to the existing methods.

Further reading

Ezekiel, M. and Fox, K.A. *Methods of Correlation and Regression Analysis*. John Wiley and Sons, New York 1996.

Framji, K.K. Determination of stage–discharge relations for estimation of discharge from gauges. Contribution to *Field Methods and Equipment Used in Hydrology and Hydrometeorology*. United Nations Secretariat, New York 1962.

Gerhard, H. Extrapolation von Abflusskurven den Geschwindigkeitflachen–Die Wasserwirtschaft 9 Stuttgart 1971.

Guy, H.P. Fluvial Sediment Concepts. *Techniques of water resources investigations of the U.S.G.S. Book 3 Chapter C1* 1970.

ISO 7066/2. *Uncertainty in linear calibration relations ISO Geneva* 1997.

ISO 1100/2. *Liquid Flow Measurement in Open Channels: Determination of the Stage–Discharge Relation.* ISO, Geneva, Switzerland 1998.

ISO 7066/2. *Uncertainty in Non-Linear Calibration Curves.* ISO, Geneva 1988.

Kennedy, E. J. Discharge Ratings at Gauging Stations. *US Geological Survey, Book 3, Ch. A10* Washington, DC 1984.

Lambie, J. C. Measurement of flow–velocity–area methods. In *Hydrometry: Principles and Practices* (ed. R. W., Herschy). John Wiley and Sons, Chichester 1978.

Marsh, T. J. Aspects of river flow measurement and hydrometric data interpretation in the United Kingdom – with particular reference to low flows. *Flow Meas. Instrum.,* 4(1), 39–45 1993.

Otnes, J. Instruments, methods of observation and processing of basic data and water levels and streamflow. Contribution to *Hydrometeorolocal Observations and Networks in Africa.* WMO 1969.

Rantz, S.E. *Charactertistics of Logarithmic rating curves.* U.S.G.S. Water Supply Paper 1892 1968.

Schenck, Jr H. *Theories of engineering experimentation.* McGraw-Hill, New York 1968.

Tilrem, O. A. Manual on Procedures in Operational Hydrology *Ministry of Water, Energy and Minerals, of Tanzania and the Norwegian Agency for International Development* 1979.

World Meteorological Organization. Manual on Stream Gauging. WMO 519 1980.

Chapter 5

Special problems in streamflow measurement

5.1 Depth corrections for sounding line and weight

When current meter measurements are made with a cable-suspended current meter in deep swift water, the current meter and the sounding weight will be carried downstream for a certain distance before the weight touches the bottom. This is often the case when measurements are made from a bridge or cableway which is at a vertical distance above the water surface. In such cases, corrections have to be applied in order to determine the correct depth of water and the depth to which the current meter should be lowered.

Figure 5.1(a) shows the position assumed by the sounding line as the weight, just off the bed of the stream, is supported by the line only. It is seen that from the length of line af, the distance ae and the difference between the length of ef and bc must be deducted in order to determine the depth bc, assuming the stream bed cf is horizontal. Both these corrections are functions of the vertical angle θ and are given in Tables 5.1 and 5.2.

The values in Tables 5.1 and 5.2 are based on the assumptions that the drag force on the weight in the comparatively still water near the bottom can be neglected and that the sounding line and weight are designed to offer little resistance to the flow of water. The uncertainties in these assumptions are such that significant errors may be introduced if the vertical angle is more than 30°. If the direction of flow is not normal to the measuring cross-section, the corrections as shown in the tables will be too small.

The same conditions that cause errors in sounding the depth of the stream also cause errors in placing the current meter at the selected depths in the vertical. The correction tables are not strictly applicable to the problem of placing the current meter because of the increased drag force on the current meter assembly caused by the higher velocities encountered as the current meter assembly is raised from the streambed. A current meter placed in deep swift water by the ordinary methods for observations at selected fractions of the depth will therefore be too high in the vertical. The use of the correction tables will tend to eliminate this error in placing the current meter and, although not strictly applicable, their use has become general for this purpose.

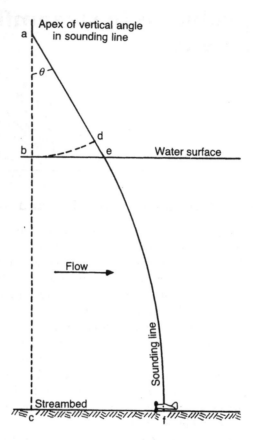

Figure 5.1(a) Assumed position taken up by sounding line and weight in deep swift water. Corrections to be made to determine depth, *bc* are air-line correction, *de* and wet-line correction as a percentage of *ef*.

The routine procedure for applying depth corrections is as follows (Fig. 5.1(a)):

(a)   Measure the vertical distance *ab* from the guide pulley on the gauging reel to the water surface. This will give the vertical distance to be used with the air-line correction table (Table 5.1).

(b)   Place the bottom of the weight at the water surface and set the depth counter on the gauging reel to read zero.

(c)   Lower the sounding weight to the bed of the stream. Read and record the sounded depth *df* and the vertical angle θ of the cable when the weight is at the bed of the stream, but entirely supported by the cable.

(d)   With the aid of the correction tables compute and record:

    (i)   the air correction *de* as a percentage of *ab* (Table 5.1);

(ii) the wet-line depth $ef = df - de$;

(iii) the wet-line correction as a percentage of ef (Table 5.2);

(iv) add both corrections together and subtract them from the sounded depth df; this will give the revised depth bc

(v) if the $0.2D + 0.8D$ method (See Chapter 2 Section 2.4) is used to locate the current meter for velocity measurement, raise the current meter from the sounding position at the streambed a distance equal to 0.2 of the wet-line depth ef minus the distance from the current meter to the bottom of the weight; this places the current meter approximately at the 0.8 depth position;

(vi) raise the current meter to the surface of the water and set the depth counter to read ae then lower the current meter until it is at a distance equal to ae plus 0.2 of the wet-line depth ef; this places the current meter approximately at the 0.2 depth position.

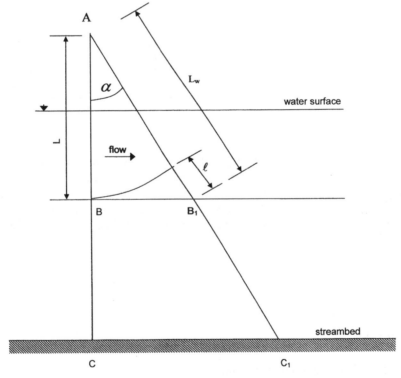

Figure 5.1(b) Position of current meter affected by velocity of water.

Table 5.1 Air-line correction

Vertical angle (degrees)	Correction (%)	Vertical angle (degrees)	Correction (%)
4	0.24	18	5.15
6	0.55	20	6.42
8	0.98	22	7.85
10	1.54	24	9.46
12	2.23	26	11.26
14	3.06	28	13.26
16	4.03	30	15.47

Table 5.2 Wet-line correction

Vertical angle (degrees)	Correction (%)	Vertical angle (degrees)	Correction (%)
4	0.06	18	1.64
6	0.16	20	2.04
8	0.32	22	2.48
10	0.50	24	2.96
12	0.72	26	3.50
14	0.98	28	4.08
16	1.28	30	4.72

Examples

In a gauging measurement in a deep swift stream using cableway, the total depth by sounding line and weight was found to be 7.55 m. The depth from the guide pulley to the surface was measured as 3.00 m and the vertical angle was 20°. Find the true depth and the positions for the current meter using the two point (0.2D and 0.8D) method. The distance from the centre line of the current meter to the bottom of the weight was 0.3 m.

From Fig. 5.1 (a)

$af = 7.55$ m

$ab = 3.00$ m

$\theta = 20°$.

Therefore $df = 7.55 - 3.00 = 4.55$ m.
Air-line correction, de, from Table 5.1 for 20° = 6.42%,
therefore air-line correction = 6.42% × 3.00 = 0.19 m.
Therefore wet-line depth ef = df − de = 4.55 − 0.19 = 4.36 m.
Wet-line correction, from Table 5.2 for 20° = 2.04%,

therefore wet-line correction = 2.04% × 4.36 m = 0.09 m.
Therefore true depth bc = 4.36 − 0.09 = 4.27 m.
   To place the current meter at the 0.8D position in the vertical

   $0.2 \times$ wet-line depth $= 0.2 \times 4.36 = 0.87$ m.

Therefore the current meter should be raised a distance of (0.87 − 0.3) m = 0.57 m to locate it at 0.87 m from the bed.
   To place the current meter at 0.2D from the surface, raise the meter to the surface and set the depth counter to read ae then lower the meter until it is at a distance ae plus 0.2 of the wet-line depth ef, that is,

   $(3.00 + 0.19 + 0.2 \times 4.36)$ m $= 4.06$ m.

Another method of obtaining depth and locating the current meter may be used where the cross-section is stable. The bed cross-section is surveyed and the water levels related to the reference gauge. At any stage therefore the depth at each vertical may be taken either from the cross-sectional drawing or from a table prepared from that drawing. To locate the meter for velocity measurement the actual depth bc is related to the reading ef (Fig. 5.1(a)). In the above example

$$\frac{ef}{bc} = \frac{4.36}{4.27} = 1.021.$$

Then

   $0.2 \times 4.27 \times 1.021 = 0.87$ m.

Therefore the current meter is located at 0.87 m from the surface and 0.87 m from the bed as before. The actual operation in placing the meter is carried out as detailed above.
   Figure 5.1(b) shows a typical situation where the position of a current meter is affected by the velocity of the water. The required position for the current meter is at point B a distance L from the point of suspension. The velocity of the water will cause the current meter to be pulled downstream so that the angle of suspension is a.
   From Fig 5.1(b)

$$\cos a = \frac{L}{L_w}$$

$$and\ L_w = L + \ell$$

$$\cos a = \frac{L}{L + \ell}$$

$$\therefore L + \ell = \frac{L}{\cos a}$$

$$\therefore \ell = \frac{L}{\cos a} - L.$$

If the cross-section is pre-surveyed then the value of L is either known or can be measured on site. The angle a can be measured and the correction factor ℓ_1 calculated. The process may be iterative as the angle a will alter as the current meter is lowered into the water. This method assumes that the bed is horizontal between points C and C1. If the cross-section is not pre-surveyed then the current meter can be lowered to the bed at point C1 and the distance AC1 measured. At the same time the angle a can be measured and thus the distance AC can be calculated

$$\cos a = AC/AC1.$$

Echo sounders or ultrasonic devices are also used to determine the actual depths of flow at each vertical. These devices may also be incorporated in the sounding weight as shown in Fig. 2.16. The location of the meter, however, follows the same procedure whether the cable suspension is from cableway, boat or bridge (Fig 5.2).

It is evident that the procedure outlined above requires to be carried out with care in order to avoid magnifying the errors involved. There are also two main difficulties involved in the measurement from a cableway: (a) employing a sufficiently heavy sounding weight to maintain the suspension in an almost vertical position, and (b) measuring the vertical angle θ. The main uncertainties involved in the measurement are:

(a) Unless the cross-sections within the area of measurement are substantially similar there will be an additional error in depth because the sounding weight is dragged downstream. The depth measured therefore will not be that at the measuring cross-section. This uncertainty can be disregarded if an echo sounder is used to measure depth.
(b) The location of the meter is approximate because of the sounding weight being dragged downstream. The meter will therefore be unavoidably located higher in the vertical than it should. In addition the velocity will be measured downstream of the measuring cross-section.
(c) The measurement of angle θ.
(d) The assumptions made in developing the air-line and wet-line correction tables (Tables 5.1 and 5.2).

5.2 Oblique flow

The angle of the current, as applied to stream gauging, is the difference between the normal to the measuring section and the angle made by the current with the

Figure 5.2 Direct depth measurement by sounding weight.

measuring section. To eliminate errors introduced by such angles, it is necessary to obtain the component of the velocity normal to the cross-section. To make the correction for oblique flow for a rod-supported meter the current meter is held rigidly in line with the direction of flow (Fig. 5.3(a)). For a current meter suspended from a cable the meter will automatically take up the direction of the current. The velocity in the direction of the current and the angle of deflection are measured. The measured velocity when multiplied by the cosine of the angle of the current gives the velocity component normal to the measuring cross-section. For small angles, say less than 8°, the correction is negligible.

Propeller-type current meters with a component propeller measure the velocity component normal to the measuring section in oblique flow up to an angle of 45° and no correction is therefore necessary if the current meter is held rigidly at right angles to the measuring section (Fig. 5.3(b)).

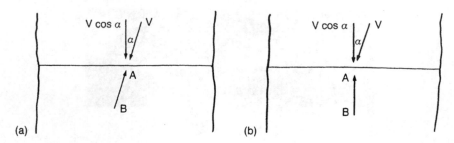

Figure 5.3 Correction for oblique flow (*a*). (a) Current meter is held rigidly in direction of flow AB; required velocity component normal to measuring section is *V* cos *a*. (b) Component propeller-type current meter is held in direction AB normal to measuring section; meter gives velocity component *V* cos *a* automatically.

The assumption is made that the point velocity corrections apply to the whole vertical or between velocity points in the vertical.

5.3 Stilling well lag and draw-down

During periods of rapid change of stage, water levels in the stilling well may lag behind those in the stream because of head loss in the intake system. This phenomenon is known as **stilling well lag** and may occur on a rising or falling stage. In addition, the protruding end of the intake pipe may cause a disturbance of flow past the pipe producing a reduction of the water level in the stilling well. This effect is known as **draw-down**.

Stilling well lag

The stilling well lag is a measure of the head loss in the intake system. Referring to Fig. 5.4 which shows the falling stage case:

Let h_L = lag or head loss between the water level in the well and the water level in the stream (m);

A = area of stilling well (m^2);

a = area of intake pipe (m^2);

λ = a non-dimensional coefficient dependent on the relative roughness and Reynolds number (see Section 16.2);

D = diameter of stilling well (m);

d = diameter of intake pipe (m);

L = length of intake pipe (m);

$\dfrac{\Delta h}{\Delta t}$ = rising velocity in stilling well (m s^{-1});

n = number of intake pipes;

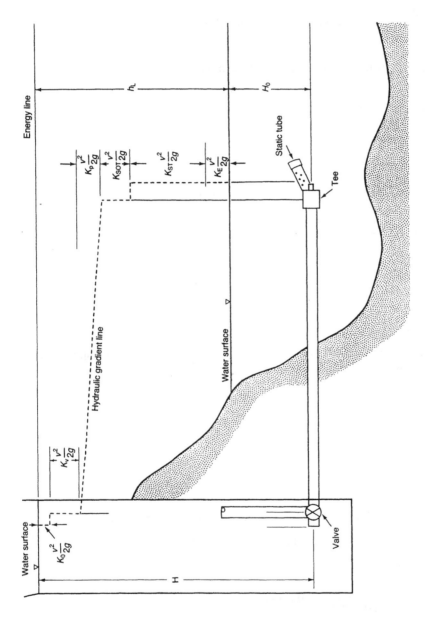

Figure 5.4 Schematic view of stilling well and intake system to determine stilling well lag; diagram shows case of lag due to falling stage.

$\dfrac{0.5v^2}{2g}$ = head loss due to entry or exit (intake pipe) (m);

$\dfrac{2.0v^2}{2g}$ = head loss due to a valve if fitted (m);

$\dfrac{v^2}{2g}$ = head loss due to static tube if fitted (m).

Then the velocity in the intake pipe is

$$v = \frac{A}{a} \frac{\Delta h}{\Delta t} \frac{1}{n} \text{ m s}^{-1}.$$

From Bernoulli's theorem[1] and equation 10.4 (Chapter 10)

$$h_L = \left(\frac{0.5v^2}{2g} + \frac{2.0v^2}{2g} + \frac{v^2}{2g} + \frac{\lambda L v^2}{2gd} \right) \tag{5.1}$$

$$= \frac{v^2}{2g} (0.5 + 2.0 + 1 + \lambda L/d). \tag{5.2}$$

Example

Determine the stilling well lag on a rising stage given the following information (there are no valves or static tubes):

$L = 10$ m

$D = 1.00$ m

$d = 0.076$ m

$\dfrac{\Delta h}{\Delta t} = 5$ mm s^{-1}

$k = 0.03$ mm (effective roughness) (Chapter 15, Section 15.2)

$v = 1.14 \times 10^{-6}$ (kinematic viscosity) (Chapter 15)

$n = 1$.

$\dfrac{k}{d} = \dfrac{0.03}{1000 \times 0.076} = 0.0004,$

---

1   Bernoulli was one of the foremost scientists during the 18th century and his theorems greatly accelerated the beginning of hydraulic and hydrostatic studies.

$$v = \frac{A}{a}\frac{\Delta h}{\Delta t}\frac{1}{n} = \frac{0.785}{0.0045} \times 0.005 = 0.872 \text{ m s}^{-1}.$$

Therefore

$$R_e = \frac{0.872 \times 0.076}{1.14 \times 10^{-6}} = 0.6 \times 10^5$$

and from Fig. 15.8 $\lambda = 0.02$ (Chapter 15).
From equation (5.1)

$$h_L = \text{lag} = \frac{0.872^2}{19.62}\left[0.5 + \frac{0.02 \times 10}{0.076}\right]$$

$$= 0.122 \text{ m}.$$

It can be seen from the above that the lag varies inversely with the square of the number of intakes for a given rate of change of stage and in direct proportion to the square of the ratio of well area to intake area. Thus the lag for a given well would be reduced by a factor of four if two intakes were used instead of one and by a factor of nine if three intakes were used.

As an aid to design, curves may be constructed by plotting different values of lag h_L against the rate of change of stage $\Delta h/\Delta t$, h_L against A and h_L against a.

Draw-down

Unless the stream end of the intake pipe is protected from the dynamic effects of the water flowing past it, there may be a draw-down (rising stage) or a building up of the height of water in the well (falling stage) as compared with the height of water in the river channel at the end of the pipe. These differences in height may be as much as 0.3 m, and they vary not only for different stages but also for the same stage of the river. Stilling wells attached to bridge piers or abutments are generally provided with small openings whereby the water is admitted directly into the wells, without the use of intake pipes. Under these circumstances the dynamic effects of the water may be even more pronounced than those experienced where the connection is made through intake pipes. Except in rivers where heavy silt loads make the use of intake pipes impracticable, it is sometimes possible to make intake connections to stilling wells on bridge piers and abutments similar to that used with stilling wells in river banks. Various devices attached to the ends of intake pipes are used in an attempt to eliminate the so-called 'draw-down', or difference in the heights of the water in the well and in the river.

Static tubes

The static tube appears to be the most satisfactory device for eliminating draw-down at intakes to gauge wells.

The static tube device may be made in various designs and in any size corresponding to the size of the intake pipe with which it is to be used. A typical design of static tube showing dimensions and arrangement of openings is given in Fig. 5.5. A suitable arrangement of openings to provide the required area for a 100 mm diameter static tube may be obtained by using six rings of holes with six holes in each ring, the rings being about 25 mm apart and the holes staggered in alternate rings as shown in Fig. 5.5.

Generally the minimum length of static tube should be about four times the diameter of the intake pipe and the total area of the holes should be about 20–25% greater than the cross-sectional area of the pipe, in order that deposits of mud and silt may be effectively removed by flushing of the intake. The outer end of the tube may be threaded for a standard pipe coupling and closed by means of a standard plug, or it may be closed by a plug specially fitted to the inside of the pipe. If preferred, a cap may be used instead of a coupling and plug.

The static tube may be connected to the intake pipe by means of a 90° elbow or by a standard tee. If the intake pipe points upstream from an artificial control, the connection may be made as shown in Fig. 5.6. The tube is fixed securely in a horizontal position, and provision is made so that it will not tilt downward by the turning of the coupling under its own weight or any additional weight that may be placed upon it. In using a standard tee connection the static tube is attached to the leg of the tee, the run of the tee opposite the well being closed with a plug. Figure 5.7 shows a typical installation of a static tube.

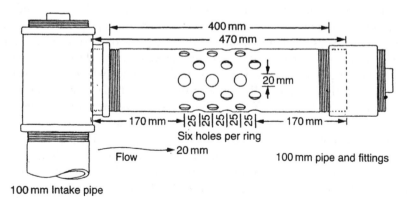

Note: Edges of holes are free of burrs but not rounded.
The tube is fixed in a horizontal position.

Figure 5.5 Typical design of static tube to mitigate draw-down.

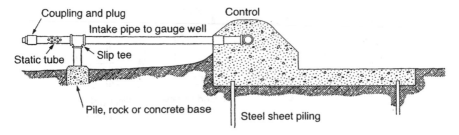

Figure 5.6 Static tube connected to intake pipe from an artificial control.

Figure 5.7 Typical installation of intake pipe fitted with static tube.

Baffles

A baffle piece inserted in the end of the intake pipe may be used instead of a static tube to mitigate draw-down if the direction of flow is at an angle of 90° to the intake. However, the baffle is not so effective if the direction of flow is at an angle considerably greater or less than 90°.

A typical baffle for a 100 mm diameter intake is shown in Fig. 5.8. The length of baffle for a 50 mm diameter intake may be taken as about 100 mm and for a 75 mm diameter intake as about the same as for the 100 mm diameter intake.

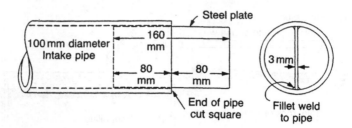

Baffle to be fixed in a vertical position normal to the current

Figure 5.8 Baffle inserted in end of intake pipe as an alternative device to a static tube.

5.4 Rapidly changing discharge

At stream gauging stations located in a reach where the slope is very flat, the stage–discharge relation is frequently affected by the superimposed slope of the rising and falling limb of a passing flood wave. During the rising stage, the velocity and discharge are normally greater than they would be for the same stage under steady flow conditions. Similarly, during the falling stage the discharge is normally less for any given gauge height than it is when the discharge is constant.

The method used in developing rating curves at single-gauge stations is, as discussed in Chapter 4, to draw a median curve through a scatter of plotted discharge measurements. This procedure gives a correct result when all discharge measurements are made under steady or nearly steady flow conditions. In fact, if each plotted measurement had been tagged as to whether it had been measured on a rising or falling stage, the curve might have taken the shape of a loop (Fig. 5.9(a)). This effect is especially noticeable for larger rivers having very flat slopes with channel control extending far downstream. For smaller rivers having section control or steeper slopes, and the measuring section not too far from the site where the stage is observed, the looping effect is seldom of such a magnitude as to have any practical consequence. The looping effect is due to several causes. The first is channel storage. If a discharge measurement is made at some distance from the station control during a period of rising or falling stage, the discharge passing the measuring section will not be the same as the discharge at the control: a correction for the channel storage has to be applied to the measured discharge by adding or subtracting from the measured discharge a quantity equal to the product obtained by multiplying the water surface area between the measuring section and the control by the average rate of change in stage in the same river reach. If the measurement is made above the control, the correction will be positive for falling stages and negative for rising stages. If made below, it will be the converse.

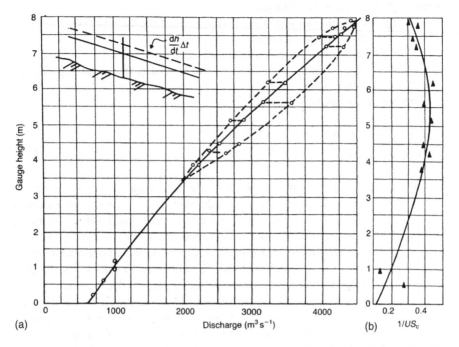

Figure 5.9 A typical loop rating (a) and the US_c rating curve (b) during conditions of rapidly changing discharge.

Example

A measurement is made 1000 m upstream from the control; average width of channel in the reach is 100 m; average rate of rise of water surface in the reach during measurement is 0.15 m h^{-1}; measured discharge is 120 m^3 s^{-1}.

Then, the rate of change of storage in the reach is given by

$$ds = 1000 \times 100 \times 0.15 = 15\,000 \text{ m}^3 \text{ h}^{-1}$$
$$= 4.2 \text{ m}^3 \text{ s}^{-1}.$$

The discharge measurement is plotted as $(120 - 4)$ m^3 s^{-1} = 116 m^3 s^{-1} (rounded) since this is the discharge passing the control to which the mean gauge height during the measurement corresponds.

The second reason for the looping of rating curves is the variation in surface slope which occurs as a flood wave passes a river gauging station. Discharge measurements taken on either side of a flood wave may be corrected to the theoretical steady-state condition by application of the following equation

$$\frac{Q_m}{Q_r} = \left(1 + \frac{1}{US_c}\frac{dh}{dt}\right)^{1/2} \tag{5.3}$$

where Q_m = measured discharge;
        Q_r = estimated steady-state discharge from the rating curve;
        U = wave velocity (celerity);
        S_c = energy slope for steady-state flow;
        dh/dt = rate of change of stage, positive for rising stage and negative for falling stage.

Rearranging equation (5.3) gives

$$\frac{1}{US_c} = \frac{(Q_m/Q_r)^2 - 1}{dh/dt}. \tag{5.4}$$

If a sufficient number of measurements have been made at a gauging station during both rising and falling stage and under steady-state conditions, equation (5.4) may be solved for Q_r by a graphical method, the so-called Boyer method, without having to compute the energy slope and the velocity of the flood wave.

The discharge measurements are plotted in the usual manner and a rating curve is drawn as a median curve through the uncorrected values (Fig. 5.9(a)). The steady-state discharge Q_r is estimated from this median curve. Q_m and dh/dt have been measured and are therefore known quantities; by substituting in equation (5.4), the term $1/US_c$ is obtained for each discharge measurement. This term is plotted against stage and a mean curve fitted to the plotted points (Fig. 5.9(b)). From the $1/US_c$ against stage relation, new smoothed values are obtained and inserted in equation (5.4) in order to obtain the steady-state Q_r. The new values of Q_r are then plotted against stage to obtain a corrected steady-state rating curve.

Example

Referring to Fig. 5.9(a) on a rising stage, the measured discharge is 3550 m³ s⁻¹, the stage measurement is 5.60 m and the rate of change of stage is 0.11 m s⁻¹. Calculate the steady-state value of discharge (equation (5.3))

$$\frac{Q_m}{Q_r} = \left(1 + \frac{1}{US_c}\frac{dh}{dt}\right)^{1/2}.$$

From Fig. 5.9(b), US_c is 0.42 for a stage value of 5.60. Then

$$\frac{3550}{Q_r} = \left(1 + \frac{1}{0.42} \times 0.11\right)^{1/2}$$

$$Q_r = \frac{3550}{1.123}$$

$$= 3160 \text{ m}^3 \text{ s}^{-1}$$

which is the steady-state value of discharge through which the steady-state Q_c rating curve is drawn. It will be noted that to determine the discharge from the Q_c rating curve for known values of stage and rate of change of stage, the above procedure is reversed, the required discharge being Q_m.

For gauging stations situated in tidal reaches with significant unsteady flow, calculation of the discharge is generally carried out by special methods such as the method of cubature and unsteady flow mathematical modelling.

The method of cubature is based on the law of continuity. The rate of rise and fall of the water surface is used to determine the rate of gain and loss of channel storage in a reach. The discharge at the downstream end of the channel reach is calculated from the known inflow to the reach and the computed gain or loss in channel storage during the time required for the water surface to rise and fall.

Unsteady flow mathematical models are based on assumptions of moderately unsteady, homogeneous and one-dimensional flows and prismatic channel geometry. On these assumptions, a system of unsteady flow equations can readily be set up to describe the tidal flow. Initial and boundary conditions are determined by field measurements. The actual computation of discharge is conveniently performed by digital computer.

5.5 Shifting control

General

Shifts in the control features occur especially in alluvial sand-bed streams. However, even in stable stream channels shifts will occur, particularly at low flow because of weed growth in the channel, or as a result of debris caught in the control section.

In alluvial sand-bed streams, the stage–discharge relation usually changes with time, either gradually or abruptly, due to scour and silting in the channel and because of moving sand dunes and bars. These variations will cause the rating curve to vary with both time and the magnitude of flow, and observations and measurements have to be carried out in the best way possible.

Characteristics of sand-bed channels

In sand-bed channels, the configuration of the bed varies with the magnitude of the flow. The bed configurations occurring with increasing discharge are ripples, dunes, plane bed, standing waves, anti-dunes and chute and pool (Fig. 5.10). The bed forms are associated with a particular mode of sand movement and with a particular range of resistance to the flow of water. The resistance to the flow is greatest in the dunes' range. When the dunes are washed out and the sand is rearranged to form a plane bed, there is a marked decrease in bed roughness and resistance to the flow causing an abrupt discontinuity in the stage–discharge relation.

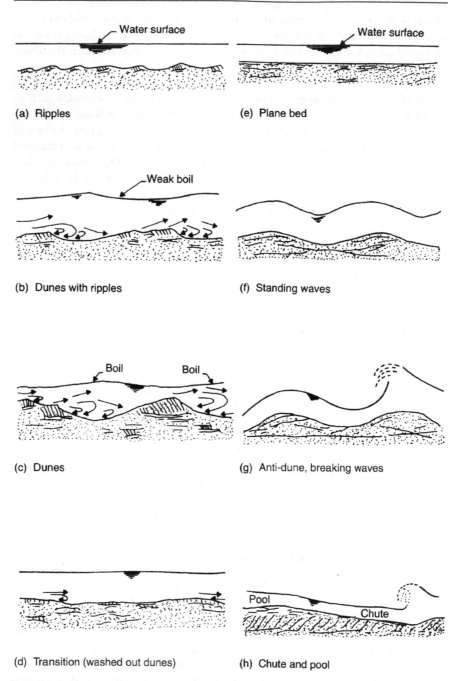

Figure 5.10 Bed and surface configurations found in sand-bed channels.

The sequence of bed configurations shown in Fig. 5.10 is arranged as developed by increasing discharge. The bed configurations are grouped into two regimes. The lower regime, a–c, occurs with lower discharges, the upper regime, e–h, with higher discharges; an unstable discontinuity, d, in the depth–discharge relation appears between these more stable regimes.

Fine sediment present in the water influences the configuration of the sand-bed and thus the resistance to flow. A concentration of fine sediments in the order of 40 000 mg litre^{-1} may reduce the resistance to flow in the dune regime by as much as 40%. Thus the stage–discharge relation for a stream may vary with the sediment concentration if the water is heavily loaded with fine sediments.

Changes in water temperature may also alter the bed form, and hence roughness and resistance to flow in sand-bed channels. The viscosity of the water will increase with lower temperatures and thereby the mobility of the sand will increase.

Discharge rating of sand-bed channels

For sand-bed streams where neither bottom nor sides are stable, a plot of stage against discharge will very often scatter widely and thus be indeterminate (Fig. 5.11). By changing variables, however, a hydraulic relationship will become

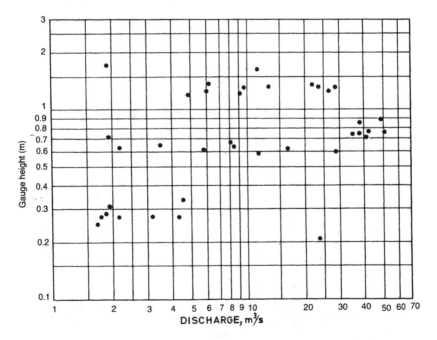

Figure 5.11 Plot of discharge against stage for a sand-bed channel with indeterminate stage–discharge relation.

apparent. The effect of variation in bottom elevation is eliminated by replacing stage by mean depth (hydraulic radius). The effect of variation in width is eliminated by using mean velocity instead of discharge.

Plots of mean depth against mean velocity are very useful in the analysis of stage–discharge relations, provided the measurements are referred to one and the same cross-section. These plots will identify the bed-form regime associated with each individual discharge measurement (Fig. 5.12). Thus only the measurements associated with the upper flow regime are used to define the upper part of the rating curve and, similarly, only measurements identified with the lower flow regime are used to define the lower part. Measurements made in the transition zone will scatter widely and may not be taken as representing shifts in the more stable parts of the rating.

Knowledge of the bed-form which existed at the time of the individual discharge measurements is helpful in developing discharge ratings. Indication of bed-forms may be obtained by visual observation of the water surfaces.

A very smooth surface indicates a plane bed, large boils and eddies indicate dunes, standing waves indicate smooth bed waves in phase with surface waves, and breaking waves indicate anti-dunes. The visual observations of the water surfaces should be recorded on the note sheet when making discharge measurements in sand-bed channels.

The upper part of the stage–discharge relation associated with the upper flow regime for a sand-bed channel is usually comparatively stable. The middle part

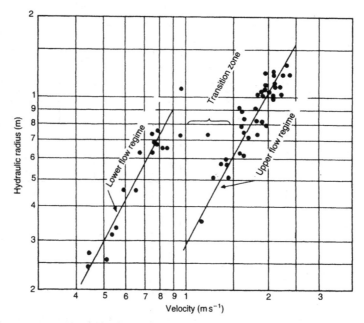

Figure 5.12 Relation of mean velocity to hydraulic radius of channel in Figure 5.11.

of the stage–discharge relation associated with the transition zone between the upper and lower regimes varies almost randomly with time, and frequent discharge measurements are necessary in order to define this part of the relation.

At low flow when the water is not covering the whole width of a sand-bed channel, the flow tends to meander in the course of time. Under this condition, it is not possible to observe a systematic gauge height and a record of the discharge can therefore not be obtained.

A continuous definition of the stage–discharge relation for a sand-bed stream at low flow is difficult. If at all feasible, a permanent control structure for the lower flow may be considered in these cases.

The Stout method

For making adjustment for shifting control the Stout method is commonly used. In this method, the gauge heights corresponding to discharge measurements taken at intervals are corrected so that the discharge values obtained from the established rating curve may be the same as the measured values. From the plot of these corrections against the chronological dates of measurements, a gauge height correction curve is made. Corrections from this curve are applied to the recorded gauge heights for the intervening days between the discharge measurements.

A staff gauge is established at the best available site on the river and readings taken at appropriate intervals, say once a day. Discharge measurements are made as often as found necessary, and may be required as often as once or twice a week. How often discharge measurements need to be taken depends on several factors, such as the hydraulic conditions in the river, the accuracy and the feasibility based on economic and other factors.

The measurements are plotted against observed gauge heights on arithmetic graph paper and a median curve is fitted to the points. Most of the subsequent discharge measurements will deviate from the established curve. For points lying above the curve, a small height, Δh, is subtracted from the observed gauge height in order to make these points lie on the curve. That is, minus corrections are applied to all points above the curve and plus corrections are applied to points lying below the curve (Fig. 5.13(a)).

Next, a correction graph is made as shown in Fig. 5.13(b). The plus and minus corrections are plotted on the date of measurement and the points connected by straight lines or a smooth curve. Gauge height corrections for each day are now obtained directly from this correction graph, remembering that the parts of the graph below the abscissa axis give minus corrections and the parts above give plus corrections.

When discharge measurements plot within 5% of the rating curve, with some plus and some minus deviations, it is acceptable to use the curve directly without adjustment for shifting control. For computation purposes special forms may be made.

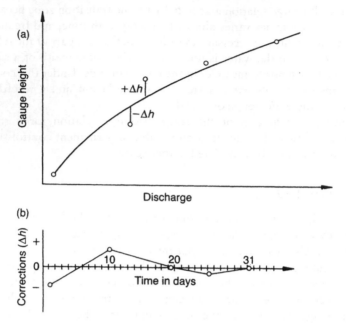

Figure 5.13 The Stout method for correcting stage readings when control is shifting.

It is not too important how the median curve is drawn between the measurements. Different curves will give different corrections and the final result will be approximately the same. Extrapolation of the curve, however, has to be done with care.

A rating of this type requires much work in order to obtain good results. The accuracy depends on the hydraulic conditions in the river and on the number and accuracy of the discharge measurements and the gauge height readings. The reliability is much less than for a station with a permanent control.

The Stout method presupposes that the deviations of the measured discharges from the established stage–discharge curve are due only to a change or shift in the station control, and that the corrections applied to the observed gauge heights vary gradually and systematically between the days on which the check measurements are taken.

In fact, the deviation of a discharge measurement from an established rating curve may be due to:

(a)   gradual and systematic shifts in the control;
(b)   abrupt random shifts in the control; and
(c)   error of observation and systematic errors of both an instrumental and a personnel nature.

The Stout method is strictly appropriate for making adjustments for the first type of error only. If the check measurements are taken frequently enough, fair adjustments may be made for the second type of error also. However, the drawback of the Stout method is that the error of observation and the systematic errors are disregarded as such and simply mixed with the errors due to shift in control, although, at times, the former errors may be of a higher magnitude than the latter. This means that 'corrections' may be applied to a discharge record when in reality the rating is correct. The apparent error is not due to shifting control but to faulty equipment or careless measuring procedure.

Erosion

A shift in control may often be caused by erosion (scour). When this occurs it will appear in the discharge rating curve as a shift to the right of the previous curve indicating an increase in the channel conveyance over that previously found for a given stage. The more common situation is that the lower part of the rating curve may change as a result of erosion of the low water control only, while the upper part of the curve may retain its original trace because the high water elements of the control come into effect at the high stages maintaining the previous stage–discharge relation. On the other hand, the opposite situation may often occur where the low water control is essentially permanent, as in the case of a rock ledge across the stream, while the high water elements of the control, such as alluvial banks, are subject to erosion during high floods.

If a radical change has affected the controlling reach downstream from the station in streams with channel control, the variation in the stage–discharge relation may extend over the entire range of stage and discharge. In such cases, some degree of parallelism may exist between the new and the old discharge rating curves.

Deposition

The effect of deposition (fill) on the controlling section and in the controlling reach is opposite to that of erosion and the discharge rating curve representing the new stage–discharge relation will be positioned to the left of the previous curve.

The effect of deposition on a low water control will be drowned out when full submergence occurs and the downstream high water controlling elements take effect. Usually, deposits on a low water section control will be washed away at high water. Typically, low water controls subject to erosion and deposition produce a series of discharge rating curves spreading out fanwise at the lower end and converging for stages above the beginning of submergence of the low water control.

Shifts in channel above the station control

Shifts and changes in the stream channel above the station control may significantly affect the velocity of approach at the gauging site as a result of changes in the slope or the cross-sectional area of the channel. If scouring of the channel takes place, the greater capacity of the forebay upstream from the control results in a lower velocity of approach and thus a decrease of the discharge for a given stage. If deposition occurs, an increase in the velocity of approach and a consequent increase in the discharge for a given stage will result.

In Chapter 4, Fig. 4.5 shows a complex stage–discharge curve for the Yangtze River due to deposition and scour among other factors.

Weed growth in the stream channel

Weed growth in the stream channel will affect both the roughness and the effective cross-sectional area of the channel and thereby also the stage–discharge relation. The growth is generally greatest in streams polluted by organic wastes.

Weed growth on and at the control may have greater effects at low water than at high water, and except for the seasonal characteristics, the effects will resemble those due to scour and deposition. Floods may remove part of the effect from weed growth by flattening them out.

In a measuring structure weed growth on the crest or throat may cause significant errors in discharge. Weed growth causes a head rise and a consequent overestimate of the discharge. The remedy is to maintain the structure in good order by frequent brushing or scraping during the weed-growing season.

5.6 Extrapolation of rating curves

General

Extrapolation of the rating curve in both directions is often necessary. If the point of zero flow has been obtained, the curve may be interpolated between this point and the lowest discharge measurements without much error. But if the point of zero flow is not available, it is not advisable to extrapolate far in this direction.

In the upper part of the curve extrapolation is almost always necessary. Only in a few cases are discharge measurements available at about the highest flood peak observed.

Logarithmic extrapolation has proved to be a reliable method for shorter extensions (Chapter 4). If, however, extended extrapolations have to be made, special methods must be used.

The stage–velocity–area method

The best method to use is the extension of the stage against the mean velocity curve. A plot with stage as the ordinate and the mean velocity as the abscissa gives a curve which, if the cross-section is fairly regular and no bank overflow occurs, tends to become asymptotic to the vertical at higher stages. That is, the rate of increase in the velocity at the higher stages diminishes rapidly and this curve can therefore be extended without much error. Further, by plotting the stage–area curve (stage as ordinate, area as abscissa) for the same cross-section as that from which the mean velocity was obtained, the area can be read off at any stage desired. Multiplication of the area by the mean velocity gives the discharge (Fig. 5.14).

The area is obtained by a field survey up to the highest stage required and is therefore a known quantity.

The Manning equation method

The uniform flow equation as developed by Manning, expressed as

$$Q = 1/n \, AR^{2/3} \, S^{1/2} \qquad (5.5)$$

where n = a constant,
   A = area of cross-section,
   R = hydraulic radius,

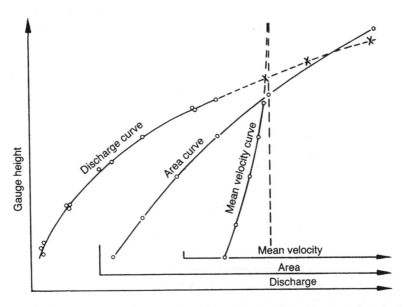

Figure 5.14 The stage–area–velocity method for extrapolating a stage–discharge curve.

S = slope of water surface and
Q = discharge,

may be used for extrapolation of rating curves. In terms of mean velocity the equation may be written (Chapter 8)

$$v = \left(\frac{1}{n}\right) R^{2/3} S^{1/2} \tag{5.6}$$

where $\dfrac{1}{n} S^{1/2}$ is a constant K.

For the higher stages, the factor $nS^{1/2}$ becomes approximately constant. Equations (5.5) and (5.6) can therefore be rewritten as

$$Q = KAR^{2/3} \tag{5.7}$$

and

$$v = KR^{2/3}. \tag{5.8}$$

By using various values of v from the known portion of the stage against the mean velocity curve and the corresponding values of R, values of K can be computed by equation (5.8) for the range in stage for which the velocity is known. By plotting these values of K against the gauge height, a curve is obtained that should asymptotically approach a vertical line for the higher stages (Fig. 5.15). This K-curve may then be extended without much error and values of K obtained from it for the higher stages. These high stage values of K combined with their respective values of A and $R^{2/3}$ using equation (5.7) will

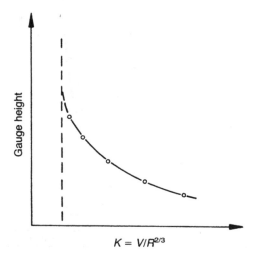

$$K = V/R^{2/3}$$

Figure 5.15 The Manning equation method for extrapolating a stage–discharge curve.

give values of the discharge Q which may be used to extrapolate the rating curve. A and R are obtained by field surveys and are known for any stage required.

The Stevens method

The so-called Stevens method is a variation of the method described above. It is based on the Chezy formula for uniform flow (Chapter 8)

$$Q = AC(RS)^{1/2}. \qquad (5.9)$$

For shallow streams with a relatively small depth–width ratio, the mean depth D does not differ much from the hydraulic radius R. Then, by substituting D for R, equation (5.9) may be written

$$Q = CS^{1/2}AD^{1/2}. \qquad (5.10)$$

At higher stages, the slope S in most cases may be considered constant. Then, by plotting $AD^{1/2}$ against Q in equation (5.10), an approximately straight line is obtained which is readily extended.

As illustrated in Fig. 5.16, values of $AD^{1/2}$ are plotted against both gauge height h and discharge Q, and the latter curve extended up to the higher stages.

Both A and D are obtained by field surveys and are therefore known factors.

5.7 Overflow

Streams with large overflow or out-of-bank flow present many complications in streamflow measurement and in the determination of the stage–discharge relation, particularly during rising and falling stage. It is frequently practicable

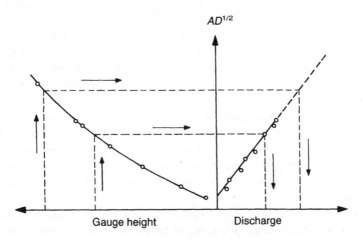

Figure 5.16 The Stevens method for extrapolating a stage–discharge curve.

to establish separate discharge rating curves for the flow in the main channel and in the overflow area, the total discharge being the sum of these. Bridging over the overflow area is a practical solution, where possible, and gaugings can be made from the bridge.

5.8 Current meter measurements from ice cover

Current meter measurements made under ice cover require special equipment for cutting holes in the ice through which to suspend the current meter. The development of power drills has facilitated what was formerly a laborious procedure and 150 mm holes can now be cut with ease and speed.

When holes are cut in ice the water, which is usually under pressure because of the weight of ice, rises in the hole. To determine the effective depth of flow, ice measuring sticks are used to measure the distance from the water surface to the bottom of the ice. The ice measuring stick consists of a bar about 1.2 m long, graduated in 10 mm divisions and having an L-shaped projection at the lower end. The horizontal part of the L is held on the underside of the ice and the depth to that point is read at the water surface on the graduated part of the ice stick. The horizontal part of the L is at least 100 mm long so that it may extend beyond any irregularities on the underside of the ice.

In order to measure the depth of flow and to suspend the current meter, a special assembly may be used because a normal sounding weight will not fit through the hole. The weight and meter are placed in a framework that will pass through the drilled hole.

A special vane ice meter is used in the United States for use under ice cover (Fig. 5.17) because the vanes do not become filled with slush ice as the cups of a cup-type meter often do. Further, the vane meter will fit into the hole made by the ice drill, and the yoke and ice rod can serve as an ice measuring stick.

In Russia equipment has been developed which enables the current meter to be lowered vertically into the hole and then rotated into the horizontal position.

The measurement of streamflow follows the same principles as those described in Chapter 2. At least 20 holes are made for the measurement, spaced so that no segment has more than 10% of the total discharge (Fig. 5.18). On narrow streams, however, it may often be simpler to remove all the ice in the cross-section.

The SonTek ADCP River Surveyor 100 mm diameter transducer head is used as a hand-held current meter for measuring under ice cover (Fig. 5.19). The meter has a user-friendly software package and is custom-designed for both velocity and discharge measurements using the mid-section method. While measuring under ice cover, the electronics may be conveniently carried by backpack.

Because of the roughness of the underside of the ice cover, the location of the filament of maximum velocity is some distance below the underside of the ice. In making a discharge measurement, the 0.2 and 0.8 depth method is preferred

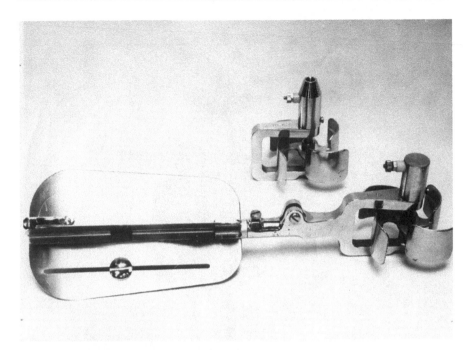

Figure 5.17 Vane ice current meter (upper) and vane meter with cable suspension yoke (lower).

for depths greater than 0.75 m and the 0.6 depth method is recommended for depths less than 0.75 m. It is also recommended that several vertical velocity curves be defined when ice measurements are made in order to determine whether any coefficients are necessary to convert the velocity obtained by the above point measurements to the mean velocity. When measuring the velocity, the meter is kept as far upstream as possible to avoid any effect that the vertical pulsation of water in the hole might have on the current meter. The current meter is kept free of ice when the velocity is being observed and it is therefore important to eliminate, as far as possible, the exposure of the meter to the cold air during the measurement. This can often be done by placing the meter in a bucket of river water or in a hot-air chamber during transportation between verticals.

Partial ice cover

When partial ice cover is observed at a gauging station discharge measurements are carried out in the ice-covered portion as described above. In the ice-free part of the stream normal gauging methods are employed using wading, footbridge, cableway, boat or floats. Floating ice can often serve as floats (Chapter 7). In Russia, ice floes distributed over the channel are photographed from aircraft in

Figure 5.18 Drilling holes in ice (foreground); ice rod being used to support current meter for a velocity measurement under ice cover (background).

order to determine the discharge in large ice-melt streams or special floats are released from aircraft and photographed to determine the discharge of the ice-free portion of a stream.

Multi-layered ice conditions

When several layers of ice have formed at the measuring site, it is usually more convenient to obtain a better site and have the discharge measurement correlated to the permanent station. If this is impractical, several holes may be drilled at the measuring site to ascertain the location of the main portion of flow within the layers. The discharge is measured by separately determining the effective depths and velocities between the ice layers.

Figure 5.19 The SonTek ADCP River Surveyor 100 mm diameter transducer head being used as a hand-held current meter during a measurement under ice in the Amur River in China. Note software electronics being carried on site in operator's backpack.

Water above the ice

If flow is over the ice surface, discharges both above and under the ice are measured separately. In this case, depth measurements under the ice cover may usually be omitted but it is necessary to measure the depth from the water surface to the upper surface of the submerged ice.

Channel completely frozen

If the channel freezes to the bed in winter and is completely filled with ice and snow, it is sometimes practical to cut a ditch about 0.5 to 1.0 m wide and at least 20 m long to convey the first flow of spring meltwater when discharge measurements can commence.

Safety measures

The danger of working on ice-covered streams can never be underestimated. As a general rule, measurements are not conducted on ice unless the thickness of the ice is at least 100 mm and the air temperature no higher than − 5°C. Normally, testing the actual strength of the ice is made by solid blows using a sharp ice chisel. Ice thickness may be irregular, especially late in the season when a thick layer of snow may act as an insulator or when water appears on the surface of the ice or new layers of ice are formed. Vehicle speeds on ice should be low since the hydrometric wave increases pressure on the ice and observers making measurements should wear non-slip footwear.

5.9 Streamflow in arid regions

Streamflow measurement is made difficult in arid catchments because of the fact that runoff only occurs when conditions are favourable and sometimes only on a few days per year. Intense rainfall may produce catastrophic floods of a fairly short duration and the problem of measurement is intensified. Further, many streams may have very low flows for some of the time but yield significant discharges during sudden floods. The problem of access and transport is made more difficult because of poor roads and observers are scarce because of the sparseness of the population. These problems and others are often only alleviated by improvisation.

Discharge measurements are generally carried out by the velocity–area method. Natural controls are rare however, and if possible bed controls are installed to stabilise the stage–discharge relation. Gabions are frequently used in this connection, consisting of baskets, each of about 3 m^3, filled with boulders (Fig. 5.20). These gabions require to be anchored to the alluvial bed by iron bars.

Stations in arid zones require to be designed to measure flows from zero to major floods. These floods are often accompanied by velocities of the order of 6 m s^{-1}. Measurement by current meter then becomes impracticable and the measurements are made by floats (Chapter 7).

It is more convenient to use cableways under normal flow conditions rather than footbridges which may be unsafe or swept away in floods.

Where access is difficult, for one reason or another, during floods crest–stage gauges may be installed and discharge computed later by the slope–area method (Chapter 8). Alternatively, the surface slope during floods may be obtained by refuse marks or traces left by the flood. Historical floods may often be estimated by ancient traces in uninhabited areas. In inhabited regions the historical floods of wadis may be estimated from the results of making enquiries amongst local villagers. The date of such floods however, may only be obtained by their having coincided with some notable event in the village, the birth of a child for example, or something of note which has occurred in village life, and extreme caution has to be exercised as to the reliability of witnesses.

Figure 5.20 Installation of bed control using gabions in arid-zone streams.

In arid regions the bed is usually unstable and is therefore susceptible to scouring and deposition during floods. However, if the bed geometry is altered with each flood the mean depth often remains the same, except for low flows, and the stage–discharge relation remains substantially unchanged. In many wadis, however, a violent flood may gouge out the bed by several metres and it may then be necessary to wait some years in order to establish the same cross-sectional profile which existed prior to the flood.

In order to obtain the actual level of the bed at the time of a flood peak it is sometimes convenient to bury a pile of bricks in the bed prior to the flood. The layers of top bricks are washed away in the flood and even if deposition occurs as the flood subsides, the minimum bed level may be obtained by uncovering the bricks. A better method makes use of a sounding weight and chain which are buried in the bed to a depth greater than is expected during scouring. The free end of the chain should just reach the bed level prior to the flood and should be installed vertically. For recovery of the chain, it is lengthened by means of a rope to which a float is attached. Several such devices are installed in the cross-section. The section of chain which is cleared by the flood by removal of the surrounding alluvium follows the direction of the current and will take up a position at right angles to the remaining buried portion of the chain. The entire chain is buried once again with alluvium as the flow subsides, and the chain can be exposed by digging and the minimum bed level obtained.

5.10 Non-standard indirect methods for measuring discharge and peak flow

General

There are occasions when a measurement of flow is required where there are no gauging stations available. In these circumstances resort has often to be made to non-standard methods as such a measurement may be considered 'better than no measurement at all'. This situation may occur in estimating flood peaks where very few large floods have been measured directly.

There are numerous methods that can be used either for direct or indirect measurements or for post- or historic measurements (palaeofloods). A selection of these methods that could be interrelated follows.

Manning's 'n' and Chezy's 'C'

Discharges, especially floods, may be estimated by the Chezy or Manning equations and the slope–area method by noting flood trash marks or flood wrack, from which slope may be estimated, and an estimation of roughness coefficients. (see Chapter 8). At the present state of knowledge, the selection of roughness coefficients (Manning's 'n' or Chezy's 'C') remain an art and often developed from experience. Generally 'n' may be selected by 'visual' estimation or based on the following semi-empirical equations:

Darcy-Weisbach:
$$n = (fR^{1/3})^{1/2}/(8g)^{1/2}$$
where 'f' is the Darcy-Weisbach friction factor and R is the hydraulic radius.

Jarrett:
$$n = 0.32S^{0.38} R^{-0.16}$$
where S is between 0.002 and 0.04.

Limerinos:
$$n = \frac{0.113R1/6}{1.16 + 2 \log_{10} (R/D_{84})}$$
where D_{84} is the 84th percentile of particle b axis size distribution.

Strickler:

$$n = 0.0151 \ (D_{50}^{1/6})$$

where D_{50} (mm) is the 50th percentile of particle b axis size distribution expressed as $D = (21.1n)^6$.

In some cases hydrologists have recorded 'n' or 'C' values at gauging stations as part of normal gaugings so that a series of these values may be made available for estimating current or historic floods or other discharges in the catchment or basin. By estimating roughness coefficients during gaugings, seasonal values may also be obtained.

In the UK, an example in the application of the slope–area method to estimate a previous flood was the Glen Ample flood in Scotland in 2004. No gauging stations existed and slope was estimated from trash lines left by the flood. Three 'n' values were chosen in the cross-section, 0.030 for the bed section, 0.080 for the banks and 0.035 for the bank overflow on one side (Fig. 5.21) (see Further reading).

A study in the UK to estimate 'n' values found values in the range 0.041 to 0.785 in one day in one river, the latter value being under extreme weedy conditions. This demonstrates the difficulty in choosing 'n' values (see Further reading).

In the US, 50 different rivers were examined for 'n' values. These rivers were stable but with a wide range of hydraulic conditions from boulder-strewn mountain rivers of the western US to the heavily vegetated flat-sloped rivers of the south (see Further reading).

In the US study, 'n' values ranged from 0.024 to 0.075 and coloured photographs of each river studied were presented (Figs 5.22 and 5.23). In general, in the US study, 'n' values increased with discharge.

ISO 1070 gives suggested values of 'n' and 'C' ranging from 0.016 for gravel beds to 0.200 for rivers with dense willow trees in mid-summer; 'C' values are

Figure 5.21 Estimating the Glen Ample flood in 2004, showing a diagrammatic view of the cross-section with the selection of three 'n' values (Cargill, SEPA UK).

in the range 5 to 91. and these are presented for various values of 'R' (the hydraulic mean depth). Fig. 5.24 shows an envelope of the world's maximum floods, many of which have been determined in this way (see Further reading).

It can be seen from Fig. 5.24 the equation of the world envelope is

$$Q = 500 \, A^{0.43}$$

and for catchments under 100 km^2

$$Q = 100 \, A^{0.8}.$$

(For Europe the envelope has the equation

$$Q = 230 \, A^{0.43}).$$

Floats (see Chapter 7)

Floats can be used to good effect and especially if flood debris is being carried in the flow from which velocity can be estimated. Generally an uncertainty of

Figure 5.22 Indian Fork below Atwood Dam, near New Cumberland, Ohio. $n = 0.026$.

Figure 5.23 Rock Creek near Derby, Mon. $n = 0.075$.

some 15–20% may be expected but ISO 748 may be referred to in this connection.

Weirs and flumes (see Chapter 10)

Existing non-standard river weirs, or flumes, may be used employing the simple basic equation $Q = Cbh^{3/2}$ where Q is discharge, C the coefficient of discharge, b the length of crest or width of flume and h the head. The units in this equation may be metric or U.S.

In Chapter 10, simplified equations are presented for spot measurements of discharge of standard and other weirs that may be used for river weirs of similar geometry. The United States Geological Survey 'Manual on the measurement of peak discharge over river dams by indirect methods' using the above equation (see Further reading) presents 14 weirs of unusual shape from which a selection is made of a comparative weir. Note that in the tables accompanying the

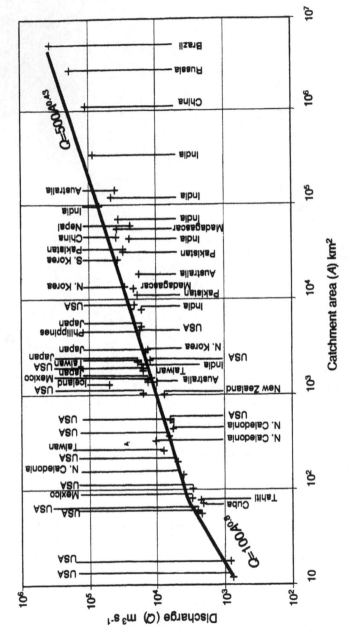

Figure 5.24 Plot of discharge (Q) against catchment area (A) for the world's maximum floods.

manual, the units are US units. Simon has presented a series of 48 broad-crested weir shapes with their coefficients of discharge determined from laboratory calibration (see Further reading).

Width contractions

Measuring peak discharges at width contractions, notably under road or railway bridges is presented in the United States Geological Survey 'Manual on the measurement of peak discharge at width contractions by indirect methods' (see Further reading).

The equation used is

$$Q = CA \left(2g \left(\Delta h + aV^2/2g - h_f\right)\right)$$

where

Q is discharge,

C is coefficient of discharge,

A is area of section between abutments,

Δh is the difference in elevation of the water surface between upstream and downstream of the flow through the contraction,

$AV^2/2g$ is the velocity head,

h_f is the head loss due to friction between the upstream and downstream of the flow through the contraction.

The units are US units.

Culverts

The flow through culvert has been studied by the United States Geological Survey and reported in their 'Manual on the measurement of peak discharge at culverts by indirect methods'.

The equation given for any shape of culvert is

$$Q = A^{3/2} (gd)^{1/2} \text{ in US units}$$

where

A is the cross-section of flow,

d is the maximum depth of water in the critical flow section.

For rectangular sections the equation is given as

$$Q = 5.67 \, bd^{3/2} \text{ in US units}$$

where b is the width of section.

For circular culverts the equation given is

$$Q = CD^{5/2}$$

where D is the inside diameter of the circular section.

Runoff from small watersheds

Peak rates of runoff were studied by the US Bureau of Public Roads for small watersheds with areas of 25 square miles or less (65 square kilometres) in 1961 (see Further reading).

The objective was to estimate the volume of streamflow to be expected at peak periods in the design of bridges and culverts.

The research consisted of examining some 246 ungauged watersheds and 96 gauged watersheds over five geological and physiographical zones east of the 105th meridian. Runoff from these zones was predicted from the area of the watersheds, the rainfall, topographic indexes and drainage density.

From the research, a procedure was presented to estimate the peak rate of runoff from a small basin. The results showed that 68% of the estimated values of Q from ungauged watersheds had a difference between the estimated values and the true values of Q of less than 20%, and for 95% of the ungauged watersheds, the difference was 40%.

The measurement of velocity from standing waves

An indirect measurement of velocity can be made in rockbed rivers, which are essentially free from sediment, by measuring the wavelength of the standing wave train. Wavelengths between 0.5 m and 7m have been studied so far (see Further reading).

In rocky rivers with beds in either eroded bedrock or very coarse to boulders, standing wave trains are quite common at virtually all stages of flow. The wave trains indicate critical flow conditions and normally stand in marked contrast to subcritically flowing water (Froude number less than unity) immediately adjacent which is either slower moving slackwater, or often a backwater. The standing wave appears stationary to an observer on the bank because it is moving upstream at the same velocity as the stream as the flow is moving downstream.

The minimum wavelength λ of stationary waves is related to mean velocity from

$$\lambda = 2\pi v^2/g$$

where v is the velocity of the wave.

Therefore, solving for v gives

$$v = 1.2495\sqrt{\lambda}.$$

It is also possible to estimate velocity from the travel time of floats from wave crest to wave crest through a wave train. From the above equations it can be seen that wavelength increases as the square of velocity and therefore travel time, t, will increase. Dividing travel distance from crest to crest by velocity gives

$$t = 0.8003\sqrt{\lambda}$$

where t is in seconds.

Velocity is therefore related by the ratio 1.2495/0.8003 to give

$$v = 1.5613t.$$

Thus, taking travel time from crest to crest along a wave train using a stopwatch an estimate of velocity may be made and, if area is known, a measurement of discharge can be made.

Estimation of ungauged peak discharge in bedrock rivers

Using bedrock blocks an equation may be established of the form

$$Q = 3.06.10^{-1}Ad^{-2/9}S^{-1/6}n^{-1}$$

where

Q is the discharge (m^3/s)

A is the cross-section area of flow (m^2)

d is the mean boulder diameter (m)

n is Manning's n.

Another method for estimating Q in a bedrock river having a 'plunge pool' at the downstream end of a reach of a drop structure in a flat bedrock river. Normally this method is used to design outfalls where the discharge is given. However provided that the approach channel and pool sections are of uniform width and depth and floored by bedrock, the standard equation for drop structures takes the form

$$Q = 9.4wy^{1.5}$$

where

w is the width of the approach channel

y is the critical depth (m)

Q is the discharge (m^3/s).

The use of boulders to estimate flood peak

The use of boulders to measure streamflow measurement relates to the 'sixth power law' which states that the weight of a rock particle is proportional to the sixth power of channel velocity (see Further reading).

The use of unit stream power, w, and critical stream power, w_c, are used to estimate the flow needed to initiate boulder movement where

$$w = \rho Qs/wb$$

where

w = unit stream power $(kg\ m^{-1}\rho s^{-1})$

ρ = density of water $(kg\ m^{-3})$

Q = discharge $(m^3 s^{-1})$

s = water surface slope $(m\ m^{-1})$

w = width of channel (m).

Boulders of a given size will move when

$$w_c = 290\ D\ 1.5\ \log\ (12Y)/D\ (kg\ m^{-1}s^{-1})$$

D = median diameter of boulders (m)

Y = mean depth of flow (m).

In a major flood in Cornwall, UK in August 2004, the above equation was used to estimate the flow in what is now classed as an historical flood.

The median boulder diameter, D, used was 0.76m giving a peak discharge of 310 $m^3 s^{-1}$.

Estimating discharge from superelevation in bends

A fundamental characteristic of streamflow measurement is the deformation of the free surface in a bend because of the action of centrifugal force. The water surface rises on the concave or outside bank of the bend and lowers along the convex or inside bank of the bend. The difference in water surface elevation between the banks is the superelevation. Superelevation varies with angular distance in the bend because of acceleration of the flow entering and leaving the curve and because of the varying curvature of streamlines within the bend.

The discharge is given by the following equation

$$Q = A\ (gh/k)^{1/2}$$

where

Q = discharge (m³/s)

A = average radial cross-section in the bend (m²)

g = acceleration of gravity (9.81 ms⁻²)

h = superelevation (maximum difference in water surface elevation measured along a radius of the bend between inner and outer banks of the bend (m)

K = superelevation coefficient.

The value of K is determined from the equation

$K = 5/4 \tanh (r_i \theta/b) \ln (r_o/r_i)$.

Note: tan$h$ is the hyperbolic tangent.

Figure 5.25 shows a plan of a bend and the symbols used in the above equations. It should be noted that the method only applies to (1) bends having no overbank flow and (2) superelevations greater than 0.08 m (mainly because of the uncertainties in the high water marks on the bank.

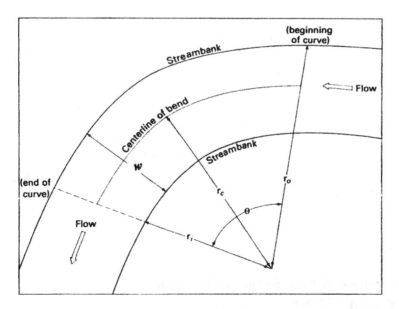

Figure 5.25 Plan showing water surface superelevation at outer and inner banks (from USGS WSP 2175).

An alternative approach by applying Newton's second law of motion gives

$$se = v^2\, w/gr_c$$

where

se is the superelevation

v is the mean velocity
w is the width of channel and

r_c is the radius of curvature at the centre of the bend.

Modelling the transverse surface profile as logarithmic curve, the equation becomes

$$se = 2.30\, v^2/g \log r_0/r_1$$

where

r_0 and r_1 are the outer and inner radii respectively.

Other empirical flood formulae (see Further reading)

Dickens formula for moderate sized basins in North and Central India

$$Q = CA^{0.75}$$

where

$C = 11\text{--}14$ where average annual rainfall is 60–120 cm.

Ryve formula for South India

$$Q = CA^{0.66}$$

where C varies between 6 and 10 depending on location.

Ingles formula for fan shaped catchments of Bombay State (Maharashtra)

$$Q = 124\, A/(A + 10.4)^{1/2}.$$

Myers formula

$$Q = 175\, A^{1/2}.$$

Ali Nawab Jang Bahadur formula for Hyderabad State

$$Q = CA^{(0.993 - 1/14 \log A)}$$

where C varies from 48 to 60.

Fuller's formula (1914)

$$Q = CA^{0.8} (1 + 0.8 \log T_T) (1 + 2.67 A^{-0.8})$$

where constants derived from USA basins. C varies from 0.026 to 2.77 and T is the recurrence interval in years.

Greager's formula for USA

$$Q = C(0.386A)^{0.894(0.386A) \text{ to the power} - 0.048}$$

where C is 130 (140.5 for areas subject to large floods).

Burkii Ziegler formula for USA

$$Q = 412 \, A^{0.75}.$$

Rational formula

$$Q = CiA$$

where C is the runoff coefficient $(0.2 - 0.8)$ and i is the intensity of rainfall corresponding to the time of concentration for the catchment in cm h^{-1}.

In the above formulae Q is in m^3/s and A is in km^2.

Uncertainty

It is difficult to place an uncertainty in any of the above methods. Each may have a place to be used under special circumstances: especially in floods when no gauging stations are available and they may be the only methods available.

5.11 Safety measures in streamflow measurement

The safety record in streamflow measurement over the past 100 years has been a good one. Nevertheless, fatal accidents have occurred, notably drowning, which with adequate safety measures could have been avoided. The main hazard is of course the proximity to water and the main considerations towards safety measures are to be found in BS 3680 Part 3Q to which reference may be made.

Further reading

Ackers, P. Stage discharge functions for two-stage channels: the impact of new research. *Journal IWEM*, 7(1), 52–61 1993.

Archer, D. *Personal communication* 2007.

Bagnold, R.A. *An empirical correlation of bedload transport rates in flumes and natural rivers*. Proc. R. Soc. London, A 372 453–473 1980.

Barnes, H.H. Roughness characteristics of natural channels. *Water Supply Paper 1849 U.S. Geological Survey.* 1967.

Benson, M.A. Measurement of peak discharge by indirect methods. *World Meteorological Organization. Technical Note No. 90* 1968.

Black, A., et al. Extreme Precipitation and runoff. SW Perthshire, Scotland, August 2004. *Abstracts volume Royal Geographical Society (with the Institute of British Geographers) Annual Conference* 2005.

Bodhaine, G.L., Measurement of peak discharge at culverts by indirect methods, *United States Geological Survey in Techniques of water resources investigations Book 3* 1968.

BS 3680 Part 3Q. *Guide for Safe Practice in Stream Gauging.* HMSO, London 1993.

Calenda, G., Calvani, L. and Mancini, C.P. Simulation of the great flood of December 1870 in Rome. *Proc. Instn. Civ. Engrs. Water and Marit. Engrs. 156, 305–312* 2003.

Carling, P.A. The Noon Hill flash floods: July 17th 1983. Hydrological and geomorphological aspects of a major formative event in an upland landscape. *Trans.Inst Br Geogr N.S. 11: 105–118 ISSN: 0020–2750* 1986.

Carling, P.A. and Grodek, T. Indirect estimation of ungauged peak discharges in a bedrock channel with reference to design discharge selection. *Hydrological Processes vol. 8 497–511 (1994)* 1993.

Clark, C. Flood risk assessment using hydrometeorology and historic flood events. *International Water Power and Dam Constr. 59 (4) 22–30* 2007.

Clark, C. Catastrophic floods: magnitude and frequency investigations. *International Water Power and Dam Construction* 2005.

Costa, J.E. Paleohydraulic reconstruction of flash flood peaks from boulder deposits in the Colorado Front Range. *Geol. Soc. Amer. Bull. 94, 986–1004* 1983.

Ge, We, Luo, X. and Tang, P. The method for uniformizing the stage–discharge relations of stable river beds, and its application. *IAHS Symposium, Exeter Publication No. 134* 1982.

Hulsing, H. Measurement of peak discharge at dams by indirect method. *United States Geological Survey in Techniques of water resources investigations Book 3* ISO 1070 1992 Slope area method 1967.

ISO 9210. *Measurement of Discharge in Meandering Rivers.* ISO, Geneva, Switzerland 1992.

ISO 11332. *Measurement of Flow in Unstable Channels and Ephemeral Streams.* ISO, Geneva, Switzerland 1998.

Manual of British Water Engineering Practice, Third Edition 1961, Fourth Edition (1969), The Institution of Water Engineers.

Matthai, H.F. Measurement of peak discharge at width contractions by indirect methods, *United States Geological Survey in Techniques of water resources investigations Book 3* 1967.

National Environment Research Council (NERC), Flood Studies Report (five volumes) 1975.

Pierce, C. H. Investigation of Methods and Equipment used in Stream Gauging. *Part 2: Intakes for Gauge Wells.* US Geological Survey Water Supply Paper 868B 1941.

Potter, W.D. Peak rates of runoff from small watersheds. *U.S. Department of Commerce, Burea of Public Roads* 1961.

Powell, K.E.C., Weed growth – a factor of channel roughness. In *Hydrometry: Principles and Practices*, (ed. R. W. Herschy) John Wiley and Sons Chichester 1978.

Smith, W., Hanson, R. L. and Cruff, R. W. *Study of Intake Lag in Conventional Stream Gauging Stilling Wells*. US Geological Survey Open File report 1965.

Tilrem, O. A. *Manual on Procedures in Operational Hydrology*. Ministry of Water, Energy and Minerals of Tanzania and the Norwegian Agency for International Development 1979.

Tinkler, K.J. Indirect velocity measurement from standing waves in rockbed rivers, *Journal of Hydraulic Engineering, October pp. 918–921* 1997.

Williams, G.P. Paleohydrological methods and some examples from Swedish fluvial environments. In cobble and boulder deposits. Geogr. Ann. 63A, 227–243 1983.

World Catalogue of Maximum Observed Floods. 2003. IAHS Pub. 284 (Compiled by Reg Herschy) 2003.

Chapter 6

The ADCP method of streamflow measurement

6.1 General

The acoustic Doppler current profiler is a velocity area method using a boat, launch or towed platform to traverse the river to make a single measurement of discharge normally for plotting on the stage–discharge relation. The ADCP collects measurements of velocity depth and position as it passes across the measuring section. The velocity is measured by the Doppler principle and the area is measured by tracking the bed to provide river depth and boat position. The method has therefore similarities to the moving boat method (Chapter 2) where a current meter is used instead of the acoustic Doppler profiler.

In the Doppler principle, the reflection of sound waves (backscattering) from a moving particle of sediment or air bubbles in the flow, causes an apparent change in frequency. This difference in frequency between the transmitted and reflected sound waves is a measure of the relative velocity of flow in both magnitude and direction and known as Doppler shift (Fig. 6.1).

An ADCP is basically a cylinder with a transducer head on the end being a ring of three or four transducers with their faces angled to the horizontal and at right angles to each other.

6.2 Depth cells

In applying the Doppler principle, the ADCP subdivides the water column being sampled by each of three beams into depth cells or bins, ranging from 0.01m to 1m or more. A centre-weighted radial velocity is measured for each depth cell in each beam. Using trigonometric relations, a three-dimensional water velocity is determined and assigned to a given depth cell in the water column (Fig. 6.2). The water set-up parameters are adjusted to optimise the system for geometric characteristics of the river cross-section being measured. These parameters include the depth cell size, the number of depth cells, the number of pulses and velocity reference commands. It should be noted that when combining 'along-beam' velocity data from different beams, the velocity seen at a given depth cell is the same for all beams (which are measuring in

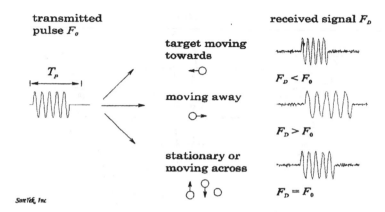

transmitted
pulse F_o

received signal F_D

target moving
towards

$F_D < F_o$

moving away

$F_D > F_o$

stationary or
moving across

$F_D = F_o$

T_p

SonTek, Inc

Figure 6.1 Doppler shift: reflection of sound waves by a moving particle resulting in apparent change in frequency of the sound waves; this frequency shift is known as the Doppler effect.

different physical locations). This is substantially true in open water conditions but may not be so in conditions near bridges. In its simplest form the equipment may be deployed by the use of a tethered boat and tow rope. Two ropes are required that will stretch across the river. Two operators are needed, one who is able to cross the river with the end of a rope. It may be possible to set up a pulley system with a single loop of rope. If a bridge is available it may be possible to use a single rope. These methods are normally suitable for small rivers.

6.3 ADCP moving boat method

In applying the ADCP moving boat method, the Doppler profiler is mounted on the boat (Fig. 6.3) and the boat moved across the river perpendicular to the flow. Water velocities are measured when the acoustic Doppler profiler transmits pulses along three beams for vertical current profiling (two or four beams are available for special studies). The pulses are transmitted at a constant frequency of between 75 and 3000 kHz. The beams are positioned at precise horizontal angles from each other at 120° for a three-beam instrument and 90° for a four-beam instrument. The beams are directed at a known angle from the vertical, typically 20° or 30° (Fig. 6.2). (These beams may be upward looking or downward looking depending on the method of measurement employed.) The instrument processes echoes throughout the water column along each beam. The difference in the frequency (shift) between transmitted pulses and received echoes, known as the Doppler effect, is used to measure the relative velocity between the instrument and suspended material in the water that reflect the pulses back to the instrument.

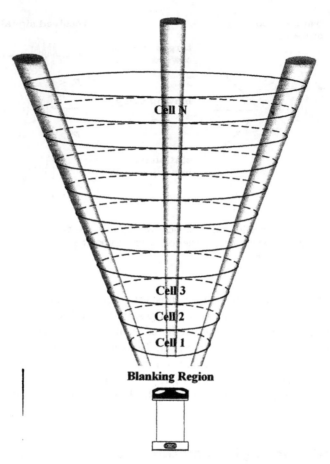

Figure 6.2 Schematic arrangement of ADCP depth cells or bins. The ADCP sub-divides the water column by each beam into depth cells ranging from 0.01 m and 1 m or greater. A centre-weighted radial velocity is measured for each depth cell in each beam. With these results and using trigonometric relations a 3-dimensional velocity is estimated and assigned to a given depth cell in the water column. This is analogous to a velocity–area vertical from a point-velocity current meter. The system may be down-looking or up-looking (as in the Figure) depending on the method and instrument used.

If the system has a built-in magnetic compass the route taken across the river does not need to be straight or perpendicular to the bank. The instrument collects measurements of velocity, depth and position as it proceeds across the river (Fig. 6.3). If no magnetic compass system is available it is important to ensure that the ADCP is deployed perpendicular to the flow avoiding any movement of the instrument as is the case with an ordinary current meter gauging.

Acoustic Doppler current profiler (ADCP)

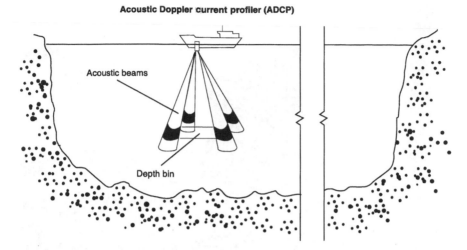

Figure 6.3(a) Diagrammatic view of ADCP mounted on a boat crossing the measuring section and having four beams and depth bins.

Figure 6.3(b) ADCP, in its simplest form for discharge measurement being moved across the river by an operator on each bank.

Source: Hydro-Logic.

Figure 6.3(c) An ADCP mounted on a motor launch on the Huanpu River in China.
Source: SonTek.

6.4 Bottom tracking – determining boat velocity, depth and distance travelled

The flow velocity determinations incorporate both flow velocity and boat velocity, the latter being measured by using the Doppler shift of separate acoustic pulses reflected from the bed. This technique is known as bottom tracking. In addition to measuring boat speed, the flow depth is determined by analysing the bottom track echoes. Bottom tracking also incorporates determination of movement and distance along the cross-section. As an alternative, the Differential Global Positioning System (DGPS) may be used to measure the boat velocity.

Bottom tracking measurements are similar to water velocity measurements but separate pulses are used. Bottom tracking pulses are longer and are used to measure depth. The sound pulses are reflected from the channel bed and used to calculate the velocity of the instrument relative to the bed. The bed is then assumed stable as seen by the instrument. The on-board magnetic compass can combine this data with bottom tracking data to determine speed and direction (see also 6.14 and 6.15). Compasses used by most ADCPs include a calibration facility to correct for ambient magnetic fields (see also 6.16).

For bottom tracking to be reliable, it should be noted that the channel bed does need to be stationary (not being moved by the water velocity).

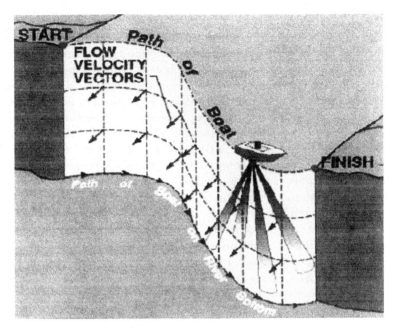

Figure 6.3(d) Schematic of typical ADCP route across river demonstrating that path need not be straight or perpendicular to the bank.
Source: United States Geological Survey.

6.5 Measurement of discharge

An ADCP determines the velocity in each depth cell and knowing the depth cell size and distance between successive profiles ('verticals'), the discharge for that cell can be calculated. When the ADCP is used to determine discharge, a series of acoustic pulses are transmitted, known as pings. Pings for measuring the flow velocity are known as water pings, and pings for measuring the boat velocity are known as bottom tracking pings. A group of these pings are referred to as an ensemble and an ensemble is analogous to a vertical in the conventional velocity–area method (Chapter 2) whereby current meter velocity measurements are taken at one or more points in the vertical. An ADCP measurement in a river, say 5 m deep for example, may make up to 18 averaged velocity measurements spaced at 0.25 m cells in the vertical which are then used to compute a discharge for each cell. Because velocity information cannot be collected near the water surface or near the bed (typically 7–10% of total depth) the velocity cells are used to produce a vertical velocity curve which estimates discharge within the unmeasured sections (top and bottom) of the profile. An alternative sometimes used is to estimate discharge on the top and bottom based on a constant extrapolation from the first and last cell for the top and bottom discharge estimates, respectively.

6.6 Equipment

Making a discharge measurement with the ADCP, requires three main pieces of equipment: the boat, launch or platform for mounting the instrument and the acoustic instrument–transducer assembly; because the system software produces and displays a large amount of data, a laptop computer with a minimum 200 MHz processor and greater than 64 MB of RAM memory is recommended. The computer screen display should be visible in direct and diffuse sunlight. The instrument includes a pressure case that contains most of the electronics, a transducer assembly and normally a DGPS on board. The electronics include a CPU and firmware, a flux-gate compass and a pitch-roll sensor.

The choice of appropriate instrument depends on the size of river and characteristics of the site and the type of mounting to be used. Small units available are less than 30 cm tall and weigh only a few kilograms.

A communication cable is connected to the pressure case that enables users to configure the ADCP software running on the laptop computer. This arrangement enables the system to transmit data to the computer for post-processing. When ADCPs, however, are run from tethered platforms, wireless communication is normally used (e.g. radio modems, Bluetooth).

6.7 Making an ADCP measurement

A discharge measurement is made by traversing the river cross-section with the boat or platform (e.g. catamaran) (Figs 6.4 and 6.5). A single traverse, called a transect, consists of a collection of ensembles. A typical transect may contain thousands of velocity measurements collected continuously across the river width compared to a conventional current meter measurement of, for example, 20 verticals. Usually four or more transects are made under relatively steady flow conditions and averaged, but under high water or flood conditions it may only be possible to make one transect. In order to arrive at an acceptable average and to discard outliers, if the measured discharge on any given transect is, say, for example, more than 5% of the mean discharge, a further set of transects is best made.

An ADCP measurement consisting of 4 transects or more typically takes up to ten times less than that of a conventional measurement by current meter.

6.8 Calibration

Like the ultrasonic method of discharge measurement (Chapter 12), the ADCP does not have independent calibration and verification facilities at the present time and resort is made to a check by current meter where necessary. However work is proceeding to discover a suitable special rating tank. Many comparative measurements, however, both in the USA, China and other countries have

Figure 6.4 A side-mounted ADCP on a large boat in China.
Source: SonTek.

shown that differences between conventional methods and the ADCP are in general agreement.

In the USA for example, comparison measurements at 12 sites across the country, in rivers ranging in width from 43 to 1100 m were made. Measured discharges varied from 21 to 1690 m^3s^{-1} and all 31 of the measurements made with the ADCP differed from measurements made by conventional current meter methods by less than 8% and 26 of the measurements were within 5%.

In China, comparison measurements were made on nine gauging stations on the Changjiang (Yangtze) River with DGPS, magnetic compass and bottom tracking functions. Discharges ranged from 5,000 m^3s^{-1} to 65,000 m^3s^{-1} and velocities from 0.45 ms^{-1} to 3.62 ms^{-1}. Velocities were measured by current meters in 537 verticals in over 82 cross-sections at a total of 1,537 measuring points in the verticals. Some 337 ADCP measurements were made at 153,700 points in 53,700 verticals. The depths of flow were 6.0–70 m.

The comparison results between current meters and the ADCP for a single discharge measurement showed a relative standard deviations of 6.5%–7%. Using DGPS and magnetic compass, comparison errors were within an average relative standard deviation of 2% to 5%. This difference between the bottom tracking and DGPS referenced measurements may be the result of bias introduced into the bottom tracking measurements due to a moving bed. DGPS compensates for the bias if used carefully with a calibrated compass.

Figure 6.5(a) A small ADCP catamaran mounted on a small motor launch on the Vaal River at Gladdedrif.

Source: Republic of South Africa.

6.9 Methods of deployment and mountings

(a) motorised launch (Fig. 6.6)
(b) floating platform (eg catamaran) (Fig.6.7)
(c) tethered deployment on a tow rope (Fig. 6.8, 6.13)
(d) tethered deployment from a cableway (Fig.6.9)
(e) tethered from a cable (Fig.6.7)
(f) tethered from a bridge (Fig 6.10)
(g) ADCP side-mounted on gauging launch (Fig 6.11)
(h) customised off-road trailer accommodating ADCP catamaran and equipment (Fig. 6.12)
(i) measurement under ice cover (see Chapter 5).

Deployment of the ADCP on (a) or (b) above or on a remote-controlled platform are the preferred methods where there is no cableway or cable and no way for the operator to cross the river. In the remote-controlled mode, it is advisable

Figure 6.5(b) An ADCP gauging on the Vaal River at Gladdedrif.
Source: Republic of South Africa.

that a light line is attached in case of failure of the motor or motor control device for recovery of the ADCP.

6.10 Stationary operation

The ADCP can be used effectively in place of the conventional current meter, on a cableway or from a bridge for example, and its horizontal position identified similarly to conventional flow measurement using verticals. The stationary method is somewhat simpler for the operator with few chances of error and may avoid a variety of problems such as determining boat motion and moving bed.

6.11 Number of transects for discharge measurements

A minimum of four transects, two in each direction, are usually taken, preferably under steady flow conditions, and averaged.

6.12 Size and frequency

The size and frequency of the ADCP to be used depends on the characteristics of the river being measured and the platform used for deployment. In large deep rivers, a large unit mounted on a motor gauging launch is used (Fig. 6.4). The

Figure 6.6(a) A small catamaran with aluminium jib-mounting for small boats.
Source: Republic of South Africa.

Figure 6.6(b) An ADCP gauging in operation.
Source: Republic of South Africa.

Figure 6.7(a) Details of an ADCP catamaran on a small motor vessel tethered from a cable on the Vaal River at Goose Bay Canyon.
Source: Republic of South Africa.

depth of the river determines the frequency of the system because the higher the frequency of the transmitted signal, the greater the attenuation of the acoustic signal and so the shorter the usable range. A 300 kHz unit may be used for a depth of, say, 130 m, while a 1200 kHz ADCP may be used to a depth of 20 m. The maximum depth for a given frequency will vary depending on the amount of material in the water column.

6.13 Small rivers with shallow depths

Small rivers with tranquil flow may be measured with the ADCP installed in small powered launches or tethered boats or catamarans. Models of 1200 and 3000 kHz are capable of measuring rivers with depths as shallow as 0.3 m. Flow disturbances can become significant in these minimum depths and care should be made in order to reduce them as much as possible (using smaller transducers or a hydrodynamic float).

6.14 Moving bed conditions

The number of suspended particles in the water can affect the range of the velocity and depth measurement. Too few particles may result in a limited number of return signals from each cell depth of the ensembles, although this

Figure 6.7(b) Details of the mounting of an ADCP with a motorised catamaran on the Vaal River at Goose Bay Canyon.

Source: Republic of South Africa.

Figure 6.7(c) ADCP gauging on the Vaal River at Goose Bay Canyon.

Source: Republic of South Africa.

Figure 6.8 An ADCP gauging with motor launch and catamaran under tow cable on the Songhua Jiang River in China.

Source: SonTek.

Figure 6.9 ADCP gauging using current meter cableway with current meter and sounding weight to move the small catamaran across the measuring section. Aksu River, China.

Source: SonTek.

Figure 6.10 ADCP gauging downstream of bridge using small catamaran on the Vaal River
   at Blaukop.
Source: Republic of South Africa.

condition may be rare. On the other hand, too many suspended particles is a
more common occurrence, especially during floods having a sediment concentration of perhaps as much as 10 kg/m³ or more. The penetration of the acoustic
pulses to the lower depths is reduced as the suspended sediment concentration
increases. Rivers with a large part of the sediment load transported as bed-load,
or having high sediment concentrations near the bed, create additional problems. In such cases, bottom track measurements are contaminated with returns
from near-bed sediments. As a result, bottom track measurements of the boat
position may be located upstream of the true location and velocity measurements may be biased low because the ADCP measures not only the boat speed,
but also the speed of the moving sediments near the bed. The changing pattern
of the moving bed prevents the bottom tracking mode of the ADCP from
accurately measuring the traverse of the measuring platform and the width of
the river.

A DGPS may be used to measure the boat location and velocity, perhaps with
increased uncertainty depending on conditions. Alternatively stationary measurements may be made using a traditional measurement method which reference only the fixed position of the ADCP rather than a moving position based
on bottom tracking or DGPS references

The higher the bed-load speed, the larger is the ADCP deviation when

Figure 6.11(a) Gauging with a front (bow) mounted ADCP on a modern motor launch.
Source: United States Geological Survey.

compared to current meter measurements and conversely, the smaller the moving bed speed, the closer are the current meter measurements.

6.15 Moving bed test

Before conducting an ADCP measurement where moving bed conditions may be suspect, a moving bed test may be carried out. This may be performed by taking the boat to various locations on the river, often to the centre or areas where fast currents are observed. At these locations, the boat is anchored or its position is maintained by other means. This is usually done for, say, 15 minutes at each location. During this time the operator observes the boat's position on the laptop, commonly called the track position. If the boat is seen to be moving upstream, even though the boat is anchored, then a moving bed has been detected. This is the method adopted in China where moving beds may have velocities as high as 1.3 m/s.

In this case the DGPS system is used to measure the boat's motion and hence computes the water velocities. This information is keyed into the laptop in real time and merged with the ADCP velocities to compute actual velocities. Where no DGPS system is available, a satellite system is necessary using a rover and base station. Such a system was used in Zhengzhou in Henan Province in China to measure the flow in the Yellow River.

Figure 6.11(b) Gauging with a side-mounted ADCP on a modern motor launch on the Pearl River Estuary in China.

Source: SonTek.

Figure 6.12 Customised off-road trailer accommodating the ADCP catamaran and equipment.

Source: Republic of South Africa.

Figure 6.13 Details of a complete ADCP assembly of a SonTek catamaran with tow rope connection.

Source: SonTek.

Failing the use of the DGPS or satellite systems, the stationary method may be used as described in Section 6.10 using the mid-section method.

It is practice by many hydrological services to carry out a moving bed test prior to a measurement.

6.16 Interference under external magnetic field conditions

Tests in China have deployed the ADCP on boats with iron hulls and on boats with wooden hulls using the DGPS mode, the bottom tracking mode and also with a magnetic compass on board.

The object of the tests was to examine the effect of the magnetic compass on magnetic declination. It was discovered that when there is no external magnetic field interference the ADCP results were close to current meter results but that boats with wooden hulls were much closer than boats with iron hulls.

6.17 Interference from boundaries and reflection of side lobes

ADCP systems are unlikely to be able to measure velocities accurately all the way from the surface to the river bed. There are direct reflections of the pulses from the boundary and reflections of side-lobe energy that take a direct path to the boundary (Fig 6.14). Also, the ADCP cannot measure velocities near

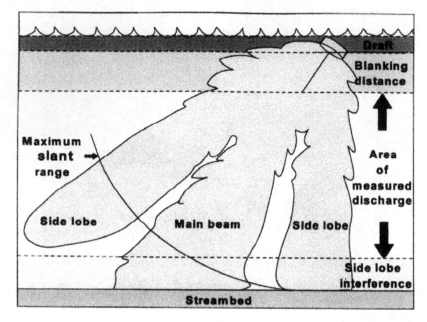

Figure 6.14(a) Diagram showing acoustic Doppler limitations for discharge measurement near the surface and near the bed; velocities near the surface are not measured because of the draft of the transducer (depth of transducer face below surface) and blanking distance of the instrument. This is typically 0.2 to 1 m and a typical blanking distance is 0.3 m. Velocities near the bed are not measured because of interference from side lobes of acoustic energy at some 30–40 angles from the main beams.

Source: United States Geological Survey.

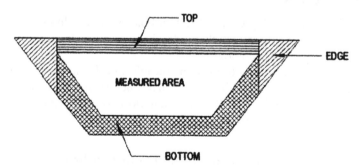

Figure 6.14(b) Schematic showing that the velocity is measured in the central area only; area hatched is estimated by extrapolation by a software package.

the edge of the river cross-section being measured. If the unmeasured discharge area is assumed to be triangular, the velocity for the unmeasured section V_e, say, may be estimated by the equation

$$V_e = 0.707 \, V_m \tag{6.1}$$

where V_m is the mean velocity at the first or last ensemble.

The assumption that the unmeasured flow area is triangular is reasonable for many situations where the bed gradually slopes upwards towards the shore (Fig 6.14b). However, sometimes the edge of the water is a vertical wall and as the ADCP approaches, the acoustic beams impinge the wall and cause a false bottom return. The distance at which the acoustic beam impinges the wall depends on the depth of water near the wall and the orientation of the transducers on the instrument relative to the wall. This distance is typically equal to the depth of the stream. Velocities for the unmeasured edge sections may then be estimated by setting $V_e = V_m$. However, since the velocity decreases to zero as the wall is approached, therefore the velocities near the walls may be estimated from

$$V_e = 0.91 \, V_m. \tag{6.2}$$

The edge discharge may then be estimated from

$$Q_e = (C V_m L d_m) \tag{6.3}$$

where

Q_e = estimated edge discharge
C = coefficient equal to 0.707 or 0.91 depending on channel shape
L = distance to the shore from the first or last ADCP measured subsection
d_m = depth at the first or last ADCP measured subsection.

The ADCP software packages, however, provide procedures to estimate the unmeasured discharge using optional algorithms for estimating unmeasured discharge near the ends of the cross-section, the unmeasured discharge due to side lobe interference and the unmeasured discharge due to blanking distance and transducer draft. The software display typically provides a comparison of the measured discharge and the estimated discharge for each of the unmeasured areas.

6.18 Site selection

For the best results of an ADCP measurement, the following should be noted:

(a) sites with excessive weed growth should be avoided;
(b) sites with excessive moving bed should be avoided;

(c) sites with excessive aeration or turbulence should be avoided;
(d) velocities should be greater than the minimum response speed of the sensor and less than the maximum;
(e) reflectors such as suspended solids etc. should be available under the full range of flows;
(f) for safety purposes the average velocity in the cross-section should not exceed 4m/s, and may be less depending on flow conditions;
(g) the faces of the transducers are susceptible to damage and measuring sections should be free of rocks;
(h) for many situations the depth should not be less than 0.2 m but reference should be made to the manufacturer's manual. It should be noted that at least three depth cells plus blanking distance, transducer depth and unmeasured area of the bed should be advised by the manufacturer;
(i) the ADCP equipment was originally designed and manufactured in the 1970s specifically for application in oceanography; research and investigation bringing the system successfully into the present day rivers application has been carried out since the early 1990s.

6.19 Training

The ADCP is a complex expensive piece of equipment and in order to get the best use of its application and use in measurement, staff training is essential. Figure 6.15 shows an ADCP training exercise being carried out on the Songhua Jiang River in China by the Bureau of Hydrology.

6.20 Advantages of the ADCP over velocity–area methods

(a) normally quicker than existing current meter methods
(b) the largest of the world's rivers may be measured as well as the smallest
(c) floods may be more conveniently measured
(d) may be used to check existing methods
(e) easily transported from site to site
(f) may be conveniently used from bridges or cableways
(g) generally less expensive than conventional methods
(h) better for measurement under ice-cover than conventional methods
(i) less expensive than existing velocity–area methods
(j) used for sediment discharge measurements
(k) used for bathymetric surveys especially in lakes.

6.21 Uncertainties

Uncertainty calculation of an ADCP measurement of discharge is more complex than that of the conventional methods of streamflow measurement (see

Figure 6.15 ADCP training exercise by the Bureau of Hydrology on the Songhua Jiang River (tributary of the Amur River) in China.
Source: SonTek.

Chapter 13) but the principle is the same in that the velocity in each of n verticals is measured and the discharge estimated from the mid-section method.

From experience so far, it is estimated that the uncertainty of an ADCP measurement under good conditions is similar to the uncertainty of a current meter measurement. For a stationary ADCP measurement the estimation of the uncertainty is similar to that of the uncertainty in a current meter measurement with minor modifications. Generally, the error sources in an ADCP measurement consist essentially of the following:

u_1 spatial resolution (velocities estimated by monastatic diverging multibeam geometry)

u_2 noise (may be large in low flows with high turbulence)

u_3 velocity ambiguity (ADCP measures phase angle difference between pulses)

u_4 side lobe interference (estimated by power curve fitting)

u_5 temporal resolution (velocity data sampled as time series at *equally* spaced intervals)

u_6 sound speed (ADCP assumes speed of sound and salinity constant)

u_7 beam angle (like u_1 is due to instrument tolerances)

u_8 boat speed (high ratios of boat-to-flow velocity may affect this error)

u_9 sampling time (may not be as critical for discharge measurements as it is for estimating mean velocity)

u_{10} near transducer (ringing waiting time-blanking period causing errors in velocities in upper bins)

u_{11} reference boat velocity (boat-mounted ADCPs measure in water column relative to boat movement)

u_{12} depth (transmit time for bottom tracking profiling and immersion depth of ADCP)

u_{13} cell positioning (maintaining top of first cell at constant position across section)

u_{14} rotation (pitch, roll, heading, attitude, and motion related to instrument configuration)

u_{15} time (needed to establish boat velocity and gating the return signal)

u_{16} edge (distance of ADCP from bank from assumed velocity distribution and discharge algorithm)

u_{17} vertical velocity profile model (depends on moving or fixed boat)

u_{18} discharge model (velocity area methods of estimation may be used in the algorithm)

u_{19} finite summation (as in velocity area the uncertainty in number of verticals may be taken)

u_{20} site selection and operation (secondary currents, aspect ratios, bed, turbulence, etc.)

Research and experience of providing values for each of the above uncertainty components is at an early stage but components u_3, u_4, u_7, u_{13} may be considered small or insignificant or may be included with other sources:

The total uncertainty may then be estimated by the root-sum-squares method as follows

$$u(Q) = (u_1^2 + u_2^2 + u_5^2 + u_6^2 + u_8^2 + u_9^2 + u_{10}^2 + u_{11}^2 + u_{12}^2 + u_{14}^2 + u_{15}^2$$
$$+ u_{16}^2 + u_{17}^2 + u_{18}^2 + u_{19}^2 + u_{20}^2)^{1/2} \tag{6.4}$$

The above component uncertainties are at the 68% confidence level and $u(Q)$ is multiplied by two to present the result at the 95% confidence level (see Chapter 13, Accuracy). Values for the component uncertainties in equation 6.4 are estimated by both the user and manufacturer of the equipment.

Further reading

Guide to the expression of uncertainties in measurement ISO, IEC, IFCC, IUPAC, IUPAP, OIML 1995.

Huhta, Craig. *ADCP Uncertainties. SonTek. private communication* 2007.

ISO 1088. Collection and processing of data for determination of uncertainties in flow measurement 2007.

ISO 5168. *Estimation of the uncertainty of a flowrate measurement* 1978.

ISO/CEN PDTS 25377. *Hydrometric Uncertainty Guide (HUG)* 2007.

ISO/PDTS 24154. Hydrometry – *Measuring river velocity and discharge with acoustic Doppler profilers*. International Standards Organization, Geneva 2005.

ISO/TR 24578 (in press). *Guide to the application of acoustic Doppler current profiler for measurement of discharge in open channels*.

Muste et al. Standardised uncertainty analysis framework for acoustic Doppler current profilers measurement. *University of IOWA, South Florida Management District* 2005.

Muste, M. ADCP Uncertainties. *The University of Iowa. Private Communications* 2005–2007.

Muste, M., and Stern, F. Proposed uncertainty assessment methodology for hydraulic and water resources engineering. *Proc. ASCE joint Conference on Water Resources Engineering and Water Resources Planning & Managemenr Minneapolis, MN* 2000.

Muste, M., Yu, K., Gonzalez-Castro, J. and Starzmann, E. Methodology for estimating ADCP measurement uncertainty in open channel flows. *Proc. World Water & Environmental Research Congress (EWRI) Salt Lake City, UT* 2004.

Muste, M., Yu, K., Pratt, T. and Abraham, D. *Practical aspects of ADCP data use for quantification of mean river flow characteristics: Part 11 Fixed-vessel measurements* J. Flow Meas. And Instr 15(1) pp. 17–28 2004.

Principles of River Discharge Measurements. SonTek/YSI. San Diego, California 6pp. inquiry@sontek.com 2003.

River discharge measurement on the Yangtze River with acoustic Doppler current profiler (ADCP). Field Measurement Verification Test and Development Study. Bureau of Hydrology, Changjiang Water Resources Commission, *The Ministry of Water Resources, Wuhan, China 29pp.* 2004.

Simpson, M.R. Discharge measurements using a broadband acoustic Doppler current profiler. *US Geological Survey Open File Report 61pp.* 2001.

Simpson, M.R. and Oltman, R.N. Discharge measurement system using an acoustic Doppler current profiler with application to large rivers and estuaries. *US Geological Survey Water Supply Paper 2395 32pp.* 1993.

SonTek/YSI, SonTeck *ADP acoustic Doppler profiler. Principles of Operation.* SonTek/YSI. San Diego, California 25pp. 2000.

US Geological Survey. Office of Surface Water Technical Memorandum *Policy and Technical Guidance on Discharge Measurements using Acoustic Doppler Current Profilers 2pp.* 2002.

Yorke, T.H. and Oberg, K.A. Measuring river velocity and discharge with acoustic Doppler profilers *Flow Measurement and Instrumentation Vol. 13 Number 5–65pp.* 2002.

Chapter 7

Measurement by floats

7.1 General

The float method is used in the measurement of streamflow where excessive velocities, depths and floating drift prohibit the use of a current meter. The method consists essentially of observing the time required for a float to traverse a course of known length and noting its position in the channel so that

$$v = \frac{L}{t} \tag{7.1}$$

where v is the float velocity (m s^{-1}), L is the distance travelled (m), and t is the time of travel over distance L (s).

For float measurements, ideally three cross-sections are selected – one at the beginning of the reach, one midway and one at the end of the reach, but depending on site conditions the midway section is sometimes omitted. The upstream and downstream cross-sections are made far enough apart so that the time, t, which the float takes to pass from one cross-section to the other can be measured accurately. A travel time of at least 20 s is recommended but a shorter time may be used for streams with high velocities when it may not be possible to find a straight reach of channel having adequate length. As a guide, the distance between the upstream and downstream cross-sections should be about four or five times the width of the midway section or the average width of the river.

The midway section, where employed, is used for checking the measurement of velocity and the location of the float in the cross-section.

7.2 Cross-sections

The selected cross-sections, normal to the direction of flow, are demarcated on the banks by clearly visible markers. Staff gauges, set to the same datum, are established at each section and are read at regular intervals during the measurement. Where there is a likelihood of a difference in water level between the banks, auxiliary staff gauges are installed on the opposite bank. The mean of the two readings is used as the mean stage. Stage data can be used as a check on float results during floods by use of the slope–area method (Chapter 8). In order to obtain the average cross-section for the reach, sufficient cross-sections are surveyed to obtain a reliable value. If the reach is fairly uniform, this may only involve a survey of the upstream (starting) and downstream (finishing) cross-sections but where the channel is irregular several of them have to be surveyed.

7.3 Floats

Surface floats

The surface float (Fig. 7.1(a)) is the simplest form of float although it is influenced by wind. Nevertheless, this type of float is valuable when quick measurements are needed.

Canister (or subsurface) floats

Canister floats (Fig. 7.1(b)) consist of a submerged canister or subsurface float connected by a thin adjustable line. The canister dimensions and its immersed depth are chosen so that the float velocity is equal to the mean velocity in the vertical.

Rod floats

Rod floats (Fig. 7.1(c)) are cylindrical rods, weighted so that they float vertically in still water with only the tip protruding above the surface. These are used to measure the mean velocity in the vertical and are designed so that they extend through as much of the stream depth as possible without the lower end touching the bed.

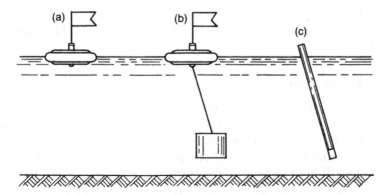

Figure 7.1 Floats: (a) surface float, (b) canister float, (c) rod float.

Drift or ice cakes

Selected pieces of drift or ice cakes will often indicate velocity more reliably than ordinary surface floats. The results are comparable to those obtained with canister or rod floats. Floating ice cakes and heavy drift comprising logs and trees that are largely submerged hold their courses well against surface disturbances and if observed at a sufficient number of points distributed across the cross-section give a fairly accurate measurement of the surface velocity.

Use of distinctive float colours

A most important aspect of measurement by floats is that the float should be easily recognisable and the use of distinctive colours is recommended. If a series of colours is used, it assists in the performance of the measurement if the colour sequence is laid down in advance. For example, it may by decided that the first batch of floats will be red, the second yellow, the third green, and so on.

7.4 Placing of floats

In making a float measurement the floats are placed in the stream so that they are distributed across the stream width. If groups of floats are used the floats are placed in accordance with the distribution of segments. The position of each float with respect to the distance from the bank or zero point is noted and the floats introduced a short distance upstream from the upper cross-section so that they will take up the speed of the current when they reach the upper cross-section. A stopwatch is used to time their travel between the end cross-sections of the reach. The estimated position of each float with respect to the bank, or zero point, is noted at the downstream cross-section and at the midway section if available. If there is no bridge or cableway from which to introduce the floats into the stream, the floats are tossed in from the bank. In a wide stream it may be impracticable to position any floats in the central part of the stream where most of the flow occurs and a boat is normally used for this purpose. In a method used in China for night gauging, using searchlights, the dispensable floats are illuminated by small disposable batteries and triggered by a launching device from the cableway.

7.5 Position fixing

The position of a float in the river may be found as follows. Two stopwatches are used, one to measure the time taken for a float to pass from the upstream cross-section to the downstream cross-section (t_1) and the second to measure the time taken for the float to cross the diagonal between the upstream reference point on the left bank and the downstream reference point on the right bank (t_2). If the upstream and downstream cross-sections are parallel, then from Fig. 7.2

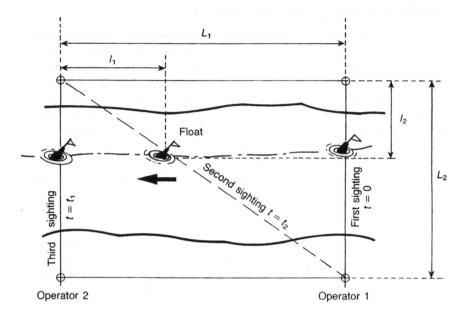

Figure 7.2 Determining the float position.

$$\frac{l_1}{L_1} = \frac{l_2}{L_2} \tag{7.2}$$

therefore

$$l_1 = \frac{L_1 l_2}{L_2} \tag{7.3}$$

and similarly

$$\frac{l_2}{L_2} = \frac{(t_1 - t_2)}{t_1} \tag{7.4}$$

therefore

$$l_2 = L_2 \frac{(t_1 - t_2)}{t_1}. \tag{7.5}$$

A temporary telephone line or walkie-talkie set is useful for communication between the upstream and downstream operators where this method is used.

7.6 Determination of mean velocity

The float velocity is determined by taking several float readings, if possible, in each segment and the mean velocity in each segment multiplied by a reduction coefficient to reduce the surface velocity to the mean velocity in the segment. The appropriate coefficient is best determined from current meter measurements which include a surface velocity measurement. If the relation between velocity at various stages in the vertical and mean velocity is plotted for all measurements, it is possible to establish a relation for stages above those actually measured by extrapolation. When determining the mean float velocity the results for floats which behave erratically or which lie, say 20%, outside the mean velocity should be discarded and a new mean derived from the remainder of the group.

If it is not possible to estimate the coefficient directly, it may be taken, in general, to vary between 0.8 and 0.9 depending on the shape of the vertical velocity distribution. The higher values are usually obtained for a smooth river bed. With canister or rod floats the coefficient may be taken as approximately 1.0 when the subsurface float is located at the 0.6 depth position. This coefficient, k say, may also be found approximately from the Manning and Chezy equations (Chapter 8) as follows

$$v = \frac{1}{n} R^{2/3} S^{1/2} \text{ (Manning)} \tag{7.6}$$

and

$$v = CR^{1/2}S^{1/2} \text{ (Chezy)} \tag{7.7}$$

where v is the velocity (ms^{-1}), R is the hydraulic radius (m), S is the surface slope, n is the Manning roughness coefficient and C is the Chezy roughness coefficient. Then

$$CR^{1/2}S^{1/2} = \frac{1}{n} R^{2/3} S^{1/2} \tag{7.8}$$

and

$$C = \frac{1}{n} R^{1/6}. \tag{7.9}$$

Therefore, estimating the value of n for the hydraulic conditions (Chapter 8), C can be found. k may then be computed from

$$k = \frac{C}{C + 6}. \tag{7.10}$$

7.7 Computation of discharge

A simple float gauging is shown diagrammatically in Fig 7.3 where the upper and lower sections are shown divided into a suitable number of segments of equal width. Generally the number of segments for a float gauging should be at least four but preferably more. If the channel is very irregular each segment should have approximately the same discharge in preference to making them equal in width. Increasing the number of floats used in each segment improves the accuracy of the determination of velocity.

By averaging, the areas of the segments can be determined. A midway section MN is drawn, as shown in Fig. 7.3, parallel to the cross-sectional lines. The starting and finishing points of each float are then plotted and joined by firm lines as shown. The surface points separating the segments of the two cross-sections are joined by dotted lines.

Where the firm lines cross the line MN, the corresponding mean velocity (float velocity multiplied by the appropriate coefficient) is plotted normal to MN and the points joined to form a velocity distribution curve. From this curve the mean velocity for each segment may be computed by planimeter or an approximate value may be adopted equal to the velocity halfway across the segment. The procedure for computing discharge is then similar to that of the velocity–area method (Chapter 2). The mean areas of corresponding segments of the upper and lower cross-sections are multiplied by the mean velocity.

When it is impracticable to obtain satisfactory movement of the floats across the whole width of the river, or if the floats move towards the centre of the flow,

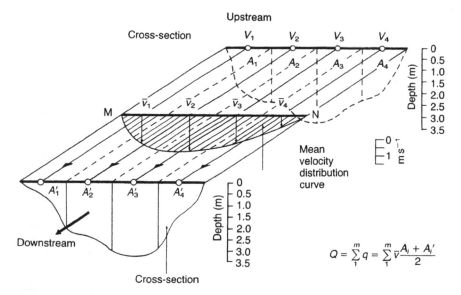

Figure 7.3 Computation of discharge from float measurement. $\bar{v}_1, \bar{v}_2 \ldots, \bar{v}_4$ are the mean velocities in each of the four panels.

an unadjusted discharge may be determined by measuring the mean of the surface velocities. The discharge is then multiplied by the appropriate coefficient determined, if possible, from the results of current meter measurements carried out simultaneously with the float measurements.

7.8 Example

In order to illustrate the procedure of a float measurement, the following example is taken from practice adopted in Russia.

1. Details of each float observation are given in Table 7.1. The mid-section is taken as the measuring cross-section and distances to crossing points of the floats are observed along this section. The total length between the upstream and downstream cross-sections is 50 m.

2. Details of segments are given in Table 7.2 and a sketch of the gauging is shown in Fig. 7.4 where the grouping of the floats is illustrated together with the average group travel times (the average of the float times). The depth verticals, 2.50 m apart, are also shown in Fig. 7.4 and it will be seen that these segments do not necessarily coincide with the float group boundaries. The final segments, therefore, are defined with respect to the nearest matching verticals.

Table 7.1 Float observation

(1) Float gauging number	(2) Crossing point of mid-section with reference to zero mark (m)	(3) Travel time Upstream section (min, s)	(4) Downstream section (min, s)	(5) Total travel time (s)	(6) Float group number	(7) Surface velocity (m s^{-1})
1	13.2	0.00	3.42	222	I	0.225
2	31.5	0.00	1.08	68	IV	0.735
3	14.0	0.00	2.59	179	I	0.279
4	17.5	0.00	2.11	131	II	0.382
5	32.2	0.00	2.24	144	IV	a
6	50.1	0.00	1.36	96	VI	0.521
7	46.1	0.00	1.24	84	V	0.596
8	42.0	0.00	1.14	74	V	0.675
9	26.2	0.00	1.22	82	III	0.610
10	24.1	0.00	1.40	100	III	0.500
11	35.0	0.00	2.19	139	–	a
12	33.7	0.00	1.17	77	IV	0.649
13	20.4	0.00	1.29	89	II	0.562
14	56.0	0.00	1.54	114	VI	0.438
15	64.5	0.00	2.59	179	VII	0.279
16	61.8	0.00	3.32	212	VII	0.236
17	41.0	0.00	1.23	83	V	0.602

[a] Discarded.

Table 7.2 Float observations and segment details[a]

(1) Serial number of depth vertical	(2) Distance from zero (m)	(3) Depth Upper section I (m)	(4) Depth Lower section II (m)	(5) Mean (m)	(6) Mean adjusted depth (m)	(7) Mean depth between verticals (m)	(8) Distance between verticals (m)	(9) Cross-sectional area between verticals (m²)	(10) Group cross-section zones (m²)	(11) Average group travel time (s)	(12) Average group surface velocity (m s⁻¹)	(13) Group discharge (m³ s⁻¹)
WELB[b]	12.2	0.00	0.00	0.00	0.00	2.80	0.45	1.26				
1	15.0	0.90	0.90	0.90	0.90	2.50	1.18	2.95	4.21	200	0.250	1.05
2	17.5	1.44	1.46	1.45	1.45	2.50	1.66	4.15				
3	20.0	1.85	1.86	1.86	1.86	2.50	2.05	5.12	9.27	110	0.454	4.21
4	22.5	2.23	2.24	2.24	2.24	2.50	2.34	5.85				
5	25.0	2.45	2.45	2.45	2.45	2.50	2.58	6.45				
6	27.5	2.72	2.73	2.72	2.72	2.50	2.68	2.70	19.00	91	0.549	10.4
7	30.0	2.65	2.64	2.64	2.64	2.50	2.72	6.80				
8	32.5	2.80	2.79	2.80	2.80	2.50	2.66	6.65	19.50	72	0.694	13.5
9	35.0	2.52	2.52	2.52	2.52	2.50	2.43	6.08				
10	37.5	2.35	2.34	2.34	2.34	2.50	2.30	5.75				

11	40.0	2.25	2.26	2.26	2.26	2.50	2.22	5.55			
12	42.5	2.17	2.18	2.18	2.18	2.50	2.12	5.30			
13	45.0	2.05	2.07	2.07	2.06	2.50	2.04	5.10	21.70	80	0.625	13.6
14	47.5	2.00	2.02	2.01	2.01	2.50	1.94	4.85			
15	50.0	1.86	1.86	1.86	1.86	2.50	1.82	4.55			
16	52.5	1.76	1.78	1.77	1.77	2.50	1.74	4.35			
17	55.0	1.70	1.70	1.70	1.70	2.50	1.65	4.12	17.9	105	0.476	8.52
18	57.5	1.60	1.61	1.60	1.60	2.50	1.48	3.70			
19	60.0	1.34	1.36	1.35	1.35	2.50	1.32	3.30			
20	62.5	1.27	1.29	1.28	1.28	2.50	1.18	2.95			
21	65.0	1.08	1.09	1.08	1.08	2.50	1.02	2.55			
22	67.5	0.95	0.96	0.96	0.96	2.50	0.85	2.12			
23	70.0	0.74	0.74	0.74	0.74	2.10	0.37	0.78	15.40	196	0.255	3.93
WERB[c]	72.1	0.00	0.00	0.00	0.00				107			55.2

[a] The average travel time for each group of floats is recorded in column 11, whilst column 12 represents the average surface velocity for each group computed by: $v = L/t$ ms^{-1} where L = distance between sections I and II, i.e. 50 m, and t = average travel time of the group.

[b] WELB: water edge left bank.

[c] WERB: water edge right bank.

Example   277

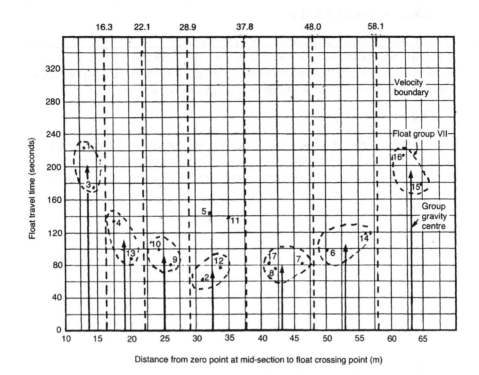

Figure 7.4 Method used in Russia to determine discharge by floats (see Tables 7.1 and 7.2).

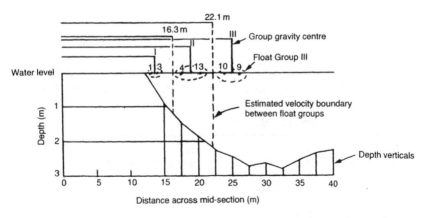

Figure 7.5 Determination of segments from float group boundaries (see Table 7.2).

This is demonstrated in Fig. 7.5 and Table 7.3. For example, the boundary between groups I and II crosses at a distance of 16.3 m (Fig. 7.5), the nearest vertical being at 17.5 m. Therefore, in column 9 of Table 7.2 the cross-sectional areas, or segment areas, between the verticals at 17.5 m are summed as

$$1.26 \ \text{m}^2 + 2.95 \ \text{m}^2 = 4.21 \ \text{m}^2$$

and this value is inserted in column 10 of Table 7.2.

3. The segment discharges summed in column 13 (55.2 m³ s⁻¹) are multiplied by the reduction coefficient which in this example was found, by current meter gaugings, to be 0.866. The discharge is then calculated as

$$55.2 \times 0.866 = 47.8 \ \text{m}^3 \ \text{s}^{-1}.$$

If no current meter gaugings are available, the procedure demonstrated in Section 7.6 may be followed, where R = hydraulic radius = 1.79 m (average depth) and n = Manning's roughness coefficient = 0.033 (approximately). Then (equation (7.9))

$$C = \frac{1}{n} R^{1/6}$$

$$= \frac{1}{0.033} \times 1.79^{1/6}$$

$$= 33.4$$

and from equation (7.10)

$$k = \text{reduction coefficient} = \frac{C}{C + 6}$$

$$= \frac{33.4}{39.4}$$

$$= 0.848.$$

4. An alternative graphical method of processing the results is demonstrated in Fig. 7.6 and Table 7.4. In this case a horizontal surface velocity distribution

Table 7.3 Float groups and segment boundaries

(1) Group number	(2) Average distance from zero point to group gravity centre (m)	(3) Semi-total of average distance (m)
I	13.6	
II	19.0	16.3
III	25.2	22.1
IV	32.6	28.9
V	43.0	37.8
VI	53.0	48.0
VII	63.2	58.1

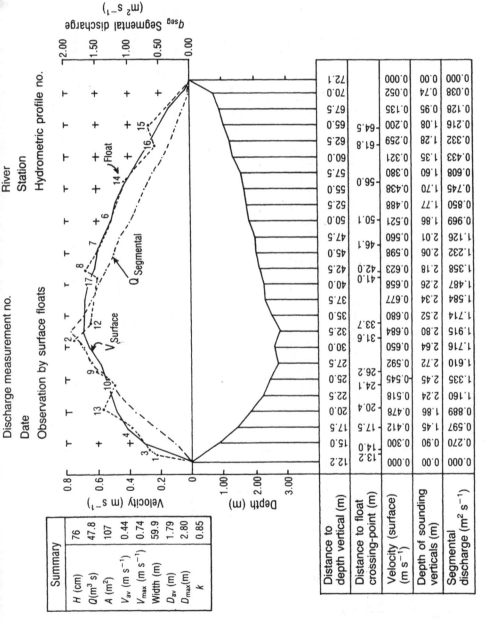

Figure 7.6 Alternative graphical method of processing the results (see Table 7.4).

Table 7.4 Alternative (graphical) method of processing float gauging data (Fig. 7.6)

Serial number	Distance from zero	Depth of vertical	Surface velocity (from Fig. 7.6)	Segmental apparent discharge at vertical	Half segmental apparent discharge between verticals	Distance between verticals	Apparent discharge between verticals
	(m)	(m)	$(m\ s^{-1})$	$(m^2\ s^{-1})$	$(m^3\ s^{-1})$	(m)	$(m^3\ s^{-1})$
WELB	12.2	0.00	0.00	0.00	0.090	2.80	0.252
1	15.0	0.90	0.30	0.270	0.433	2.50	1.08
2	17.5	1.45	0.412	0.597	0.745	2.50	1.86
3	20.0	1.86	0.478	0.893	1.030	2.50	2.58
4	22.5	2.24	0.518	1.160	1.240	2.50	3.10
5	25.0	2.45	0.545	1.320	1.460	2.50	3.55
6	27.5	2.72	0.592	1.600	1.660	2.50	4.15
7	30.0	2.64	0.650	1.720	1.820	2.50	4.55
8	32.5	2.80	0.684	1.930	1.820	2.50	4.55
9	35.0	2.52	0.680	1.710	1.650	2.50	4.12
10	37.5	2.34	0.677	1.590	1.540	2.50	3.85
11	40.0	2.26	0.658	1.490	1.420	2.50	3.55
12	42.5	2.18	0.623	1.350	1.300	2.50	3.25
13	45.0	2.06	0.598	1.240	1.180	2.50	2.95
14	47.5	2.01	0.560	1.130	1.040	2.50	2.60
15	50.0	1.86	0.521	0.967	0.908	2.50	2.27
16	52.5	1.77	0.480	0.850	0.799	2.50	2.00
17	55.0	1.70	0.438	0.748	0.678	2.50	1.70
18	57.5	1.60	0.380	0.608	0.520	2.50	1.30
19	60.0	1.35	0.321	0.432	0.382	2.50	0.955
20	62.5	1.28	0.259	0.333	0.274	2.50	0.685
21	65.0	1.08	0.220	0.216	0.175	2.50	0.438
22	67.5	0.96	0.135	0.134	0.086	2.50	0.215
23	70.0	0.74	0.052	0.038	0.012	2.10	0.025
Σ							55.6

curve is drawn by joining the float velocities shown dashed in Fig. 7.6. The velocities for each vertical are taken from this curve and entered in Table 7.4. The computation may be performed by the mid-section method (Chapter 2).

Alternatively the discharge per unit width at a vertical is

$$Aq_{vert} = vd \ \mathrm{m^2\ s^{-1}} \tag{7.11}$$

and the discharge per segment is

$$q_{seg} = \frac{q_n - 1 + q_n}{2} \times b \tag{7.12}$$

where b is the distance between verticals.

For example, in Table 7.4, serial number 2

$$Aq_{\text{vert}} = 1.45 \times 0.412 = 0.597 \ \text{m}^2 \ \text{s}^{-1}$$

and

$$q_{\text{seg}} = \frac{0.597 + 0.893}{2} \times 2.5$$

$$= 0.745 \times 2.5$$

$$= 1.86 \ \text{m}^3 \ \text{s}^{-1}.$$

7.9 The rising air float technique

Theory

The theory of the rising air float technique can be explained by considering a buoyant float released from the bed in a moving stream of water (Fig. 7.7). As it rises to the surface, the float is displaced downstream according to the flow velocity (v_x) at each successive level through which it passes. The total displacement of the float (L) in reaching the surface is a function of both depth of flow and the average velocity of flow and, providing the float rises at a constant terminal velocity (v_r), the product of depth and velocity, i.e. discharge per unit width, q, is given by

$$q = \int_0^{t_r} v_x \, v_r \, dt \tag{7.13}$$

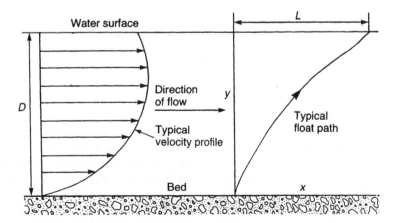

Figure 7.7 Rising bubble (air float) technique for determining discharge: schematic illustration (section).

in which t_r is the time of rise of the float which reduces to

$$q = v_r L, \tag{7.14}$$

thus the rising float essentially integrates the velocity profile, through which it passes, together with the depth.

This result can also be derived directly from a consideration of Fig. 7.7. The distance through which the float rises (depth) is given by

$$D = v_r t_r. \tag{7.15}$$

In this same period of time, the horizontal displacement is

$$L = \bar{v} t_r \tag{7.16}$$

where $\bar{v}$ is the mean flow velocity. Then

$$t_r = \frac{D}{v_r} = \frac{L}{\bar{v}}$$

giving

$$D\bar{v} = q = v_r L. \tag{7.17}$$

If we then have a number of release points across the channel the total discharge Q can be calculated as

$$Q = \int_0^b q \, db \tag{7.18}$$

where b is the channel width. In discrete form this becomes

$$Q = \sum_{i=1}^{n} q_i b_i \tag{7.19}$$

in which n is the number of release points across the section.

In the rising air float technique, the floats are a succession of air bubbles produced by a specially manufactured pipe laid across the channel bed. This produces a semi-continuous envelope of air bubbles on the water surface (Fig. 7.8). In this case

$$Q = \int_0^b q \, db = \int_0^b v_r L \, db \tag{7.20}$$

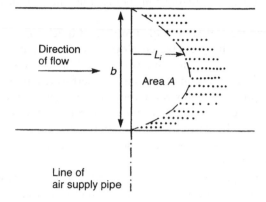

Figure 7.8 Rising bubble (air float) technique for determining discharge: schematic illustration of bubble envelope.

$$= v_r \int_0^b L \, db = v_r A \tag{7.21}$$

where A is the area formed on the surface between the bubble envelope and the line of the air supply pipe. Thus to calculate Q, v_r must be known and A must be determined for each gauging.

Laboratory experiments have found v_r to have an average value of 0.218 m s^{-1} and Δb and $\bar{L}$ (area A in Fig. 7.8) are found by photographing the water surface. The cross-section of the channel therefore need not be known.

Equipment

The equipment consists of 20 mm diameter rubber high-pressure air hose mounted in 20 m lengths on hose trolleys. Lengths are connected in the field using self-sealing snap connectors. The pipe is fitted with commercially available drip irrigation nozzles spaced at 250 mm centres in order to produce a semi-continuous bubble envelope on the water surface. The air supply is provided by a small portable petrol-engine-driven compressor unit.

Data capture and analysis

With the pipe laid across the channel and the air supply connected, a semi-continuous curve of bubbles is formed on the water surface. Upstream of the point of surfacing the surface is clear, whilst downstream a trail of bursting bubbles is formed. The front of the bubbles is then clearly visible by eye in all except very turbulent or very windy conditions.

If the displacement of the bubbles from their point of issue can be determined at a number of points across the channel, and with the rise velocity relating to the nozzle configuration being known, the discharge can be calculated.

In order to produce a method both simple and quick to use, and to provide 'hard-copy' of the resulting bubble envelope, the data are captured by photographing the water surface. One camera is used from any convenient position on the bank and the calculation of discharge is performed, preferably by microcomputer. The input data required are the camera position, the location in the same coordinate system of a single ground control point, normally on the far bank, the water level relative to the camera elevation, the camera focal length and the magnification of the projected image or print from which the measurements are taken.

Photographic data may be taken off manually but it is more convenient and quicker to use a chart digitiser which can be interfaced directly into the microcomputer. Having calculated the area on the water surface between the bubble envelope and the line of the pipe, and multiplying this area by the bubble rise velocity, the discharge is calculated.

The above technique is suitable for streams of up to about 50 m in width, but for greater widths it is necessary to subdivide the surface into two or more segments, using lenses of different focal lengths and with the advantage that the furthest segment can be magnified.

The rising air float technique has been used in UK rivers with satisfactory results and the uncertainties claimed compare favourably with those of current meter measurements. The main limitations appear to be the assumed value of the bubble rise velocity, which should be established in field tests, and the requirement for reasonably unbroken water surface. It will be noted that the method produces a measurement of instantaneous discharge and, since the cross-sectional area is not measured, does not produce a measurement of the average velocity.

A typical gauging in progress is shown in Fig. 7.9.

7.10 Float gauging using aircraft

Measuring discharge in large and inaccessible rivers in Russia has been successfully carried out using floats dropped from aircraft. Two methods are used. One measures surface velocity and the other uses the method of the rising float technique, the principle being similar to that described in Section 7.9.

Surface velocity method

In this method, floats are dropped from an aeroplane and their paths traced by aerial photography, photographs being taken twice during travel of the floats. Normally the floats contain dyes for clearer identification. Since the scale of the photographs are known, the travel distance L during time t is determined and the surface velocity found, as before (Section 7.1, equation (7.1)), from

Figure 7.9 Typical gauging by rising bubble (air float) method: note the line of bubbles on the surface.

$$v = \frac{L}{t}.$$

A reduction coefficient is applied to reduce the surface velocity to mean velocity.

Rising float method

In this method floats are dropped containing oil. On impact with the bed, oil is released and traces the distribution of velocity.

Ground markers are required to determine the photographic scale and the theory is similar to the rising bubble technique so that, from Section 7.9 (equation (7.21))

$$Q = v_r A$$

and in this case the laboratory value of v_r may be checked on site from the aerial photographs by timing from the point of ejection of the oil to its appearance on the surface.

Further reading

ISO 748 *Liquid Flow Measurement in Open Channels: Velocity – Area Methods. ISO, Geneva, Switzerland* 2007.

Kuprianov, V. V. Aerial methods of measuring river flows. In *Hydrometry: Principles and Practices* (ed. R. W. Herschy). John Wiley and Sons, Chichester 1978.

Lučševa, A. A. *Applied Hydrometry*. The Ministry of Agriculture, Leningrad 1957.

Rodier, J. and Roche, M. River flow in arid regions. In *Hydrometry: Principles and Practices* (ed. R. W. Herschy). John Wiley and Sons, Chichester 1978.

Sargent, D. M. The development of a viable method of streamflow measurement using the integrating float technique. *Proc. Instn Civ. Engrs, Part 2*, 71 1981.

Sargent, D. M. The rising air float technique for the measurement of stream discharge. *IAHS Symposium, Exeter, Publication No. 134* 1982.

Starosolszky, O. Run-off and river flow measurements. *NATO Advanced Research Workshop, Tucson, Arizona* 1993.

Chapter 8

The slope–area method of streamflow measurement

8.1 General

The most important use of the slope–area method is for the indirect determination of flood discharge, normally after the flood has passed. It may also be used, however, where the flow is affected by backwater. The method consists of the estimation of three basic factors. First, the area of the average cross-section in a longitudinal reach of channel of known length second, the slope of the water surface or the slope of the energy gradient in the same reach of channel and third, the character of the streambed so that a suitable roughness factor may be chosen.

When these factors are known, the mean velocity of the stream may be computed by either the Chezy equation or the Manning equation.

8.2 Chezy and Manning equations

Chezy: $\bar{v} = C R^{1/2} S^{1/2}$ m s^{-1} (8.1)

Manning: $\bar{v} = \dfrac{1}{n} R^{2/3} S^{1/2}$ m s^{-1} (8.2)

where $\bar{v}$ = mean velocity of stream (m s^{-1});
where $\bar{R}$ = hydraulic radius = $\bar{A}/P$ (m) is the mean cross-sectional area and P the wetted perimeter (m);
    S = slope or energy gradient;
    C = Chezy roughness coefficient;
    n = Manning roughness coefficient.

It will be recalled (equation (7.9)) that by equating equations (8.1) and (8.2), C may be related to n in the form

$$C = \frac{R^{1/6}}{n}.$$

Generally the Manning equation is preferred in practice because it is simple to apply and many years of experience in its use have shown that it produces reliable results.

The product of the mean velocity so obtained from either equation (8.1) or equation (8.2) and the area $\bar{A}$ of the average cross-section provides an estimation of the discharge.

If the mean velocity does not remain constant from section to section along the reach of channel, the surface slope may not coincide with the energy gradient, and for those conditions the energy gradient is used instead of the surface slope.

The Manning equation, like the Chezy equation, was developed for conditions of uniform flow in which the water surface slope and energy gradient are parallel to the streambed, and the area, hydraulic radius and depth remain constant throughout the reach. For lack of a better solution, it is assumed that the equations are also valid for non-uniform reaches that are often encountered in natural channels, if the water surface gradient is modified by the difference in velocity head between the cross-sections.

8.3 The energy equation

The energy equation for a reach of non-uniform channel between cross-section 1 (upstream) and cross-section 2 (downstream), shown in Fig. 8.1, is

$$h_{\mathrm{F}} = \left(Z_1 + \frac{a_1 \bar{v}_1^2}{2g} \right) - \left(Z_2 + \frac{a_2 \bar{v}_2^2}{2g} \right) \tag{8.3}$$

where h_{F} = energy loss between sections 1 and 2;
$\quad Z_1$ = elevation of water surface at section 1 above a common datum (m);
$\quad Z_2$ = elevation of water surface at section 2 above a common datum (m);
$\quad \bar{v}_1$ = mean velocity at section 1;
$\quad \bar{v}_2$ = mean velocity at section 2;
$\quad a_1$ = energy coefficient at section 1 (= 1.0);
$\quad a_2$ = energy coefficient at section 2 (= 1.0);
$\quad g_1$ = acceleration due to gravity = 9.81 m s^{-2}.

The energy slope S, therefore, to be used in equations (8.1) and (8.2) is (from equation (8.3))

$$S = \frac{(Z_1 - Z_2) + \left(\dfrac{\bar{v}_1^2}{2g} - \dfrac{\bar{v}_2^2}{2g} \right)}{L} \tag{8.4}$$

where L is the length of channel reach (m). Equation (8.4) applies to contracting reaches ($\bar{v}_2$ greater than $\bar{v}_1$).

For expanding reaches where the velocity at the downstream section is less than that in the upstream section ($\bar{v}_2$ less than $\bar{v}_1$) and where there is a

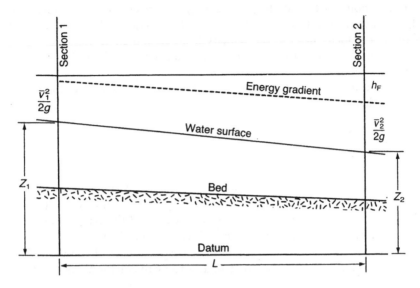

Figure 8.1 Schematic definition of slope–area reach.

tranformation of kinetic energy into potential energy, it is customary to assume the actual recovery to be 50% of the theoretical recovery.

Therefore, for $\bar{v}_2$ less than $\bar{v}_1$, the following equation is used

$$S = \frac{(Z_1 - Z_2) + 0.5\left(\dfrac{\bar{v}_1^2}{2g} - \dfrac{\bar{v}_2^2}{2g}\right)}{L}. \tag{8.5}$$

8.4 Calculation of velocity head

The velocity head, $\bar{v}^2/2g$, has to be calculated when using equations (8.4) and (8.5) where non-uniform flow conditions are encountered. The slope–area method is normally used to estimate discharge after the flood has passed or under conditions of major floods. In either situation the direct measurement of velocity is not made and $\bar{v}$ has to be calculated by an iterative procedure, similar to that shown in Chapter 10, Section 10.2, as follows.

A first approximation of the discharge is made from equation (8.1) or equation (8.2) using the actual value of the fall in the stream. Knowing the values of A_1 and A_2, $\bar{v}_1$ and $\bar{v}_2$ can be calculated from

$$\bar{v}_1 = Q/A_1 \tag{8.6}$$

$$\bar{v}_2 = Q/A_2, \tag{8.7}$$

hence $\bar{v}_1^2/2g$ and $\bar{v}_2^2/2g$ can be computed. S can then be calculated from equation (8.4), contracting reach, or equation (8.5), expanding reach.

In equations (8.4) and (8.5) the energy coefficient, or the velocity head adjustment factor, a, has been taken as 1.0. The energy coefficient a is the ratio of the true velocity head to the velocity head computed on the basis of mean velocity and for most cases its value will be unity or will have little significant variation from unity.

However, in compound channels the value of a may be greater than 1.0 and may be calculated from

$$a = \frac{\Sigma K_i^3/A_i^2}{K_T^3/A_T^2} \tag{8.8}$$

where subscript i refers to the conveyance or area of the subsections of the channel cross-section and subscript T refers to the conveyance or area of the whole cross-section. The conveyance K is a measure of the carrying capacity of the channel and has the dimensions $m^3\ s^{-1}$ (equation (8.12)).

The conveyances of each portion of the cross-section are therefore calculated separately and then added (Section 8.6) according to equation (8.8).

8.5 Estimation of Manning's n and Chezy's C

At the present state of knowledge, the selection of the roughness coefficient, n or C, remains mainly an art. There are no resistance diagrams or relationships available similar to those used in uniform pipe-flow. Consequently in view of the large range of conditions in open channels the ability to evaluate roughness coefficients is developed through experience. Otherwise, only by direct measurement of discharge, from which roughness values may be calculated, is it possible to obtain reliable values.

To the untrained beginner the selection of a roughness coefficient can be no more than a guess and different individuals may often obtain different results.

The experience necessary for the proper selection of roughness coefficients can be obtained from (a) an understanding of the factors that affect the value of the roughness coefficient, (b) an examination to become acquainted with the appearance of some typical channels whose roughness coefficients have been established from gaugings, and (c) consulting a table of typical roughness coefficients for channels of different types. Table 8.1 is such a table for suggested values of n and C (see also Chapter 5, Section 5.10).

8.6 Calculation of discharge

If the channel is straight, comparatively narrow, and the slope of the water surface essentially uniform, a reference gauge at each end of the reach is normally sufficient and discharge may be calculated from

$$Q = \bar{A}\bar{v}\ m^3\ s^{-1} \tag{8.9}$$

Table 8.1 Values of Manning's n and Chezy's C

Type of channel and description	Manning's coefficient n	Chezy's coefficient C				
		$R_h = 1\,m$	$R_h = 2.5\,m$	$R_h = 5\,m$	$R_h = 10\,m$
Excavated or dredged					
(1) Earth, straight and uniform					
a. Clean, recently completed	0.016 to 0.020	63 to 50	72 to 58	81 to 65	91 to 73
b. Clean, after weathering	0.018 to 0.025	55 to 40	64 to 46	72 to 52	81 to 59
c. With short grass, few weeds	0.022 to 0.033	45 to 30	53 to 35	59 to 40	67 to 44
(2) Rock cuts					
a. Smooth and uniform	0.025 to 0.040	40 to 25	46 to 29	52 to 33	59 to 37
b. Jagged and irregular	0.035 to 0.050	29 to 20	33 to 23	37 to 26	42 to 29
Natural streams					
Minor streams (top width at flood stage less than 30 m) on plains: clean, straight, full stage, no rifts or deep pools	0.025 to 0.033	40 to 30	46 to 35	52 to 40	59 to 44
Flood plains					
(1) Pasture, no brush					
a. Short grass	0.025 to 0.035	40 to 29	46 to 33	52 to 37	59 to 42
b. High grass	0.030 to 0.050	33 to 20	39 to 23	44 to 26	49 to 29
(2) Cultivated areas					
a. No crop	0.020 to 0.040	50 to 25	58 to 29	65 to 33	73 to 37
b. Mature row crops	0.025 to 0.045	40 to 22	46 to 26	52 to 29	59 to 33
c. Mature field crops	0.030 to 0.050	33 to 20	39 to 23	44 to 26	49 to 29

Table 8.1 Continued

Type of channel and description	Manning's coefficient n	Chezy's coefficient C				
		$R_h = 1\,m$	$R_h = 2.5\,m$	$R_h = 5\,m$	$R_h = 10\,m$
(3) Brush						
a. Scattered brush, heavy weeds	0.035 to 0.070	29 to 14	33 to 17	37 to 19	42 to 21
b. Light brush and tress (without foliage)	0.035 to 0.060	29 to 17	33 to 19	37 to 22	42 to 24
c. Light brush and trees (with foliage)	0.040 to 0.080	25 to 12	29 to 14	33 to 16	37 to 18
d. Medium to dense brush (without foliage)	0.045 to 0.110	22 to 9	26 to 10.5	29 to 12	33 to 13
e. Medium to dense brush (with foliage)	0.070 to 0.160	14 to 6.5	17 to 7.5	19 to 8	21 to 9
(4) Trees						
a. Cleared land with tree stumps, no sprouts	0.030 to 0.050	33 to 20	39 to 23	44 to 26	49 to 29
b. Same as above, but with heavy growth of sprouts	0.050 to 0.080	20 to 12	23 to 14	26 to 16	29 to 18
c. Heavy stand of timber; a few felled trees, little undergrowth, flood stage below branches	0.080 to 0.120	12 to 8.5	14 to 9.5	16 to 11	18 to 12
d. Same as above, but with flood stage reaching branches	0.100 to 0.160	10 to 6.5	12 to 7.5	13 to 8	15 to 9
e. Dense willows, in midsummer	0.110 to 0.200	9 to 5	10.5 to 6	12 to 6.5	13 to 7.5

Table 8.2 Coefficients for channels with relatively coarse bed material and not characterised by bed formations

Type of material	Size of bed material (mm)	Manning's coefficient n	Chezy's coefficient C				
			$R_h = 1\,m$	$R_h = 2.5\,m$	$R_h = 5\,m$	$R_h = 10\,m$
Gravel	4 to 8	0.019 to 0.020	53 to 50	61 to 58	69 to 65	77 to 73
	8 to 20	0.020 to 0.022	50 to 45	58 to 53	65 to 59	73 to 67
	20 to 60	0.022 to 0.027	45 to 37	53 to 43	59 to 48	67 to 54
Pebbles and shingle	60 to 110	0.027 to 0.030	37 to 33	43 to 39	48 to 44	54 to 49
	110 to 250	0.030 to 0.035	33 to 29	39 to 33	44 to 37	49 to 42

where $\bar{A}$ is the average cross-sectional area calculated from a convenient number of cross-sections and $\bar{v}$ is the mean velocity in the channel found from equation (8.1), or equation (8.2), or

$$Q = \frac{1}{n} \bar{A} R^{2/3} S^{1/2} \text{ m}^3 \text{ s}^{-1} \quad \text{(Manning)} \tag{8.10}$$

or

$$Q = C \bar{A} R^{1/2} S^{1/2} \text{ m}^3 \text{ s}^{-1} \quad \text{(Chezy)} \tag{8.11}$$

where S may be taken as the difference in the reference gauge readings, set to a common datum, divided by L.

In applying the Manning equation, however, it is normal practice to compute first the conveyance, K, for each cross-section, as

$$K = \frac{1}{n} A R^{2/3}. \tag{8.12}$$

The mean conveyance in the reach is then calculated as the geometric mean of the conveyances at the two cross-sections. This procedure is based on the assumption that the conveyance varies uniformly between sections. The discharge is then computed from

$$Q = \sqrt{(K_1 K_2 S)} \tag{8.13}$$

noting that, if the reach is contracting or expanding, S is found from equation (8.4) or equation (8.5), respectively.

Note that the conveyance in the Chezy equation is

$$K = C \bar{A} R^{1/2}. \tag{8.14}$$

Conveyance is a measure of the carrying capacity of the channel and has dimensions m^3s^{-1}. The discharge in a uniform channel is therefore equal to the product of the conveyance and the square root of the slope (equation (8.13)). One advantage of this concept is that the conveyance of separate portions of a compound cross-section can be added to obtain the total conveyance of the section. Conveyance is also used to compute the relative distribution of discharge in an approach section downstream from which division of flow occurs through separate channels.

Simplified slope–area method

In recent years, the US Geological Survey has analysed slope–area data in US rivers and suggested a simple logarithmic model for the computation of discharge in uniform channels. The analysis has indicated a relation between channel roughness and water surface slope. With the assumption that slope may replace n and that hydraulic radius is related to cross-sectional area, the Manning equation is reduced to

$$\log Q = 0.191 + 1.33 \log \bar{A} + 0.05 \log S - 0.056 (\log S)^2 \qquad (8.15)$$

where Q is in $m^3 s^{-1}$ and $\bar{A}$ is in m^2.

The model upon which equation (8.15) is based, is of the form

$$Q = aA^b S^c \qquad (8.16)$$

where S is the water surface slope, and a, b and c are constants.

The simplified slope–area method appears to give good results in channels having uniform flow with no significant difference in velocity head between the upstream cross-section and the downstream cross-section.

8.7 Selection of reach

If the slope–area method is used for a direct measurement of discharge, reference gauges, whether staff or inclined, are set to a common datum at the upstream and downstream cross-sections (primary installations). Water level recorders are normally installed at these sections. If the channel is essentially straight, the cross-sections are the same or similar and the water surface slope uniform (that is, the water surface slope is essentially the same as the bed slope), these two gauges will suffice. The slope, S, may then be calculated from the difference in the gauge readings read simultaneously.

In some reaches, however, it may be advisable to have reference gauges placed on the banks opposite the primary installations or, if information is desired regarding the uniformity of the slope or gradient line, gauges may be placed at various strategic locations along the reach. The length of reach required depends largely on channel conditions and, as a fairly uniform slope is necessary, it may be convenient to select a shorter reach than might otherwise have been chosen. Nevertheless, it is advisable to select a reach as long as practicable and in large rivers reaches of 300 m or more are desirable having at least 0.3 m difference in height of the water surface between the upstream and downstream gauges. Generally a useful rule is to estimate the reach required as being four times the channel width or 75 times the mean depth in the channel.

In determining the average area and the hydraulic radius for a reach of channel, enough cross-sections are measured to ensure representative averages. One such determination may suffice for an artificial channel of uniform cross-section, whereas in natural streams variations in the channel may necessitate several measurements of the cross-section at carefully selected points along the reach before reliable averages can be determined.

When the slope–area method is used where no reference gauges are installed and reliance has to be placed on flood marks to obtain the slope, such marks require to be carefully surveyed and interpreted.

Some good reaches are rendered useless when heavy rains follow the peak and destroy marks. Marks on the ground or those left by quiet water are usually preferable, the best types being drift, silt lines, wash lines, seed lines on trees or poles, and slit or stain lines on buildings.

Examples

1. Compute the flood discharge through a uniform, straight river reach of 200 m length given the following data

Upstream cross-sectional area
= downstream cross-sectional area = 120 m²
Fall of water surface = 0.15 m
Average width of channel = 60 m
Average cross-sectional depth = 2 m
Manning's n value = 0.030.

Then

$$R = 2 \text{ m} \quad \left(\text{or } \frac{\bar{A}}{P} = \frac{120}{60} = 2 \text{ m} \right)$$

and since the reach is straight and uniform and the cross-sectional areas similar, $\bar{v}$ is computed from equation (8.2) with $S = 0.15/200 \, n$

$$\bar{v} = \frac{1}{n} R^{2/3} S^{1/2}$$

$$= \frac{1}{0.030} 2^{2/3} \times \left(\frac{0.15}{200} \right)^{1/2}$$

$$= 52.91 \times (0.000\,75)^{1/2}$$

$$= 52.91 \times 0.0274,$$

therefore

$$\bar{v} = 1.45 \text{ m s}^{-1}.$$

Then

$$Q = \bar{A}\bar{v} \tag{8.9}$$

$$= 120 \times 1.45$$

$$= 174 \text{ m}^3 \text{ s}^{-1}.$$

Using the simplified equation (8.15) gives

$$\log Q = 0.191 + 1.33 \log \bar{A} + 0.05 \log S - 0.056(\log S)^2$$

$$= 0.191 + 1.33 \log 120 + 0.05 \log 0.000\,75 - 0.056(\log 0.000\,75)^2$$

$$= 2.2533$$

and

$$Q = 179 \text{ m}^3 \text{ s}^{-1}.$$

2. Compute the flood discharge through a river reach of 200 m length given the following data

A_1 (upstream)	$= 1231 \text{ m}^2$
A_2 (downstream)	$= 1222 \text{ m}^2$
Fall of water surface	$= 0.30 \text{ m}$
P_1 (upstream)	$= 320 \text{ m}$
P_2 (downstream)	$= 310 \text{ m}$
n	$= 0.030.$

Now

$$R_1 = \frac{A_1}{P_1} = \frac{1231}{320} = 3.85 \text{ m}$$

$$R_2 = \frac{A_2}{P_2} = \frac{1222}{310} = 3.94 \text{ m}$$

$$S = 0.30/200 = 0.001\,50.$$

Then from equation (8.12)

$$K_1 = \frac{1}{n} A_1 R_1^{2/3} = \frac{1}{0.030} \times 1231 \times 3.85^{2/3} = 100\,787$$

$$K_2 = \frac{1}{n} A_2 R_2^{2/3} = \frac{1}{0.030} \times 1222 \times 3.94^{2/3} = 101\ 603.$$

And from equation (8.13)

$$Q = \sqrt{(K_1 K_2 S)}$$
$$= \sqrt{(100\ 787 \times 101\ 603)}\ \sqrt{S}$$
$$= 101\ 194\ \sqrt{0.001\ 50}$$

giving

$$Q = 3919\ \mathrm{m}^3\ \mathrm{s}^{-1}.$$

This is a first approximation of discharge since a velocity head adjustment has to be made.

Second approximation of Q

$$\frac{\bar{v}_1^2}{2g} = \left(\frac{Q}{A_1}\right)^2 \times \frac{1}{2g} = \left(\frac{3919}{1231}\right)^2 \times \frac{1}{19.62} = 0.516\ 58\ \mathrm{m}$$

$$\frac{\bar{v}_2^2}{2g} = \left(\frac{Q}{A_2}\right)^2 \times \frac{1}{2g} = \left(\frac{3919}{1222}\right)^2 \times \frac{1}{19.62} = 0.524\ 21\ \mathrm{m}.$$

The velocity head difference is therefore $-0.007\ 63$ m, and, since $\bar{v}_1$ is less than $\bar{v}_2$, the flow is contracting. The adjusted fall is therefore (equation (8.4))

$$0.30 - 0.007\ 63 = 0.292\ 37\ \mathrm{m}$$

and S becomes

$$\frac{0.292\ 37}{200} = 0.001\ 4619$$

and

$$Q = K\sqrt{S} \quad \text{where} \quad K = \sqrt{(K_1\ K_2)}$$
$$= 101\ 194\sqrt{0.001\ 4619}$$
$$= 3869\ \mathrm{m}^3\ \mathrm{s}^{-1}.$$

Third approximation of Q

The above procedure is repeated using the new value of $Q = 3860$ m^3 s^{-1}

$$\frac{\bar{v}_1^2}{2g} = \left(\frac{3869}{1231}\right)^2 \times \frac{1}{19.62} = 0.503\ 480 \text{ m}$$

$$\frac{\bar{v}_2^2}{2g} = \left(\frac{3869}{1222}\right)^2 \times \frac{1}{19.62} = 0.510\ 924 \text{ m};$$

velocity head difference $= -0.007\ 444$ m;
adjusted fall $= 0.30 - 0.007\ 444 = 0.292\ 556$;

$$S = \frac{0.292\ 556}{200} = 0.001\ 4628$$

$$Q = 101\ 194\sqrt{0.001\ 4628}$$

$$= 3870 \text{ m}^2 \text{ s}^{-1}.$$

Fourth approximation of Q

$$\frac{\bar{v}_1^2}{2g} = \left(\frac{3870}{1231}\right)^2 \times \frac{1}{19.62} = 0.503\ 740 \text{ m}$$

$$\frac{\bar{v}_2^2}{2g} = \left(\frac{3870}{1222}\right)^2 \times \frac{1}{19.62} = 0.511\ 188 \text{ m};$$

velocity head difference $= -0.007\ 448$ m;
adjusted fall $= 0.30 - 0.007\ 448 = 0.292\ 552$ m;

$$Q = 101\ 194\left(\frac{0.292\ 552}{200}\right)$$

$$= 3870 \text{ m}^3 \text{ s}^{-1}$$

which is the same value as calculated by the third approximation. Therefore $Q = 3870$ m^3 s^{-1}.

Further reading

Barnes, H. H. Roughness Characteristics of Natural Channels. *US Geological Survey Water Supply Paper 1849 1967.*

Barnes, H. H. Indirect methods. In *Hydrometry: Principles and Practices* (ed. R. W. Herschy). John Wiley and Sons, Chichester 1978.

ISO 1070. *Liquid Flow Measurement in Open Channels: Slope – Area Method*. ISO, Geneva, Switzerland 1992.

Powell, K.E.C. Weed growth – a factor of channel roughness. In *Hydrometry: Principles and Practices* (ed. R. W. Herschy). John Wiley and Sons, Chichester 1978.

Rodier, J. and Roche, M. River flow in arid regions. In *Hydrometry: Principles and Practices* (ed. R. W. Herschy). John Wiley and Sons, Chichester 1978.

Chapter 9

The stage–fall–discharge method of streamflow measurement

9.1 General

Relatively simple relations between stage and discharge are presented in Chapter 4 and normally most streamflow records are based on this principle. By simply recording stage, therefore, and developing the stage–discharge relation, a continuous record of discharge can be obtained. Several factors, however, can cause scatter of discharge observations about the stage–discharge relation at some stations. Backwater is one of these factors whereby the velocity is retarded so that a higher stage is necessary to maintain a given discharge than would be necessary if the backwater were not present. Backwater is caused by constrictions such as narrow reaches of a stream channel or artificial structures downstream such as dams or bridges or downstream tributaries. All of these factors can increase or decrease the energy gradient for a given discharge and cause variable backwater conditions. If, however, the backwater caused by a fixed obstruction is always constant at any given stage, the discharge rating is a function of stage only. Constant backwater, as caused by section controls for example, will not adversely affect the simple stage–discharge relation. The presence of variable backwater, on the other hand, does not permit the use of simple stage–discharge relations for the accurate determination of discharge.

Regulated streams may have variable backwater virtually all of the time, while other streams will have only occasional backwater from downstream tributaries or from the return of overbank flow. A complex situation arises when all of the above factors are present plus scour and deposition of the bed. Such an example is shown in Chapter 4, Fig. 4.5 for the Yangtze River at Wuhan where stage has been plotted against current meter discharge observations taken over a period of a year. The complex rating is due to many factors including variable backwater, variable slope, scour and deposition.

Many of these sites can be operated by the so-called stage–fall–discharge method using a reference gauge (base gauge), at which stage is measured continuously and current meter measurements are made occasionally, and an auxiliary reference gauge some distance down-stream where stage is also measured continuously. When the two reference gauges are set to the same datum, the difference between the two stage records is the water surface fall and this provides a measure of water surface slope. The shorter the slope reach, the closer the relation between fall and water surface slope. On the other hand, the longer the slope reach, the smaller the percentage uncertainty in the recorded fall.

Precise time synchronisation between base gauge and auxiliary gauge is very important if stage changes rapidly, or when fall is small. Reliable records can usually be computed when fall exceeds about 0.1 m. Timing and gauge height uncertainties which may be negligible at high flows become significant at very low flows. Likewise, at low flows, poor measuring conditions or wind can contribute to measurement uncertainties.

Stage–fall–discharge ratings are established from observation of (1) stage at a base gauge, (2) the fall of the water surface between the base gauge and an auxiliary gauge downstream, and (3) the discharge.

The plotting of the discharge measurements with the fall marked against each measurement, will reveal whether the relationship is affected by variable slopes at all stages or is affected only when the stage rises above a particular level. If the relationship is affected by variable backwater at all stages, a correction is applied by the *constant-fall* method; on the other hand, however, when the relation is affected only when the stage rises above a particular value, the *normal-fall* method is applied.

In order to observe the fall, an auxiliary gauge is established a distance L downstream from the base gauge at the station and set to the same datum as the base gauge. With a difference in gauge reading of F (fall), the surface slope is approximated by

$$S \sim \frac{F}{L}. \tag{9.1}$$

The Manning uniform flow formula may be written in the form

$$Q = NAR^a S^b \tag{9.2}$$

where N = a constant
A = area of cross-section
R = hydraulic radius
S = slope of water surface
a,b = two exponents.

Substituting for S, equation (9.2) becomes

$$Q = NAR^a \left(\frac{F}{L}\right)^b. \tag{9.3}$$

The roughness factor N and the conveyance AR^d are functions of stage, then for a given stage the relation between discharge and fall can be developed, thus

$$\frac{Q_m}{Q_r} = \left(\frac{F_m}{F_c}\right)^b \tag{9.4}$$

in which Q_m and F_m are the measured discharge and fall, and Q_r and F_c are the adjusted discharge from the rating curve and a selected constant fall on which the rating curve is based.

From fluid mechanics, the exponent b in equation (9.3) would be expected to equal 0.5. However, where the slope S is only approximated by F/L, the exponent b would then not necessarily equal 0.5 and must be determined empirically; this is done by a graphical plot of Q_m/Q_r, against F_m/F_c as explained below.

The procedure of the constant-fall method in developing a stage–fall–discharge relation is as follows:

(1) Plot the discharge measurements against stage in the usual manner and indicate the observed fall beside each point. Select a constant fall for which a rating curve can be drawn through the measurements and draw the curve by visual estimation; for curve (a) in Figure 9.1 a constant fall of 0.50 m has been selected.
(2) Read the discharge Q from the rating curve for each measurement and compute the ratios Q_m/Q_r and F_m/F_c. Plot these ratios and draw a smooth curve through them, i.e. curve (b) in Figure 9.1: Note that curve (b) expresses the exponent b in equation (9.4).
(3) From the curve of relation between discharge ratios and fall ratios, curve (b) in Figure 9.1, select the smoothed discharge ratio for each measurement.

Adjust each measurement by dividing it by its smoothed discharge ratio, then replot the adjusted discharge measurements. Examine the plot. If the first approximation of the rating curve does not appear to be well-balanced among the adjusted discharge measurements, then adjust the rating curve and repeat the procedure. Usually, not more than one repeat is necessary.

A constant-fall rating is not the usual case in natural streams. However, if discharge measurements cover the whole range of flow and if the measurements conform to a constant-fall rating, the constant-fall method is sufficient and there is no need to use a more complicated technique. If the stream geometry is not too far from uniform and the velocity head increments are negligible, the relation between the discharge ratio and fall ratio should approach a single curve.

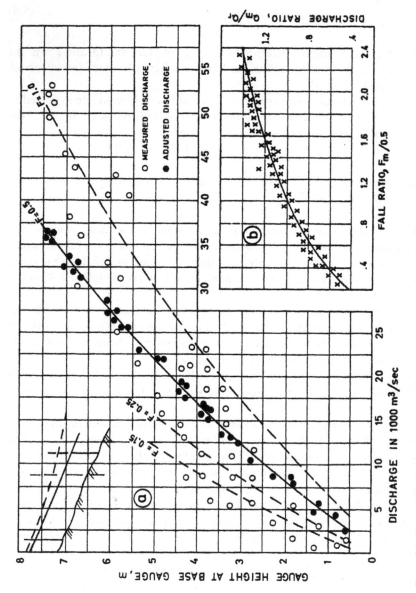

Figure 9.1 Stage–fall–discharge rating for variable backwater, nearly uniform channel.

Example

(See Table 9.1 and Figure 9.2)

14 discharge measurements are available at a twin-gauge station. In columns 2, 3 and 5 of Table 9.1, values of observed gauge height (H), measured discharge (Q_m), and measured fall (F_m) are shown. It is desired to develop a stage–fall–discharge relation for this station:

(1) Plot the discharge measurements in the usual manner indicating measured fall beside each point. Select a constant fall for which a rating curve can be drawn. Here a suitable fall would be $F_c = 0.30$m. Draw the curve (Figure 9.2a).
(2) Compute the fall ratios F_m/F_c and enter in column 6.
(3) Read the discharge Q_r from the rating curve for each measurement and enter in column 4. Compute the discharge ratio Q_m/Q_r and enter in column 7.
(4) Plot the fall ratios F_m/F_c against the discharge ratios Q_m/Q_r and draw a smooth mean curve of relation. Here it appears that the curve of relation approaches a straight line. Usually this curve bends downward (Figure 9.2b)
(5) Entering the curve of relation with the fall ratios, adjusted values of the discharge ratios are obtained and entered in column 8.
(6) Now, divide each value in column 3 by its corresponding value in column 8 obtaining adjusted values for Q_r which are entered in column 9.
(7) Replot the adjusted values of Q_r. It appears that in this case the original rating curve has a well-balanced fit to the new adjusted values of Q_r; therefore, no repeat is necessary.

Table 9.1 Developing a stage–fall–discharge relation

(1) No.	(2) H	(3) Q_m m³/sec	(4) Q_r m³/sec	(5) F_m m	(6) F_m/F_c	(7) Q_m/Q_r	(8) Adjusted Q_m/Q_r	(9) Adjusted Q_r m³/sec
1	1.62	580	550	0.33	1.10	1.05	1.05	552
2	1.64	526	560	0.29	0.97	0.94	0.99	531
3	1.85	668	642	0.32	1.07	1.04	1.04	642
4	1.86	468	645	0.16	0.53	0.73	0.75	642
5	2.02	568	715	0.18	0.60	0.79	0.79	719
6	2.26	710	810	0.24	0.80	0.88	0.90	789
7	2.75	922	1015	0.24	0.80	0.91	0.90	1024
8	2.98	1018	1115	0.24	0.80	0.91	0.90	1131
9	3.00	1035	1125	0.25	0.83	0.92	0.91	1140
10	3.92	1562	1560	0.29	0.97	1.00	0.99	1578
11	4.05	1620	1625	0.30	1.00	1.00	1.00	1620
12	4.56	1960	1935	0.30	1.00	1.01	1.00	1960
13	4.85	2130	2125	0.31	1.03	1.00	1.02	2088
14	4.85	2230	2125	0.31	1.03	1.05	1.02	2186

$F_c = 0.30$ m

The procedure of converting observed gauge height and fall to discharge by means of the stage–fall–discharge relation as developed in Figure 9.2 is as follows:

Observed gauge height:	H	=	3.40 m
Observed fall:	F_m	=	0.22 m
Selected constant fall:	F_c	=	0.30 m
Computed fall ratio:	0.22/0.30	=	0.73
Discharge ratio corresponding to a fall ratio of 0.73 (curve b):		=	0.86
Discharge of rating curve corresponding to a gauge height of 3.40m:		=	1300 m³/sec
True discharge, 1300 m³/sec × 0.86:		=	1118 m³/sec

The normal-fall method

At some stations, a simple single-gauge rating is applicable at low discharge when the surface slope is comparatively steep, while at higher discharges when the slope becomes flatter the discharge is affected by variable backwater. Critical values of the fall (or slope) dividing these two regions are termed the normal-fall. The value of normal-fall at any discharge can be defined by studying the plot of stage against discharge (Figure 9.3).

The points at which backwater has no effect will group to the extreme right. This is the simple single-gauge rating with no backwater effects. A plot of the normal-fall values from this curve is made against the corresponding stage and a curve of normal-fall obtained. This permits drawing a curve of discharge ratios against fall ratios when normal-fall is used in place of constant-fall. The rest of the procedure is similar to that of the constant-fall method (Figure 9.3).

The normal-fall procedure in developing a stage–fall–discharge rating is as follows:

(1) Plot the discharge measurements and write the fall beside each as indicated in Figure 9.3a (the fall values are not shown in the figure). A smooth curve is fitted to the measurements grouped to the extreme right. This curve is the stage–discharge relation for a condition of no backwater effects at the base gauge.

(2) The fall for the measurements used to fit the rating curve are plotted against stage as illustrated in Figure 9.3b; curve is fitted to the plotted points. This curve, the normal-fall curve, shows the fall at which backwater begins at any stage. That is, there are no backwater effects at a fall value to the right of the normal-fall curve; for fall values to the left of the curve there is backwater present.

(3) Each fall ratio F_m/F_n is plotted against its corresponding discharge ratio Q_m/Q_r and a smooth curve of relation drawn (Figure 9.3c).

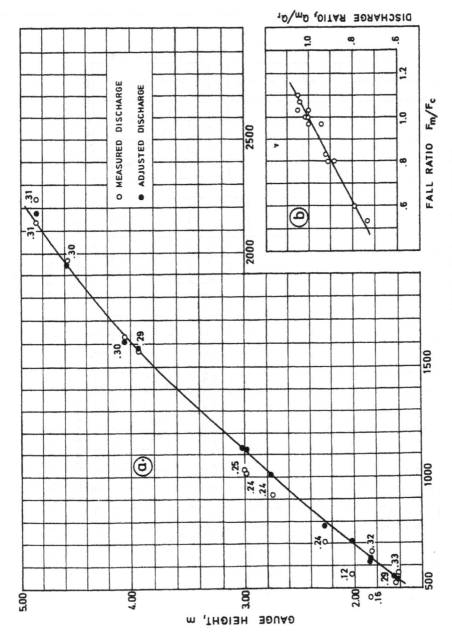

Figure 9.2 Stage–fall–discharge rating using a constant fall value of 0.30 m developed from Table 9.1.

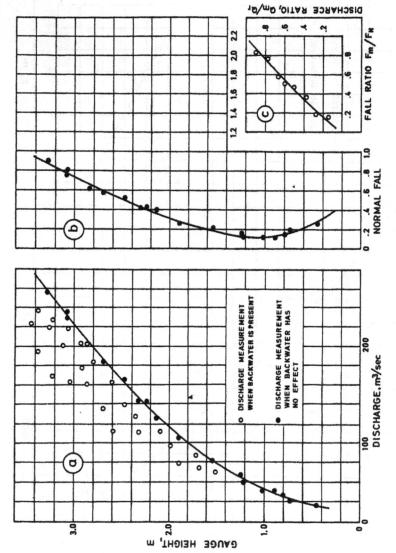

Figure 9.3 Stage–fall–discharge rating, submergence of low-flow control.

The rest of the procedure is identical to that of the constant-fall method. The only difference is that a normal-fall value varying with stage is used instead of a constant-fall value.

Further reading

Corbett, D.M. and others. Stream-gauging Procedures. *U.S. Geological Survey, Water Supply Paper 888* 1942.

ISO 9123. Liquid Flow Measurement in Open Channels: Stage–Fall–Discharge Method. *ISO, Geneva, Switzerland* 2001.

Tilrem, O.A. Manual on procedures in operational hydrology. *Ministry of Water, Energy and Minerals of Tanzania and the Norwegian Agency for International Development* 1979.

World Meteorological Organization. Manual on Stream Gauging. *WMO Report No. 13, Pub. No. 519*, Geneva 1980.

Chapter 10

Weirs and flumes

10.1 General

On small rivers it is often convenient to measure flow by means of a weir or flume. Such structures have the advantage that they are less sensitive to the downstream conditions, the channel roughness and the influence of backwater than the velocity–area method for example. A good example of the use of measuring structures is to be found in the UK where, mainly because of the small rivers and the influence of backwater, the hydrometric network of some 1400 streamflow stations contains about 750 measuring structures. Practically all of these are of a type which has been precalibrated in the laboratory, some of which have been check-calibrated in the field. In this chapter, those weirs and flumes which have been precalibrated and standardised by the International Organization for Standardization or the British Standards Institution will be discussed with an additional section on non-standard weirs.

The philosophy of the method is founded on the premise that the relation of discharge to water level is found empirically or is based on physical principles.

The water level, or head, is measured at a prescribed distance upstream of the structure. In the simple (and usual) case where the downstream water level is below some limiting condition and where it does not affect the upstream head, there is a unique relation between head and discharge. This condition is termed the free-flow or **modular** condition. If, however, the tailwater level affects the flow, the weir is said to be drowned, or submerged, and operates in the **non-modular** condition. For this condition an additional downstream measurement of head is required and a reduction factor is required to be applied to the modular or free-flow discharge equation. When the flow in the non-modular condition increases until the weir is almost or wholly submerged, the structure no longer performs as a measuring device.

There is a significant degradation in the accuracy of discharge measurements in the non-modular flow condition. This degradation is due mainly to the difficulty in measuring the downstream head because of turbulence and the uncertainty in the coefficient of discharge in the non-modular range. However, discharge equations are presented for the non-modular range for the triangular profile (Crump) weir and the Flat V weir where crest tappings are incorporated in the design. A typical design is shown in Fig. 10.20.

One of the problems of crest tappings in recent years has been the silting of the intake pipe and blocking of the tappings. A system now available is designed to remedy this situation by employing a small submersible pump with the inlet pipe connected to the crest tapping by means of a manifold. The crest tapping is kept clear by operating the pump intermittently for short periods (say periods of 20 seconds) each day as necessary. The pump operates on electrical power from the mains or from a battery charged by a solar cell.

In order to increase the range of a structure in the free-flow condition, it is sometimes convenient to raise the design height; otherwise for a weir the device may be used as a section control in the non-modular condition, and rated by a current meter in this range.

Section controls are dealt with in Chapter 4 and a weir or flume is basically a section control designed and built to specific criteria. The rating equation for a stage–discharge relation and a measuring structure, therefore, take the same form. However, because of the application of theoretical concepts of flow, the rating equations for weirs and flumes take constant exponents of the head h. This similarity can be simply demonstrated as follows.

For the velocity–area method with section control the rating equation may be expressed in the form

$$Q = Ch^n \tag{10.1}$$

and if the reference gauge is set with its zero at the level of the control, then n, the exponent of h, is constant within the flow range of the control. For a weir or flume, a simplified form of rating equation is

$$Q = Cbh^{3/2} \tag{10.2}$$

where C is the coefficient of discharge, b is the width of crest and h is the head over the weir with staff gauge zero set at crest level. For unit width of weir

$$Q = Ch^{3/2}. \tag{10.3}$$

(Note that for a V-shaped weir the exponent of h is 5/2.)

Civil engineering construction

The installation of a measuring structure in a river requires major civil engineering works and each device needs to be designed as such with due attention to foundations. Ground conditions may require sheet piling and cut-offs to prevent the seepage of water under or around the structure or risks to stability. If provision for energy dissipation downstream is not included in the construction, for example, there may be a tendency to scour depending on the ground conditions. If this becomes excessive the stability of the structure will be in danger. If the structure, however, is founded on rock, for example, such problems may not be severe but nevertheless due attention may still be required with regard to energy dissipation. The energy downstream of a structure is normally dissipated by a hydraulic jump and generally design and construction considerations involve the installation of a stilling basin and a suitable elevation for the floor of the basin.

Measuring structures are normally constructed in concrete although many have been successfully built of fibreglass, masonry or timber. To avoid abrasion or damage, steel or granite capping is often used to form the crest of triangular profile weirs and steel corners to form the edging to rectangular broad-crested weirs. The capital cost of installing a measuring structure is much greater than the capital cost of installing a velocity–area station but the operational costs of a structure are very much lower. This is due to the calibration costs in manpower in establishing the stage–discharge relation for a velocity–area station. The advantage in this respect, of a standard measuring structure is that records of discharge are obtained immediately on completion of the installation.

Theoretically there is no limit to the design width of a measuring structure. In practice, however, cost of construction usually limits the width of a weir to about 50 m. Present-day costs may be as much as US$20 000 per metre width of crest mainly due to the risks contractors have to take in dealing with water during construction. The largest triangular profile weir built in the UK has a crest width (compounded) of 52.5 m with a maximum capacity of 630 m^3 s^{-1}. The cost of construction of this weir 40 years ago was some US$6000 but at today's prices a similar weir would probably cost somewhere in the region of $1 million. Cost is therefore of considerable importance in deciding on the use of a weir for stream gauging.

Existing non-standard river weirs may be used for measuring flow but these require to be calibrated on site. The most important non-standard weir in the UK, calibrated by current meter, is on the River Thames in London (Teddington) (Fig. 10.1) where 100 years of records were completed in 1982. This weir was originally built to serve navigational needs but became a prescribed control point for the abstraction of London's water supply from the Thames. The weir is a complex one for streamflow and has 37 sluices, 34 radial gates and a 21.34 m long thin plate weir to measure low flows. The lowest flow ever recorded at the station approached zero discharge in 1976 and the highest was 1059 m^3 s^{-1} in 1894. Records are now obtained from an ultrasonic gauging station installed in 1974 just upstream of the weir.

Figure 10.1 Aerial view of River Thames at London (Teddington Weir) used for streamflow measurement 1883–1982; weir comprises 34 radial gates, 37 sluice gates and a thin plate weir for measuring the lowest flows.

10.2 Principles and theory

The flow conditions upstream of a measuring structure are governed by the geometry of the structure and the approach channel, and by the physical properties of the water. They are not affected by the flow conditions in the channel downstream or by the roughness and geometry of the channel upstream.

At the control section critical flow conditions occur, where, for a given discharge, the depth is such that the total head is a minimum. The total head is the total energy of the flow per unit weight of water and, by Bernoulli's theorem, this is the sum of the potential head, the pressure head and the velocity head. It is generally referred to the crest of the structure as a datum.

In many structures the streamlines are curved in the region of the crest and the pressure distribution within the flow is not known so it is not possible to deduce an algebraic equation to describe the head–discharge relation. The coefficients of discharge of standard measuring structures in this chapter have been determined under rigorously controlled laboratory conditions.

In presenting the theoretical analysis for the establishment of the discharge equation, the broad-crested weir is used (Fig. 10.2). Taking the critical section to occur where the flow is parallel near the end of the crest of the weir, then the Bernoulli equation gives

$$H = z + \frac{p}{g\rho} + \frac{u^2}{2g}$$
(10.4)

where H is the total head;
   z is the height of the streamline;
   p is the pressure;
   g is the acceleration due to gravity;
   ρ is the density of water;
   u is the velocity on the streamline.

As the streamlines are parallel the pressure is hydrostatic.
   Therefore

$$p = g\rho(d - z)$$
(10.5)

where d is the depth of flow, so

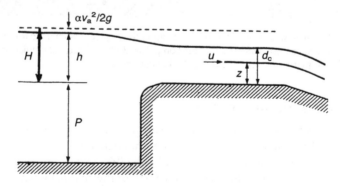

Figure 10.2 Schematic illustration of principle and theory of flow over broad-crested weir.

$$H = d + \frac{u^2}{2g}. \tag{10.6}$$

If the approach channel is deep, the total head H is the same for all stream-lines and it follows from equation (10.6) that the velocity at the critical section is constant with depth and equal to $\bar{v}$, the mean velocity. The discharge per unit width of crest is

$$q = \bar{v}d \tag{10.7}$$

and substituting from equation (10.6)

$$q = d \sqrt{[2g(H - d)]}. \tag{10.8}$$

Differentiating equation (10.8), treating H as a constant and putting $dq/dd = 0$ gives the following equation for critical flow conditions

$$d_c = \tfrac{2}{3}H \tag{10.9}$$

and substituting d_c in equation (10.8)

$$q = \tfrac{2}{3}\sqrt{(\tfrac{2}{3}\,g)}H^{3/2} \tag{10.10}$$

or

$$Q = (\tfrac{2}{3})^{3/2}b\sqrt{(g)}H^{3/2}. \tag{10.11}$$

Equation (10.11) is derived from theoretical concepts and calibration of a structure imposes on the equation a coefficient of discharge. The equation then becomes

$$Q = (\tfrac{2}{3})^{3/2}C_d b\sqrt{(g)}H^{3/2}. \tag{10.12}$$

Since the total head H cannot be measured in practice, an iterative procedure is necessary to compute the discharge from equation (10.12) (see the section on velocity of approach, below). In order to avoid this cumbersome procedure the discharge equation can be presented as

$$Q = (\tfrac{2}{3})^{3/2}C_d C_v b\sqrt{(g)}h^{3/2} \tag{10.13}$$

which is the basic equation for measuring structures where C_v is the dimensionless coefficient of velocity allowing for the velocity of approach, and C_d is the coefficient of discharge.

From an inspection of equations (10.12) and (10.13) it can be seen that

$$C_v = \left(\frac{H}{h}\right)^{3/2}.$$

(10.14)

Values of C_v for measuring structures may be obtained from Table 10.1 where C_v has been plotted against $(C_d bh)/A$, where A is the cross-sectional area of flow at the head measurement section.

Velocity of approach

In the derivation of equation (10.12) it will be noted that the total head H determines the depth and velocity at the critical section and the discharge equation is independent of the velocity in the approach channel.

In practice, the discharge is determined from a measurement of the gauged head which is converted to total head by the Bernoulli equation

$$H = h + \frac{\bar{v}^2}{2g}$$

(10.15)

where $\bar{v}$ is the average velocity in the approach channel at the head measurement section.

Now from

$$Q = A\bar{v}$$

(10.16)

where A is the cross-sectional area of flow in the approach channel at the head gauging section, and substituting $\bar{v} = Q/A$ in equation (10.15),

Table 10.1 Coefficient of approach velocity C_v for values of $C_d bh/A$, where A is the area of cross-sectional flow at the head measuring section, h is the gauged head and P is the height of the weir (or flume). For trapezoidal flumes C_d is replaced by C_s, for U-shaped flumes C_d is replaced by C_u

$\dfrac{C_d bh}{A}$	0.00	0.01	0.02	0.03	0.04	0.05	0.06	0.07	0.08	0.09
0.1	1.003	1.004	1.004	1.005	1.006	1.006	1.007	1.008	1.008	1.009
0.2	1.010	1.011	1.012	1.013	1.014	1.015	1.016	1.018	1.019	1.020
0.3	1.021	1.023	1.024	1.026	1.028	1.030	1.032	1.034	1.036	1.038
0.4	1.040	1.042	1.044	1.046	1.049	1.051	1.054	1.056	1.059	1.061
0.5	1.064	1.067	1.070	1.073	1.076	1.080	1.082	1.086	1.090	1.093
0.6	1.097	1.101	1.105	1.110	1.115	1.120	1.125	1.130	1.135	1.140
0.7	1.144	1.150	1.156	1.163	1.170	1.177	1.184	1.192	1.200	1.208
0.8	1.218	1.226	1.236	1.246	1.225					

$$H = h + \frac{Q^2}{2gA^2}.$$ (10.17)

Therefore the discharge requires to be calculated by an iterative procedure (successive approximations) as follows:

(a) determine A from a given value of h;
(b) obtain a first estimate of discharge (equation (10.12)) using the given value of h;
(c) calculate the total head from equation (10.17);
(d) recalculate the discharge from equation (10.12) using this value of the total head;
(e) using this discharge recalculate the total head from equation (10.17) and repeat the procedure until two successive values of discharge differ by less than, say, 1%. Generally four or five trials will suffice.

Although computer programmes are available to carry out the above procedure, it is now becoming practice to use the form of equations of discharge having gauged head and the coefficient of velocity factor (equation (10.13)).

Limitations on the use of discharge equations

With each of the discharge equations for measuring structures, limitations are imposed as to their use in practice. Such limitations concern the range in head under which a laboratory calibration was made, the distance upstream to the head measurement section, the limits on certain ratios such as h/P, where P is the height of the weir, and others which will be presented in Section 10.5. The values and uncertainties of the coefficients of discharge and velocity are dependent on maintaining these limitations in the field installation. If the limitations are relaxed for any reason the measuring structure may still produce a record of streamflow but the uncertainty of the measurement will be impaired. In most, if not all, of these cases no data are available to give values of uncertainties in the coefficients, and therefore in the discharge measurements where the specified limitations have been relaxed. The only resort in these circumstances is to field calibration (Section 10.4).

Coriolis coefficient

The total head is related to the gauged head by the equation

$$H = h + a\frac{\bar{v}^2}{2g}$$ (10.18)

where $\bar{v}$ is the mean velocity in the approach channel at the head measurement section, and a is a coefficient, known as the kinetic energy or Coriolis coefficient which takes account of the fact that the kinetic energy head exceeds $\bar{v}^2/2g$ if the velocity distribution across the section is not uniform. In applying the equations in this chapter, a may be taken as unity. However, if any suspicion of non-uniform flow arises at any particular station it is advisable to take a current meter measurement at the head measurement cross-section, compute $\bar{v}$ and calculate a from equation (10.18).

Froude number

The discharge coefficients of many structures vary when the velocity head becomes large in relation to the depth of flow. A dimensionless parameter which describes this ratio is the **Froude number**, Fr, which for a river of approximately rectangular geometry with sensibly parallel flow is defined as

$$\text{Fr} = \frac{\bar{v}}{\sqrt{(g\bar{d})}} \tag{10.19}$$

where $\bar{v}$ is the average velocity in the cross-section under investigation and $\bar{d}$ is the average depth of flow. When the Froude number exceeds about 0.5, measuring structures suffer a loss of accuracy and generally good design practice dictates that Fr is less than 0.5. If Fr approaches unity however, debris is liable to accumulate behind the weir, particularly on a rising stage and in practice this debris has to be cleared out, and in the worst case the weir abandoned as a measuring structure.

At the design stage therefore, it is important to determine the Froude number in the natural channel corresponding to the highest anticipated discharge to be gauged accurately and if it does not exceed a value of 0.5 it can be assumed that this value will not be exceeded after construction of the weir.

10.3 Measurement of head

Uncertainty

The measurement of head is the most important measurement in assessing the flow over a weir. Whilst much attention is rightly paid to the uncertainty in the coefficient of discharge, the uncertainty in head measurement is critical. This is particularly the case when measuring low discharges and head values are small. The percentage uncertainty in discharge under these conditions, for even a small uncertainty in head, may be significant.

The uncertainty in discharge measurement by weirs and flumes is dealt with in Chapter 13 but Table 10.2 illustrates the possible uncertainty in discharge over a triangular profile (Crump) weir, arising from uncertainties in head only, at a gauged head of 0.15 m (the percentage uncertainty in

Table 10.2 Uncertainty in discharge of a triangular profile (Crump) weir due to uncertainties in head measurement at a given head of 0.15 m

Uncertainty in head (mm)	Percentage uncertainty in discharge
2	±2
4	±4
6	±6
8	±8
10	±10

discharge being numerically equal to the uncertainty in head measurement is coincidental).

Instruments

The instruments for recording the measurement of stage are described in Chapter 3. In addition, hook and point gauges are often used at structures, particularly on small structures or on laboratory devices. Although it is customary to fix a staff gauge to the wing wall of a weir, it is usually placed on the opposite bank to the recorder and intake pipe and is therefore often difficult to read to the required accuracy. It is thus advisable to have an additional and more precise method available for the head measurement. This can be achieved, for example, using an electric-tape gauge or steel tape, marked in millimetres, from a datum plate on the wing wall immediately above the intake pipe. By this means it is possible to measure head to an uncertainty of 3 mm or better. Such measurement may be a single measurement used to compute discharge or, more usually, a measurement to set or check the recorder (Fig. 10.3).

Gauge zero

In measuring structures, the reference gauge zero is set to the crest level of weirs or the throat invert level of flumes. For V-notches or flat V weirs it is set to the lowest level (vertex) of the notch or weir. The initial setting of the reference gauge, and subsequent checking, is therefore extremely important. The setting is best performed by surveyor's optical level reading to 1 mm or better, the procedure being as described in Chapter 3. Generally it is better to level from the crest or throat to the datum on the wing wall above the intake pipe and then relate the water level and setting of the reference gauge to this datum. Because of possible movement of the structure after completion, regular checks of the zero are advisable. With careful attention to the levelling procedure it is possible to set and maintain the zero to 3 mm or better. It will be noted that whereas the head measurement contains a random uncertainty, the zero error will be

Figure 10.3 The photograph shows the electric-tape gauge (Fig. 3.8) being held against the datum plate to make a measurement of water level. As the buzzer sounds on contact being made with the water surface, the tape is read to the nearest millimetre. The distance from the datum plate to the bed of the channel being already measured by surveyors level, or other means, the depth of flow (head) is the difference between the two readings (no hump). The photograph also shows the staff gauge and an ultrasonic look-down water level gauge (see also Fig. 3.18) in the approach channel of a rectangular standing wave flume. Note the sun shade over the ultrasonic transducer to avoid faulty water level recording due to temperature change.

systematic and, if known, should be added to, or subtracted from, the head measurement before the computation of discharge. If regular checks are made (for example, once a year) the uncertainty in the zero setting may be randomised and taken as ±3 mm, or better, and added by the root-sum-square method to the uncertainty in the recorded head measurement (Chapter 13).

Location of head measuring section (gauging section)

The location of the measuring section where the head is measured upstream of the structure will be given later for each device, but generally this is specified as two to four times the maximum head to be measured. The water surface upstream of a gauging structure is not horizontal and therefore the location for the head measurement will affect the uncertainty in measurement. The designer is required to make a decision as to what he considers the most important range of flows to be measured and base the choice of location on these criteria. For example, it is better to take the design head, for the purpose of locating the gauging section, as less than the maximum anticipated gauged head if 95% of the flows are expected to be in the low or medium flow range. The design head for this purpose would be taken as that applying to the upper end of the range required to be measured accurately.

10.4 Check calibration in the field

The 'standard' measuring structures presented in this chapter are those which have been calibrated in the laboratory. The discharge equation so established is therefore used immediately after construction of the weir. The detailed design and construction of the measuring structure are required to follow the specifications laid down in International Standards.

However, it may be necessary to check the calibration of a measuring structure in the field should any suspicion arise as to its reliability. Since the laboratory calibration normally has a reliable rating, and coefficients of discharge so established are generally of the order of, typically 2–3%, it is necessary to use special equipment and methods in a check calibration in order to make field measurements of comparable accuracy.

Experience has shown that the main sources of error to receive attention in the check calibration of a measuring structure are:

(a) incorrect gauge zero resulting in a systematic error in the head measurement and therefore discharge;
(b) incorrect measurement of crest width or flume throat resulting in a systematic error in the discharge;
(c) incorrect measurement of other geometric dimensions resulting in an error in the coefficient of discharge and measurement of head;
(d) incorrect design of intake pipe and stilling well resulting in an uncertainty in head measurement;
(e) uncertainty in recording instruments.

Normally the check on discharge is carried out by current meter and, as discussed in Chapter 2, it is essential that due attention is given to the selection of the measuring cross-section, to the number of verticals used and to the number of velocity points selected in the vertical. For such a discharge measurement it is desirable that at least 30 verticals are taken and that the time of exposure of the current meter is at least one min – preferably three one min readings should be taken – if stage remains constant. If the dilution method is used (Chapter 11) an injection section is chosen which will ensure complete mixing. Multiple injection and multiple sampling techniques are also desirable.

Unfortunately river flows cannot normally be controlled during a check calibration investigation unless a regulating reservoir is sited upstream of the measuring structure. Added problems occur when the river is either rising or falling when neither head nor discharge remains constant for sufficient time to obtain a satisfactory discharge measurement which can be compared with that computed from the laboratory equation of discharge. Therefore, obtaining a series of reliable measured discharges over the full range of the structure is a time-consuming and expensive operation and resort has usually to be made to being content with a few measurements at low and medium flows.

In comparing the measured discharge with the computed discharge given by the equation of discharge, it is normally more convenient to use the form of discharge equation having gauged head (equation 10.13)) rather than total head (equation (10.12)). It will be observed that for thin plate weirs the discharge equations are already given in the form of gauged head (equations (10.20), (10.25) and (10.26)).

10.5 Types of measuring structure

Measuring structures may be divided into the following categories:

(a) thin plate weirs
(b) broad-crested weirs
(c) flumes
(d) compound measuring structures.

Thin plate weirs

Rectangular thin plate weir

A diagrammatic illustration of the basic weir form is shown in Fig. 10.4. This particular form of the thin plate rectangular weir is often referred to as a rectangular notch weir or 'contracted' weir, so called because the nappe is contracted. When $b/B = 1$, or when the width of the notch is equal to the width of the channel, the weir is termed a full width rectangular thin plate weir or sometimes a 'suppressed' weir because the nappe does not have side contractions.

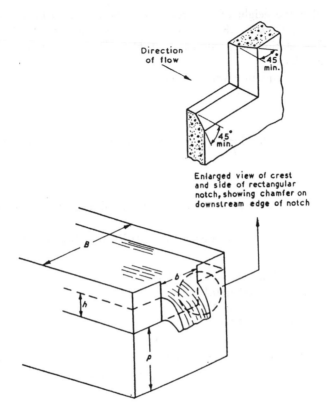

Figure 10.4 Diagrammatic illustration of thin plate weir.

Rectangular notch weir

The equation for discharge for a rectangular notch (contracted) weir is

$$Q = \tfrac{2}{3}\sqrt{(2g)}\,C_d b_e h_e^{3/2} \qquad\qquad (10.20)$$

where Q is the discharge (m³ s⁻¹);
  g is the acceleration due to gravity (9.81 m s⁻²);
  C_d is the coefficient of discharge;
  b_e is the effective breadth (m);
  h_e is the effective gauged head (m).

$$b_e = b + 0.003 \text{ m} \qquad\qquad (10.21)$$

$$h_e = h + 0.001 \text{ m} \qquad\qquad (10.22)$$

$$C_d = a + \beta\frac{h}{P} \qquad\qquad (10.23)$$

where P (m) is the height of the crest above the upstream bed, and α and β depend on the ratio b/B, where B (m) is the upstream width of the approach channel. This relationship is shown in Table 10.3 where intermediate values may be interpolated.

A more useful equation which avoids the calculation of the coefficient of discharge is

$$Q = 0.554\left(1 - 0.0035\frac{h}{P}\right)(b + 0.0025)\sqrt{(g)}(h + 0.001)^{3/2} \text{ (metric units)}.$$

(10.24)

Provided the respective limitations are observed, this equation should give flows within about 2%.

The limitations on the use of equations (10.20) and (10.24) are:

(a) h_e (or h) should not be less than 0.03 m;
(b) b should not be less than 0.15 m;
(c) P should not be less than 0.10 m;
(d) h/P should not exceed 2.5;
(e) $(B - b)/2$ should not be less than 0.10 m;
(f) the head measurement section should be located at a distance of two to four times the maximum head upstream from the weir.

Example

Calculate the discharge over a rectangular notch weir having a notch width of 4 m and an approach channel width of 8 m under a measured head of 1 m. The height of the notch (P) is 0.75 m.

(a) Check limitations
    h is greater than 0.03 m;
    b is greater than 0.15 m;
    P is greater than 0.10 m;

Table 10.3 Relationship between b/B, α and β

	b/B									
	0.1	0.2	0.3	0.4	0.5	0.6	0.7	0.8	0.9	1.0
α	0.588	0.589	0.590	0.591	0.592	0.593	0.594	0.596	0.598	0.602
β	−0.002	−0.002	0.002	0.006	0.011	0.018	0.030	0.045	0.064	0.075

$$\frac{h}{P} = \frac{1}{0.75} = \frac{4}{3}, \text{ therefore less than 2.5;}$$

$$\frac{B-b}{2} = \frac{8-4}{2} = 2, \text{ therefore greater than 0.10 m.}$$

(b) From Table 10.3 $b/B = 4/8 = 0.5$.
Therefore $a = 0.592$
and $\beta = 0.011$.

(c) $C_d = a + \beta \dfrac{h}{P} = 0.592 + (0.011 \times 1.333) = 0.607$.

(d) From the equation of discharge (equation (10.20))

$$Q = \tfrac{2}{3}\sqrt{(2g)}C_d b_e h_e^{3/2}$$
$$= \tfrac{2}{3}\sqrt{(9.81 \times 2)} \times 0.607 \times 4.003 \times 1.001^{3/2}$$
$$= 7.180 \text{ m}^3 \text{ s}^{-1}.$$

Using equation (10.24) gives

$$Q = 0.554 \left(1 - 0.0035 \times \frac{1}{0.75}\right) \times 4.0025 \times 3.132 \times 1.0015$$

$$= 6.923 \text{ m}^3 \text{ s}^{-1}.$$

Full width rectangular thin plate weirs Fig. 10.9 (b) (see p. 334)

The discharge equation for a full width thin plate weir is

$$Q = C_d \tfrac{2}{3}\sqrt{(2g)}bh_e^{3/2} \tag{10.25}$$

in which

$$C_d = 0.602 + 0.083h/p$$
$$h_e = h + 0.001 \text{ m}.$$

An alternative equation is

$$Q = 0.564 \ (1 + (0.150h/p) \ b\sqrt{(g)}(h + 0.001)^{3/2} \text{ (metric units).} \tag{10.25a}$$

Practical limits on the use of the above equations are

(a) h/p should not exceed 4.0
(b) h should be between 0.03 and 1.0 m

(c) b should not be less than 0.30 m
(d) p should not be less than 0.06 m
(e) the head measurement section should be located at a distance upstream from the weir of 2 to 4 times the maximum head to be measured.

Example

Calculate the discharge over a rectangular full width thin plate weir having a crest breadth 4 m under a measured head of 1 m with a p value of 0.75.

(a) check limitations

  b is 1 m
  b is greater than 0.30 m
  p is greater than 0.06 m
  b/p is $1/0.75 = 4/3 = 1.333$ therefore less than 4.

(b) discharge

$$Q = (0.602 + 0.083 \times 1.333) \tfrac{2}{3}\sqrt{(2 \times 9.81)} \times 4 \times 1.001^{3/2}$$
$$= 8.41 \text{ m}^3\text{s}^{-1}$$

from equation 10.25(a)

$$Q = 0.564 (1 + 0.150 \times 1.333) \times 4 \times 3.13 \times 1.001^{3/2}$$
$$= 8.47 \text{ m}^3\text{s}^{-1}.$$

Triangular (V-notch) thin plate weir

A diagrammatic illustration of the basic weir form is shown in Fig. 10.5. The three sizes of V-notches commonly used are (Fig. 10.6):

(a) A 90° notch in which the dimension across the top is twice the vertical depth ($\tan \theta/2 = 1$). This is the most common type of V-notch.
(b) A half 90° notch ($\theta = 53°8'$) in which the dimension across the top is equal to the vertical depth ($\tan \theta/2 = 0.5$).
(c) A quarter 90° notch ($\theta = 28°4'$) in which the dimension across the top is half the vertical depth ($\tan \theta/2 = 0.25$).

Notches (b) and (c) above nominally deliver a half and a quarter of the discharge, respectively, of the 90° notch.

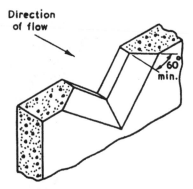

Enlarged view of V notch,
showing chamfer on down-
stream edge of notch

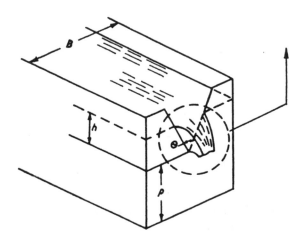

Figure 10.5 Diagrammatic illustration of V-notch thin plate weir.

The BSI equation of discharge is

$$Q = \tfrac{8}{15}\sqrt{(2g)}\,C_d \tan\frac{\theta}{2}\,h^{5/2} \qquad (10.26)$$

and the experimentally determined values of C_d may be found from Fig. 10.7
Tabulated values of C_d and Q for a 90° notch are given in Table 10.4.

Practical limitations applicable to the use of equation (10.26) are:

(a) h/P should not exceed 0.4;
(b) h/B should not exceed 0.2;
(c) h should be between 0.05 and 0.38 m;
(d) P should be not less than 0.45 m;

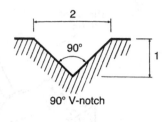

90° V-notch

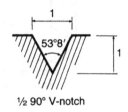

½ 90° V-notch

¼ 90° V-notch

Figure 10.6 Commonly used V-notches.

(e) *B* should be not less than 1 m;
(f) the head measurement section should be located at a distance upstream of the notch of 4 to 5 times the maximum head.

It should be noted that the maximum head to be used with equation (10.26) is 0.38 m. If larger heads are to be measured, a loss of accuracy in the discharge measurement has to be accepted. There is, however, insufficient experimental data available to give guidance on this aspect. However, if a 90° notch is used, and this is the most common form, the maximum head to be gauged may be increased provided the h/P ratio is within the range 0.2 to 2.0, P is not less than 0.09 m and P/B is between 0.10 and 1.0. When B is large compared with P ($P/B = 0.1$), the coefficient C_d is substantially constant and equal to 0.578.

For the above conditions for the 90° notch, h in equation (10.26) is replaced by $(h + 0.001)$ m. This adjustment compensates for the combined effects of viscosity and surface tension and may be an advisable adjustment at very low heads but becomes unnecessary at large heads.

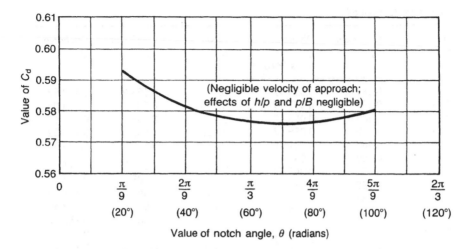

Figure 10.7 V-notch thin plate weir: coefficient of discharge C_d related to notch angle θ.

Table 10.4 Discharge of water over a 90° V-notch

Head h (m)	Coefficient C_d	Discharge Q $(m^3\ s^{-1} \times 10^{-1})$	Head h (m)	Coefficient C_d	Discharge Q $(m^3\ s^{-1} \times 10^{-1})$
0.060	0.603 2	0.012 57	0.085	0.595 5	0.029 61
0.061	0.602 8	0.013 09	0.086	0.594 8	0.030 48
0.062	0.602 3	0.013 62	0.087	0.594 5	0.031 36
0.063	0.601 9	0.014 17	0.088	0.594 2	0.032 25
0.064	0.601 5	0.014 73	0.089	0.594 0	0.033 16
0.065	0.601 2	0.015 30	0.090	0.593 7	0.034 09
0.066	0.600 8	0.015 88	0.091	0.593 5	0.035 03
0.067	0.600 5	0.016 48	0.092	0.593 3	0.035 98
0.068	0.600 1	0.017 10	0.093	0.593 1	0.036 96
0.069	0.599 8	0.017 72	0.094	0.592 9	0.037 95
0.070	0.599 4	0.018 36	0.095	0.592 7	0.038 95
0.071	0.599 0	0.019 01	0.096	0.592 5	0.039 97
0.072	0.598 7	0.019 67	0.097	0.592 3	0.041 01
0.073	0.598 3	0.020 35	0.098	0.592 1	0.042 06
0.074	0.598 0	0.021 05	0.099	0.591 9	0.043 12
0.075	0.597 8	0.021 76	0.100	0.591 7	0.044 20
0.076	0.597 5	0.022 48	0.101	0.591 4	0.045 30
0.077	0.597 3	0.023 22	0.102	0.591 2	0.046 41
0.078	0.597 0	0.023 97	0.103	0.591 0	0.047 54
0.079	0.596 7	0.024 73	0.104	0.590 8	0.048 69
0.080	0.596 4	0.025 51	0.105	0.590 6	0.049 85
0.081	0.596 1	0.026 30	0.106	0.590 4	0.051 03
0.082	0.595 8	0.027 10	0.107	0.590 2	0.052 22
0.083	0.585 5	0.027 92	0.108	0.590 1	0.053 44
0.084	0.595 3	0.028 76	0.109	0.588 9	0.054 67

Table 10.4 Continued

Head h (m)	Coefficient C_d	Discharge Q $(m^3\ s^{-1}\times 10^{-1})$	Head h (m)	Coefficient C_d	Discharge Q $(m^3\ s^{-1}\times 10^{-1})$
0.110	0.589 8	0.055 92	0.155	0.585 9	0.130 93
0.111	0.589 7	0.057 19	0.156	0.585 9	0.133 04
0.112	0.589 6	0.058 47	0.157	0.585 8	0.135 17
0.113	0.589 4	0.059 77	0.158	0.585 8	0.137 32
0.114	0.589 2	0.061 08	0.159	0.585 7	0.139 50
0.115	0.589 1	0.062 42	0.160	0.585 7	0.141 69
0.116	0.589 0	0.063 77	0.161	0.585 7	0.143 91
0.117	0.588 9	0.065 14	0.162	0.585 6	0.146 14
0.118	0.588 8	0.066 53	0.163	0.585 6	0.148 40
0.119	0.588 6	0.067 93	0.164	0.585 5	0.150 67
0.120	0.588 5	0.069 35	0.165	0.585 5	0.152 97
0.121	0.588 3	0.070 79	0.166	0.585 5	0.155 29
0.122	0.588 2	0.072 24	0.167	0.585 4	0.157 63
0.123	0.588 1	0.073 72	0.168	0.585 4	0.159 99
0.124	0.588 0	0.075 22	0.169	0.585 3	0.162 37
0.125	0.588 0	0.076 73	0.170	0.585 3	0.164 77
0.126	0.587 9	0.078 27	0.171	0.585 3	0.167 19
0.127	0.587 8	0.079 82	0.172	0.585 2	0.169 64
0.128	0.587 7	0.081 39	0.173	0.585 2	0.172 10
0.129	0.587 6	0.082 98	0.174	0.585 1	0.174 59
0.130	0.587 6	0.084 58	0.175	0.585 1	0.177 09
0.131	0.587 5	0.086 21	0.176	0.585 1	0.179 63
0.132	0.587 4	0.087 85	0.177	0.585 1	0.182 19
0.133	0.587 3	0.089 51	0.178	0.585 1	0.184 78
0.134	0.587 2	0.091 19	0.179	0.585 1	0.187 38
0.135	0.587 2	0.092 89	0.180	0.585 1	0.190 01
0.136	0.587 1	0.094 61	0.181	0.585 1	0.192 65
0.137	0.587 0	0.096 34	0.182	0.585 0	0.195 31
0.138	0.586 9	0.098 10	0.183	0.585 0	0.198 00
0.139	0.586 9	0.099 87	0.184	0.585 0	0.200 71
0.140	0.586 8	0.101 67	0.185	0.585 0	0.203 45
0.141	0.586 7	0.103 48	0.186	0.585 0	0.206 21
0.142	0.586 7	0.105 32	0.187	0.585 0	0.208 99
0.143	0.586 6	0.107 17	0.188	0.585 0	0.211 80
0.144	0.586 6	0.109 04	0.189	0.585 0	0.214 63
0.145	0.586 5	0.110 93	0.190	0.585 0	0.217 48
0.146	0.586 4	0.112 84	0.191	0.585 0	0.220 34
0.147	0.586 3	0.114 76	0.192	0.584 9	0.223 22
0.148	0.586 2	0.116 71	0.193	0.584 9	0.226 12
0.149	0.586 2	0.118 67	0.194	0.584 9	0.229 06
0.150	0.586 1	0.120 66	0.195	0.584 9	0.232 03
0.151	0.586 1	0.122 67	0.196	0.584 9	0.235 01
0.152	0.586 0	0.124 71	0.197	0.584 9	0.238 02
0.153	0.586 0	0.126 76	0.198	0.584 9	0.241 06
0.154	0.585 9	0.128 83	0.199	0.584 9	0.244 11

Table 10.4 Continued

Head h (m)	Coefficient C_d	Discharge Q ($m^3 s^{-1} \times 10^{-1}$)	Head h (m)	Coefficient C_d	Discharge Q ($m^3 s^{-1} \times 10^{-1}$)
0.200	0.584 9	0.247 19	0.245	0.584 6	0.410 34
0.201	0.584 9	0.250 28	0.246	0.584 6	0.414 54
0.202	0.584 8	0.253 39	0.247	0.584 6	0.418 77
0.203	0.584 8	0.256 52	0.248	0.584 6	0.423 02
0.204	0.584 8	0.259 69	0.249	0.584 6	0.427 30
0.205	0.584 8	0.262 88			
0.206	0.584 8	0.266 10	0.250	0.584 6	0.431 60
0.207	0.584 8	0.269 34	0.251	0.584 6	0.435 93
0.208	0.584 8	0.272 61	0.252	0.584 6	0.440 28
0.209	0.584 8	0.275 90	0.253	0.584 6	0.444 66
			0.254	0.584 6	0.449 07
0.210	0.584 8	0.279 21	0.255	0.584 6	0.453 50
0.211	0.584 8	0.282 54	0.256	0.584 6	0.457 96
0.212	0.584 8	0.285 88	0.257	0.584 6	0.462 45
0.213	0.584 7	0.289 24	0.258	0.584 6	0.466 96
0.214	0.584 7	0.292 64	0.259	0.584 6	0.471 50
0.215	0.584 7	0.296 07			
0.216	0.584 7	0.299 53	0.260	0.584 6	0.476 06
0.217	0.584 7	0.303 01	0.261	0.584 6	0.480 65
0.218	0.584 7	0.306 51	0.262	0.584 6	0.485 27
0.219	0.584 7	0.310 04	0.263	0.584 6	0.489 91
			0.264	0.584 6	0.494 58
0.220	0.584 7	0.313 59	0.265	0.584 6	0.499 28
0.221	0.584 7	0.317 17	0.266	0.584 6	0.504 00
0.222	0.584 7	0.320 77	0.267	0.584 6	0.508 76
0.223	0.584 7	0.324 39	0.268	0.584 6	0.513 53
0.224	0.584 7	0.328 03	0.269	0.584 6	0.518 34
0.225	0.584 6	0.331 68			
0.226	0.584 6	0.335 35	0.270	0.584 6	0.523 17
0.227	0.584 6	0.339 07	0.271	0.584 6	0.528 02
0.228	0.584 6	0.342 82	0.272	0.584 6	0.532 91
0.229	0.584 6	0.346 59	0.273	0.584 6	0.537 82
			0.274	0.584 6	0.542 76
0.230	0.584 6	0.350 39	0.275	0.584 6	0.547 72
0.231	0.584 6	0.354 21	0.276	0.584 6	0.552 72
0.232	0.584 6	0.358 06	0.277	0.584 6	0.557 74
0.233	0.584 6	0.361 93	0.278	0.584 6	0.562 82
0.234	0.584 6	0.365 82	0.279	0.584 7	0.567 94
0.235	0.584 6	0.369 74			
0.236	0.584 6	0.373 69	0.280	0.584 7	0.573 06
0.237	0.584 6	0.377 66	0.281	0.584 7	0.578 19
0.238	0.584 6	0.381 66	0.282	0.584 7	0.583 35
0.239	0.584 6	0.385 68	0.283	0.584 7	0.588 53
			0.284	0.584 7	0.593 75
0.240	0.584 6	0.389 73	0.285	0.584 7	0.598 99
0.241	0.584 6	0.393 80	0.286	0.584 7	0.604 25
0.242	0.584 6	0.397 90	0.287	0.584 7	0.609 55
0.243	0.584 6	0.402 02	0.288	0.584 7	0.614 87
0.244	0.584 6	0.406 17	0.289	0.584 7	0.620 23

Table 10.4 Continued

Head h (m)	Coefficient C_d	Discharge Q $(m^3 s^{-1} \times 10^{-1})$	Head h (m)	Coefficient C_d	Discharge Q $(m^3 s^{-1} \times 10^{-1})$
0.290	0.584 7	0.625 60	0.336	0.585 0	0.904 48
0.291	0.584 7	0.631 01	0.337	0.585 1	0.911 28
0.292	0.584 7	0.636 45	0.338	0.585 1	0.918 11
0.293	0.584 7	0.641 95	0.339	0.585 1	0.924 91
0.294	0.584 8	0.647 48	0.340	0.585 1	0.931 75
0.295	0.584 8	0.653 03	0.341	0.585 1	0.938 62
0.296	0.584 8	0.658 58	0.342	0.585 1	0.945 51
0.297	0.584 8	0.664 16	0.343	0.585 1	0.952 44
0.298	0.584 8	0.669 76	0.344	0.585 1	0.959 40
0.299	0.584 8	0.675 39	0.345	0.585 1	0.966 38
0.300	0.584 8	0.681 06	0.346	0.585 1	0.973 40
0.301	0.584 8	0.686 75	0.347	0.585 1	0.980 45
0.302	0.584 8	0.692 46	0.348	0.585 1	0.987 53
0.303	0.584 8	0.698 21	0.349	0.585 1	0.994 71
0.304	0.584 8	0.703 98	0.350	0.585 2	1.001 92
0.305	0.584 8	0.709 80	0.351	0.585 2	1.009 12
0.306	0.584 8	0.715 68	0.352	0.585 2	1.016 33
0.307	0.584 9	0.721 59	0.353	0.585 2	1.023 56
0.308	0.584 9	0.727 50	0.354	0.585 2	1.030 82
0.309	0.584 9	0.733 41	0.355	0.585 2	1.038 12
0.310	0.584 9	0.739 36	0.356	0.585 2	1.045 45
0.311	0.584 9	0.745 34	0.357	0.585 2	1.052 80
0.312	0.584 9	0.751 35	0.358	0.585 2	1.060 19
0.313	0.584 9	0.757 38	0.359	0.585 2	1.067 67
0.314	0.584 9	0.763 44	0.360	0.585 3	1.075 19
0.315	0.584 9	0.769 54	0.361	0.585 3	1.082 73
0.316	0.584 9	0.775 66	0.362	0.585 3	1.090 24
0.317	0.584 9	0.781 81	0.363	0.585 3	1.097 78
0.318	0.584 9	0.788 02	0.364	0.585 3	1.105 36
0.319	0.585 0	0.794 28	0.365	0.585 3	1.112 97
0.320	0.585 0	0.800 57	0.366	0.585 3	1.120 63
0.321	0.585 0	0.806 85	0.367	0.585 3	1.128 37
0.322	0.585 0	0.813 14	0.368	0.585 4	1.136 15
0.323	0.585 0	0.819 47	0.369	0.585 4	1.143 91
0.324	0.585 0	0.825 83	0.370	0.585 4	1.151 67
0.325	0.585 0	0.832 22	0.371	0.585 4	1.159 47
0.326	0.585 0	0.838 63	0.372	0.585 4	1.167 30
0.327	0.585 0	0.845 08	0.373	0.585 4	1.175 16
0.328	0.585 0	0.851 55	0.374	0.585 4	1.183 10
0.329	0.585 0	0.858 06	0.375	0.585 5	1.191 11
0.330	0.585 0	0.864 59	0.376	0.585 5	1.199 14
0.331	0.585 0	0.871 16	0.377	0.585 5	1.207 12
0.332	0.585 0	0.877 75	0.378	0.585 5	1.215 15
0.333	0.585 0	0.884 38	0.379	0.585 5	1.223 20
0.334	0.585 0	0.891 03	0.380	0.585 5	1.231 28
0.335	0.585 0	0.897 72	0.381	0.585 5	1.239 40

Note: In the above table each discharge requires to be divided by 10, e.g. for h = 0.381, discharge = 0.1239 $m^3 s^{-1}$.

Design of thin plate weirs

For small installations (e.g. up to 2 m wide) the plate is generally manufactured of stainless steel sheet 6 mm thick. The profile of the notch is normally chamfered at 45° (or 60° for a V-notch) and the crest is normally made 1–2 mm thick and carefully machined (Fig. 10.8). For large installations the plate is made thicker but not normally more than 25 mm. However the crest should not exceed 2 mm even at this thickness. Crests should not be allowed to become rounded with time and if the rounding is significant the plate is best replaced. Slight rounding of the upstream edge of the crest may increase the discharge by as much as 2%.

The fabrication of V-notches to any specified angle requires strict tolerances to be observed. The angle θ should therefore be measured after fabrication of the notch (top width divided by vertical height) and that value, if different from the specified value, should be entered into equation (10.26).

For V-notches the uncertainty in the coefficient of discharge may be taken as ± 1.0%. This makes the V-notch potentially the most accurate of all measuring devices in open channel measurement.

Site installations of rectangular thin plate weirs and a V-notch weir are shown in Figs 10.9 and 10.10.

Simplified discharge equations for design or spot measurements

For design purposes or for spot measurements, the following discharge equations may be used:

Rectangular notch (contracted thin plate weir)

$$Q = 1.73bh^{3/2} \text{ m}^3 \text{ s}^{-1}. \tag{10.27}$$

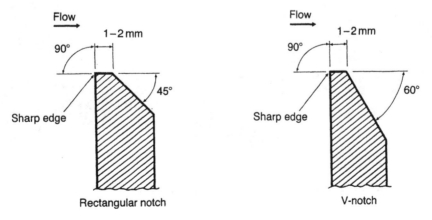

Figure 10.8 Details of the crest sections of thin plate weirs and orifices.

Figure 10.9(a) Rectangular thin plate notch (contracted) weir.

Figure 10.9(b) Rectangular thin plate full width (suppressed) weir.

Figure 10.10 V-notch thin plate weir.

Full width rectangular thin plate weir

$$Q = 1.766(1 + 0.150h/p)bh^{3/2}. \tag{10.28}$$

V-notch thin plate weir (90°)

$$Q = 1.365h^{5/2} \ \mathrm{m^3 \ s^{-1}}. \tag{10.29}$$

V-notch thin plate weir ($\frac{1}{2}$ 90°)

$$Q = 0.682h^{5/2} \ \mathrm{m^3 \ s^{-1}}. \tag{10.30}$$

V-notch thin plate weir ($\frac{1}{4}$ 90°)

$$Q = 0.347h^{5/2} \ \mathrm{m^3 \ s^{-1}}. \tag{10.31}$$

Broad-crested weirs

The triangular profile (Crump) weir

This form of weir is the one most used in United Kingdom rivers over the past 50 years and has given good operational service providing reliable records.

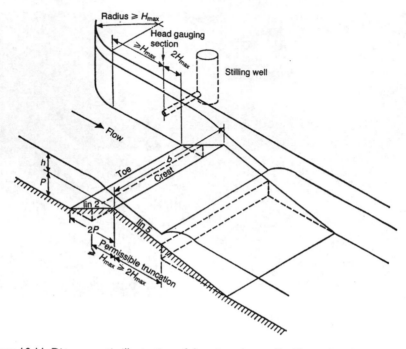

Figure 10.11 Diagrammatic illustration of the triangular profile (Crump) weir.

A diagrammatic illustration of the basic weir form is shown in Fig. 10.11. The weir has a slope of 1:2 (one vertically to two horizontally) on the upstream face and 1:5 on the downstream face, a geometry that tests have shown to give an essentially constant coefficient of discharge and a high modular limit (75%).

The equation of discharge is

$$Q = C_v C_d \sqrt{(g)} b h^{3/2} \qquad (10.32)$$

or

$$Q = 0.633 C_v \sqrt{(g)} b h^{3/2}. \qquad (10.33)$$

In terms of total head, H, the discharge equation becomes

$$Q = 0.633 \sqrt{(g)} b H^{3/2}. \qquad (10.34)$$

Note:

(a) In the form of equation (10.13), equation (10.33) would become

$$Q = (\tfrac{2}{3})^{3/2} \times 1.16 C_v \sqrt{(g)} b h^{3/2} \qquad (10.35)$$

since

$$\left(\tfrac{2}{3}\right)^{3/2} \times 1.16 = 0.633.$$

(b) When computing C_v from Table 10.1 the coefficient 1.16 should be used to calculate $(C_d b h)/A$.

(c) Since $A = b(h + P)$ at the head measuring section

$$\frac{C_d b h}{A} \text{ becomes } C_d \frac{h}{h + P} = 1.16 \frac{h}{h + P}.$$

This assumes that the width of river at the head measuring section is the same as the width of the weir (see Fig. 10.11).

Practical limitations applicable to the use of equations (10.33) and (10.34) are:

(a) h should not be less than 0.03 m (for a crest of smooth metal or equivalent);
(b) h should not be less than 0.06 m (for a crest of fine concrete or equivalent);
(c) P should not be less than 0.06 m;
(d) b should not be less than 0.3 m
(e) h/P should not be exceed 4.0;
(f) b/h should not be less than 2.0;
(g) the Fr number should not be greater than 0.5;
(h) the head measuring section should be located at a distance of twice the maximum head ($2H_{max}$) from the crest-line of the weir;
(i) h_2/h should not be greater than 0.75, where h_2 is the downstream gauged head (modular, or free flow, limit).

Example

Calculate the discharge for a single crest triangular profile weir for the following conditions

crest breadth $(b) = 15.24$ m

weir height $(P) = 0.61$ m

gauged head $(h) = 1.219$ m.

(a) Check limitations

h is greater than 0.03 m;
P is greater than 0.06 m;
b is greater than 0.3 m;

h/P is less than 4;

b/h is greater than 2.

(b) To find C_v

$$\frac{h}{h + P} = \frac{1.219}{1.219 + 0.61} = 0.666$$

$1.16 \times 0.666 = 0.773$ (notes 2 and 3 above)

and C_v from Table 10.1 is 1.194. Equation (10.32) gives

$$Q = C_v C_d \sqrt{(g)} b h^{3/2}$$

$$= 1.194 \times 0.633\sqrt{(9.81)} \times 15.24 \times 1.219^{3/2}$$

$$= 48.553 \text{ m}^3 \text{ s}^{-1}.$$

Alternatively, the discharge may be calculated by the successive approximations method using the total head equation as follows

(a) Cross-sectional area of flow at the head measuring section (A) is

$$A = (1.219 + 0.61)15.24 = 27.874 \text{ m}^2.$$

(b) First approximation of discharge is (equation (10.34))

$$Q = 0.633\sqrt{(g)} b H^{3/2}$$

$$= 0.633\sqrt{(9.81)} \times 15.24 \times 1.219^{3/2}$$

$$= 30.215 \times 1.219^{3/2}$$

$$= 40.666 \text{ m}^3 \text{ s}^{-1}.$$

(c) Second approximation

$$\bar{v} = \frac{Q}{A} = \frac{40.666}{27.874} = 1.459 \text{ m s}^{-1}$$

$$h = \frac{\bar{v}^2}{2g} = \frac{1.459^2}{2 \times 9.81} = 0.109 \text{ m}$$

$$H = 1.219 + 0.109 = 1.328 \text{ m}$$

$$Q = 30.215 \times 1.328^{3/2}$$

$$= 46.240 \text{ m}^3 \text{ s}^{-1}.$$

(d) Third approximation

$$\bar{v} = \frac{46.240}{27.874} = 1.659 \text{ m s}^{-1}$$

$$h = \frac{1.659^2}{2 \times 9.81} = 0.140 \text{ m}$$

$$H = 1.219 + 0.140 = 1.359 \text{ m}$$

$$Q = 30.215 \times 1.359^{3/2}$$

$$= 47.869 \text{ m}^3 \text{ s}^{-1}.$$

(e) Fourth approximation

$$\bar{v} = \frac{47.869}{27.874} = 1.717 \text{ m s}^{-1}$$

$$h = \frac{1.717^2}{2 \times 9.81} = 0.150 \text{ m}$$

$$H = 1.219 + 0.150 = 1.369 \text{ m}$$

$$Q = 30.215 \times 1.369^{3/2}$$

$$= 48.398 \text{ m}^3 \text{ s}^{-1}.$$

Difference in successive values of $Q = 1.1\%$.

The flat V weir

The flat V weir has the same section as the triangular profile weir, 1:2 upstream slope and 1:5 downstream slope, but in elevation has a crest slope of 1:10, 1:20 or 1:40 so that it takes the form of a shallow V when viewed in the direction of flow. With this geometry, it is therefore sensitive to low flows but at the same time has a wide flow range.

A diagrammatic illustration of the basic weir form is shown in Fig. 10.12. The discharge over a flat V weir may be within the V, above the V and within vertical side walls, or above the V and within trapezoidal side walls. To allow for these conditions, a shape factor, Z, is included in the discharge equation. The equation of discharge is

$$Q = \tfrac{4}{5}C_d C_v \sqrt{(g)} mZh^{5/2}. \tag{10.36}$$

In terms of total head the discharge equation is

$$Q = \tfrac{4}{5}C_d \sqrt{(g)} mZH^{5/2} \tag{10.37}$$

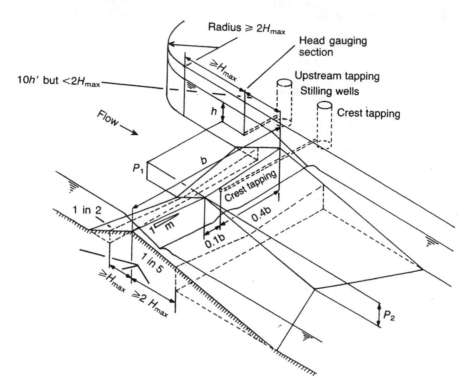

Figure 10.12 Diagrammatic illustration of the flat V weir.

where m is the cross-slope 1:10, 1:20 or 1:40. Other cross-slopes, less than 1:10, are unusual and undesirable (1 vertical: m horizontal).

To calculate Z, if h' is the height of the V (difference in crest elevation), then for the V full condition

$$\frac{h}{h'} = 1 \quad \text{and} \quad Z = 1.0;$$

when h is greater than h' (flow above the V full condition) within vertical side walls

$$Z = [1.0 - (1.0 - h'/h)^{5/2}]; \tag{10.38}$$

within trapezoidal side walls

$$Z = \left\{ h^{5/2} - \left[1 - \frac{n}{m} \left(h - \frac{b}{2m} \right)^{5/2} \right] \right\} \tag{10.39}$$

where n is the channel side slope (1 vertical: n horizontal). The discharge equations then become

for flow within vertical side walls

$$Q = \tfrac{4}{5}C_dC_v\sqrt{(g)}mh^{5/2}[1.0 - (1.0 - h'/h)^{5/2}]\tag{10.40}$$

and

for flow within trapezoidal side walls

$$Q = \tfrac{4}{5}C_dC_v\sqrt{(g)}m\left[h^{5/2} - \left(1 - \frac{n}{m}\right)\left(h - \frac{b}{2m}\right)^{5/2}\right].\tag{10.41}$$

Note that when $n = 0$ (vertical side walls), equation (10.41) reduces to equation (10.40) and when $n = m$ (flow within the V) equation (9.41) reduces to equation (10.36).

Limitations

The approach flow to a flat V weir is three-dimensional whilst all of the other broad-crested weirs presented in this chapter are two-dimensional. In addition, the crest slopes may vary and for these reasons the limitations and coefficients of discharge are presented only for weirs having crest slopes of 1:10, 1:20 and 1:40. These are given in Table 10.5.

In addition the coefficient of velocity, C_v, is related to the coefficient of discharge C_d, the ratio h/P and the ratio h/h'. Values of C_vZ are given in Table 10.6 and of Z in Table 10.7.

The head measuring section should be set at 10 h' upstream of the crest. However, if $10h' < 2H_{max}$, the tapping should be set at $2H_{max}$ upstream of the crest.

Example

Calculate the discharge for a flat V weir for the following conditions

crest slope (m) = 1:20.3

crest breadth (b) = 36.0 m

crest height at lowest point (vertex) (P) = 0.82 m

(a) for a head of 0.621 m;
(b) for a head of 2.000 m (within vertical side walls).

Limitations

The h/P ratios are less than 2.5 for (a) = 0.76 and for (b) = 2.44.

Table 10.5 Flat V limitations and uncertainties in the coefficient of discharge (note P_2 = height of weir at downstream end)

Flat V weir	Crest cross-slope		
	1:40 or less	1:20	1:10
Flow within the V, $h/h' \leqslant 1$			
C_d	0.625	0.620	0.615
$\dfrac{h}{P} \leqslant$	2.5	2.5	2.5
$\dfrac{h}{P_2} \leqslant$	2.5	2.5	2.5
Upstream tapping from crest line	10h'	10h'	10h'
Modular limit	0.70	0.70	0.70
Uncertainty in C_d (%)	3.0	3.0	3.0
Flow above the V within vertical or trapezoidal side walls, $h/h' > 1$			
C_d	0.630	0.625	0.620
$\dfrac{h'}{P} \leqslant$	2.5	2.5	2.5
$\dfrac{h'}{P_2} \leqslant$	8	8	4
Upstream tapping from crest line	10h'	10h'	10h'
Modular limit	0.75	0.75	0.75
Uncertainty in C_d (%)	3.0	3.0	3.0

(a) The flow is within the V, therefore the discharge equation (10.36) is

$$Q = \tfrac{4}{5}C_d C_v \sqrt{(g)} m Z h^{5/2}$$

where $m = 20.3$. From Table 10.5, $C_d = 0.620$. Now

$$h' = \frac{b}{2m} = \frac{36}{2 \times 20.3} = 0.887,$$

then

$$\frac{h}{h'} = \frac{0.621}{0.887} = 0.70$$

and

$$\frac{h'}{P} = \frac{0.887}{0.82} = 1.08.$$

Table 10.6 Evaluation of C_vZ in terms of h/P and h/h' for flat V weirs

h/h'	h/P												
	0.2	0.4	0.6	0.8	1.0	1.2	1.4	1.6	1.8	2.0	2.2	2.4	2.6
0.05	1.000	1.000	1.000	1.000	1.000	1.000	1.000	1.000	1.000	1.000	1.000	1.000	1.000
0.10	1.000	1.000	1.000	1.000	1.000	1.000	1.000	1.000	1.000	1.000	1.000	1.000	1.000
0.15	1.000	1.000	1.000	1.000	1.000	1.000	1.000	1.000	1.000	1.000	1.000	1.000	1.000
0.20	1.000	1.000	1.001	1.000	1.000	1.000	1.000	1.000	1.000	1.000	1.000	1.000	1.000
0.25	1.000	1.000	1.000	1.000	1.000	1.000	1.000	1.000	1.000	1.001	1.001	1.001	1.001
0.30	1.000	1.000	1.000	1.000	1.000	1.000	1.001	1.001	1.001	1.001	1.001	1.001	1.001
0.35	1.000	1.000	1.000	1.000	1.001	1.001	1.001	1.001	1.001	1.002	1.002	1.002	1.002
0.40	1.000	1.000	1.000	1.001	1.001	1.001	1.002	1.002	1.002	1.002	1.003	1.003	1.003
0.45	1.000	1.000	1.001	1.001	1.002	1.002	1.002	1.003	1.003	1.004	1.004	1.004	1.005
0.50	1.000	1.001	1.001	1.002	1.002	1.003	1.003	1.004	1.004	1.005	1.005	1.006	1.006
0.55	1.000	1.001	1.001	1.002	1.003	1.004	1.005	1.005	1.006	1.007	1.007	1.008	1.008
0.60	1.000	1.001	1.002	1.003	1.004	1.005	1.006	1.007	1.008	1.009	1.009	1.010	1.011
0.65	1.000	1.001	1.003	1.004	1.005	1.006	1.008	1.009	1.010	1.011	1.012	1.013	1.013
0.70	1.001	1.002	1.003	1.005	1.007	1.008	1.010	1.011	1.012	1.013	1.015	1.016	1.017
0.75	1.001	1.002	1.004	1.006	1.008	1.010	1.012	1.013	1.015	1.016	1.018	1.019	1.020
0.80	1.001	1.003	1.005	1.008	1.010	1.012	1.014	1.016	1.018	1.020	1.021	1.023	1.024
0.85	1.001	1.004	1.007	1.009	1.012	1.015	1.017	1.020	1.022	1.024	1.025	1.027	1.029
0.90	1.001	1.004	1.008	1.011	1.015	1.018	1.021	1.023	1.026	1.028	1.030	1.032	1.034
0.95	1.002	1.005	1.009	1.014	1.017	1.021	1.024	1.027	1.030	1.033	1.035	1.037	1.039
1.00	1.002	1.006	1.011	1.016	1.020	1.025	1.028	1.032	1.035	1.038	1.040	1.043	1.045
1.05	1.002	1.007	1.013	1.018	1.023	1.028	1.032	1.036	1.039	1.042	1.045	1.048	1.050
1.10	1.001	1.006	1.012	1.019	1.024	1.029	1.034	1.038	1.042	1.046	1.049	1.052	1.054
1.15	0.997	1.004	1.011	1.017	1.024	1.029	1.034	1.039	1.043	1.047	1.050	1.053	1.056
1.20	0.993	1.000	1.007	1.015	1.021	1.028	1.033	1.038	1.042	1.047	1.050	1.054	1.057
1.25	0.986	0.994	1.003	1.011	1.018	1.024	1.030	1.036	1.040	1.045	1.049	1.052	1.056

Table 10.6 Continued

h/h'	h'/P												
	0.2	0.4	0.6	0.8	1.0	1.2	1.4	1.6	1.8	2.0	2.2	2.4	2.6
1.30	0.979	0.988	0.997	1.005	1.013	1.020	1.026	1.032	1.037	1.042	1.046	1.050	1.053
1.35	0.971	0.980	0.990	0.999	1.008	1.015	1.022	1.027	1.033	1.038	1.042	1.046	1.050
1.40	0.962	0.972	0.983	0.992	1.001	1.009	1.016	1.022	1.028	1.033	1.037	1.041	1.045
1.45	0.953	0.963	0.974	0.985	0.994	1.002	1.009	1.016	1.022	1.027	1.031	1.036	1.040
1.50	0.943	0.954	0.966	0.976	0.986	0.995	1.002	1.009	1.015	1.020	1.025	1.030	1.034
1.55	0.932	0.944	0.957	0.968	0.978	0.987	0.995	1.001	1.008	1.013	1.018	1.023	1.027
1.60	0.922	0.934	0.947	0.959	0.969	0.978	0.987	0.994	1.000	1.006	1.011	1.016	1.020
1.65	0.911	0.924	0.938	0.950	0.961	0.970	0.978	0.986	0.992	0.998	1.004	1.008	1.013
1.70	0.900	0.914	0.928	0.940	0.952	0.961	0.970	0.977	0.984	0.990	0.996	1.001	1.005
1.75	0.889	0.904	0.918	0.931	0.942	0.952	0.961	0.969	0.976	0.982	0.988	0.993	0.997
1.80	0.878	0.893	0.908	0.922	0.933	0.943	0.953	0.960	0.968	0.974	0.980	0.985	0.989
1.85	0.867	0.883	0.898	0.912	0.924	0.935	0.944	0.952	0.959	0.966	0.971	0.977	0.981
1.90	0.856	0.873	0.889	0.903	0.915	0.926	0.935	0.943	0.951	0.957	0.963	0.968	0.973
1.95	0.845	0.863	0.879	0.893	0.906	0.917	0.926	0.935	0.942	0.949	0.955	0.960	0.965
2.00	0.835	0.852	0.869	0.884	0.896	0.908	0.917	0.926	0.933	0.940	0.946	0.952	0.957
2.05	0.824	0.842	0.859	0.874	0.887	0.899	0.909	0.917	0.925	0.932	0.938	0.944	0.949
2.10	0.814	0.833	0.850	0.865	0.878	0.890	0.900	0.909	0.916	0.923	0.930	0.935	0.940
2.15	0.804	0.823	0.841	0.856	0.869	0.881	0.891	0.900	0.908	0.915	0.921	0.927	0.932
2.20	0.794	0.813	0.831	0.847	0.861	0.872	0.883	0.892	0.900	0.907	0.913	0.919	0.924
2.25	0.784	0.804	0.822	0.838	0.852	0.864	0.874	0.883	0.891	0.899	0.905	0.911	0.916
2.30	0.774	0.795	0.813	0.830	0.843	0.855	0.866	0.875	0.883	0.891	0.897	0.903	0.908
2.35	0.764	0.785	0.804	0.821	0.835	0.847	0.856	0.867	0.875	0.883	0.889	0.895	0.900
2.40	0.755	0.776	0.796	0.812	0.827	0.839	0.850	0.859	0.867	0.875	0.881	0.887	0.893
2.45	0.746	0.768	0.787	0.804	0.819	0.831	0.842	0.851	0.860	0.867	0.874	0.880	0.885
2.50	0.737	0.759	0.779	0.796	0.811	0.823	0.834	0.843	0.852	0.859	0.866	0.872	0.878

2.55	0.870	0.863	0.859	0.852	0.844	0.836	0.826	0.815	0.803	0.788	0.771	0.751	0.728
2.60	0.863	0.857	0.851	0.844	0.837	0.828	0.819	0.808	0.795	0.780	0.763	0.742	0.720
2.65	0.856	0.850	0.844	0.837	0.829	0.821	0.811	0.800	0.787	0.772	0.755	0.734	0.711
2.70	0.849	0.843	0.837	0.830	0.822	0.814	0.804	0.793	0.780	0.765	0.747	0.726	0.703
2.75	0.842	0.836	0.830	0.823	0.815	0.806	0.797	0.785	0.772	0.757	0.740	0.719	0.695
2.80	0.835	0.829	0.823	0.816	0.805	0.799	0.790	0.778	0.765	0.750	0.732	0.711	0.687
2.85	0.828	0.822	0.816	0.809	0.801	0.792	0.783	0.771	0.758	0.743	0.725	0.703	0.679
2.90	0.822	0.816	0.809	0.802	0.795	0.786	0.776	0.764	0.751	0.736	0.718	0.696	0.671
2.95	0.815	0.809	0.803	0.796	0.788	0.779	0.769	0.758	0.744	0.729	0.711	0.689	0.674
3.00	0.809	0.803	0.796	0.789	0.781	0.773	0.762	0.751	0.738	0.722	0.704	0.682	0.657
3.05	0.802	0.797	0.790	0.783	0.775	0.766	0.756	0.744	0.731	0.716	0.697	0.675	0.649
3.10	0.796	0.790	0.784	0.777	0.769	0.760	0.750	0.738	0.725	0.709	0.690	0.668	0.642
3.15	0.790	0.784	0.778	0.771	0.763	0.754	0.743	0.732	0.718	0.703	0.684	0.662	0.636
3.20	0.784	0.778	0.772	0.765	0.757	0.748	0.737	0.726	0.712	0.696	0.678	0.655	0.629
3.25	0.779	0.773	0.766	0.759	0.751	0.742	0.731	0.720	0.706	0.690	0.671	0.649	0.622
3.30	0.773	0.767	0.760	0.753	0.745	0.736	0.725	0.714	0.700	0.684	0.665	0.643	0.616
3.35	0.767	0.761	0.755	0.747	0.739	0.730	0.720	0.708	0.694	0.678	0.659	0.637	0.610
3.40	0.762	0.756	0.749	0.742	0.733	0.724	0.714	0.702	0.688	0.672	0.653	0.631	0.603
3.45	0.756	0.750	0.744	0.736	0.728	0.719	0.708	0.696	0.683	0.667	0.648	0.625	0.597
3.50	0.751	0.745	0.738	0.731	0.723	0.713	0.703	0.691	0.677	0.661	0.642	0.619	0.591
3.55	0.746	0.740	0.733	0.726	0.717	0.708	0.697	0.686	0.672	0.656	0.637	0.613	0.586
3.60	0.741	0.735	0.728	0.720	0.712	0.703	0.692	0.680	0.666	0.650	0.631	0.608	0.580
3.65	0.736	0.730	0.723	0.715	0.707	0.698	0.687	0.675	0.661	0.645	0.626	0.602	0.574
3.70	0.731	0.725	0.718	0.710	0.702	0.692	0.682	0.670	0.656	0.640	0.620	0.597	0.569
3.75	0.726	0.720	0.713	0.705	0.697	0.687	0.677	0.665	0.651	0.635	0.615	0.592	0.563
3.80	0.722	0.715	0.708	0.701	0.692	0.683	0.672	0.660	0.646	0.630	0.610	0.587	0.558
3.85	0.717	0.711	0.704	0.696	0.687	0.678	0.667	0.655	0.641	0.625	0.605	0.582	0.553
3.90	0.712	0.706	0.699	0.691	0.683	0.673	0.662	0.650	0.636	0.620	0.600	0.577	0.548
3.95	0.708	0.701	0.694	0.687	0.678	0.668	0.658	0.646	0.632	0.615	0.596	0.572	0.543
4.00	0.704	0.697	0.690	0.682	0.674	0.664	0.653	0.641	0.627	0.611	0.591	0.567	0.538

Table 10.7 Evaluation of Z in terms of h/h' for flat V weirs

h/h'	0.00	0.01	0.02	0.03	0.04	0.05	0.06	0.07	0.08	0.09
0.0 to 0.9	1.000	1.000	1.000	1.000	1.000	1.000	1.000	1.000	1.000	1.000
1.0	1.000	1.000	1.000	1.000	1.000	1.000	1.000	1.000	0.999	0.998
1.1	0.998	0.997	0.996	0.996	0.995	0.994	0.993	0.992	0.991	0.990
1.2	0.989	0.987	0.986	0.985	0.984	0.982	0.981	0.979	0.978	0.976
1.3	0.974	0.973	0.971	0.969	0.968	0.966	0.964	0.962	0.960	0.958
1.4	0.956	0.954	0.952	0.950	0.948	0.946	0.944	0.942	0.940	0.938
1.5	0.936	0.934	0.932	0.929	0.927	0.925	0.923	0.921	0.918	0.916
1.6	0.914	0.912	0.909	0.907	0.905	0.903	0.900	0.898	0.896	0.895
1.7	0.891	0.889	0.887	0.884	0.882	0.880	0.877	0.875	0.873	0.871
1.8	0.868	0.866	0.864	0.861	0.859	0.857	0.855	0.852	0.850	0.848
1.9	0.846	0.843	0.841	0.839	0.837	0.834	0.832	0.830	0.828	0.825
2.0	0.823	0.821	0.819	0.817	0.814	0.812	0.810	0.808	0.806	0.804
2.1	0.801	0.799	0.797	0.795	0.793	0.791	0.789	0.787	0.784	0.782
2.2	0.780	0.778	0.776	0.774	0.772	0.770	0.768	0.766	0.764	0.762
2.3	0.760	0.758	0.756	0.754	0.752	0.750	0.748	0.746	0.744	0.742
2.4	0.740	0.738	0.736	0.734	0.732	0.731	0.729	0.727	0.725	0.723
2.5	0.721	0.719	0.717	0.716	0.714	0.712	0.710	0.708	0.707	0.705
2.6	0.703	0.701	0.699	0.698	0.696	0.694	0.692	0.691	0.689	0.687
2.7	0.685	0.684	0.682	0.680	0.679	0.677	0.675	0.674	0.672	0.670
2.8	0.669	0.667	0.665	0.664	0.662	0.661	0.659	0.657	0.656	0.654
2.9	0.653	0.651	0.649	0.648	0.646	0.645	0.643	0.642	0.640	0.639
3.0	0.637									

From Table 10.6, $C_vZ = 1.008$. Then

$$Q = 0.8 \times 0.620 \times 1.008 \times \sqrt{(9.81)} \times 20.3 \times 0.621^{5/2}$$
$$= 9.66 \text{ m}^3 \text{ s}^{-1}.$$

(b) The flow is above the V within vertical walls

$$C_d = 0.625 \text{ (from Table 10.5)}$$

$$\frac{h}{h'} = \frac{2.0}{0.887} = 2.25$$

$$\frac{h'}{P} = 1.08.$$

With these values Table 10.6 gives a value of C_vZ of 0.862. Then

$$Q = 0.8 \times 0.625 \times 0.862 \times \sqrt{(9.81)} \times 20.3 \times 2^{5/2}$$
$$= 155.02 \text{ m}^3 \text{ s}^{-1}.$$

Note: when computing discharge using total head a similar procedure to the example on the triangular profile weir is adopted.

The rectangular profile weir

Of all the precalibrated broad-crested weirs in operational use today there has probably been more research carried out on the rectangular profile weir than on any other. This has not been because of the merit of the weir as a gauging structure but rather due to the hydraulic considerations in the variable coefficient range. Indeed, reported installations of the weir have been few compared to other types of weir. However, there are many existing rectangular profile weirs which are used for operational purposes such as irrigation devices or for compensation water measurement from reservoirs which were built before the International Standard on the weir was published. Some of these weirs might conform, even approximately, to the limitations given later although a degradation in the uncertainty in discharge has to be accepted. The weir is one of the easiest to construct in the field, the main requisite being that it has to have sharp right-angle corners. The main disadvantages of the weir are that silt and debris collect behind the structure and it has a low modular limit. A diagrammatic illustration of the weir form is shown in Fig. 10.13.

The equation of discharge is

$$Q = (\tfrac{2}{3})^{3/2} C_d C_v \sqrt{(g)} bh^{3/2}. \tag{10.42}$$

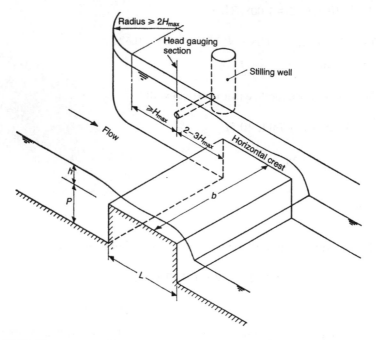

Figure 10.13 Diagrammatic illustration of the rectangular broad-crested weir.

The coefficient of discharge C_d is a constant value of 0.86 in the range

$$\frac{h}{P} \leq 0.5$$

and

$$\frac{h}{L} \leq 0.3$$

where L (m) is the length of the weir in the direction of flow.

The coefficient of velocity C_v is obtained from Table 10.1. In the variable coefficient range, C_d may be obtained from Fig. 10.14, in terms of h/P and h/L. Intermediate values may be interpolated.

For computer processing C_d may be obtained from the following equation, derived from the contour plot of Fig. 10.14

$$C_d = 0.888 - 0.093y + 0.133y^2 - 0.021y^3 - 0.151x$$
$$+ 0.102xy - 0.065xy^2 + 0.310x^2 + 0.028x^2y$$
$$- 0.102x^3 \tag{10.43}$$

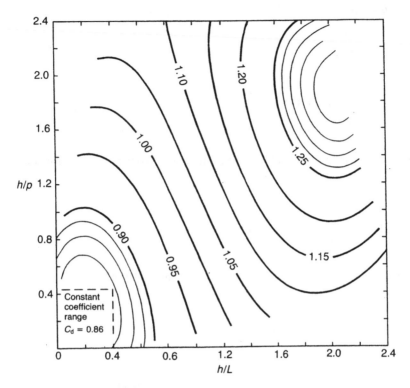

Figure 10.14 Rectangular broad-crested weir: C_d values from h/P and h/L values.

where $x = h/L$ and $y = h/P$.

Other practical limitations for the use of equation (10.42) are

h should not be less than 0.06 m;
b should not be less than 0.3 m;
P should not be less than 0.15 m;
L/P should not be less than 0.15 nor greater than 4.0;
h/P should not be greater than 1.6 (variable coefficient range);
h/L should not be greater than 1.6 (variable coefficient range).

The location of the head measurement section is at $3–4h_{max}$ upstream from the weir block.

Example

Calculate the discharge for a rectangular profile weir for the following conditions

$$h = 0.4 \text{ m}$$

$$b = 10 \text{ m}$$

$$L = 0.5 \text{ m}$$

$$P = 0.3 \text{ m}$$

$$h/P = 0.4/0.3 = 1.333$$

$$h/L = 0.4/0.5 = 0.8.$$

Therefore C_d is in the variable coefficient range. From Fig. 10.14, for the above values C_d is found to be 1.014.

Now for C_v

$$\frac{C_d bh}{A} = \frac{1.014 \times 10 \times 0.4}{10 \times (0.3 + 0.4)} = 0.579$$

and from Table 10.1 $C_v = 1.09$. Now from equation (10.42)

$$Q = (\tfrac{2}{3})^{3/2} C_d C_v \sqrt{(g)} bh^{3/2}$$

$$= 0.544 \times 1.014 \times 1.09 \times 3.132 \times 10 \times 0.253$$

$$= 4.764 \text{ m}^3 \text{ s}^{-1}.$$

Alternatively, C_d can be calculated from equation (10.43) as follows

$$\frac{h}{L} = 0.8 = x$$

$$\frac{h}{P} = 1.333 = y.$$

Then from equation (10.43)

$$C_d = 0.888 - 0.093(1.333) + 0.133(1.333)^2 - 0.021(1.333)^3$$
$$- 0.151(0.8) + 0.102(0.8)(1.333) - 0.065(0.8)(1.333)^2$$
$$+ 0.310(0.8)^2 + 0.028(0.8)^2(1.333) - 0.102(0.8)^3$$
$$= 1.014.$$

The round-nose horizontal crest weir

The standard weir comprises a level crest with an upstream rounded corner so that flow separation does not occur as it does in the rectangular profile weir. Downstream of the horizontal crest there may be:

(a) a rounded corner
(b) a downward slope
(c) a vertical face.

A diagrammatic illustration of the weir form is shown in Fig. 10.15.
  The equation of discharge is

$$Q = (\tfrac{2}{3})^{3/2} C_d C_v b \sqrt{(g)} h^{3/2}. \tag{10.44}$$

The coefficient of discharge C_d is a function of h, the crest length L in the direction of flow and the ratio h/b and is expressed by the equation

$$C_d = \left(1 - \frac{0.006L}{b}\right)\left(1 - \frac{0.003L}{h}\right)^{3/2}. \tag{10.45}$$

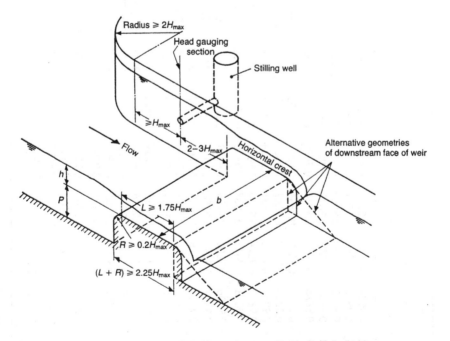

Figure 10.15 Diagrammatic illustration of round-nose horizontal crested weir.

Table 10.8 Discharge coefficients for flumes and round-nose weirs $C_d = \left(1 - \dfrac{0.006L}{b}\right)\left(1 - \dfrac{0.003L}{h}\right)^{3/2}$

$\dfrac{L}{b}$	h/L													
	0.70	0.65	0.60	0.55	0.50	0.45	0.40	0.35	0.30	0.25	0.20	0.15	0.10	0.05
0.2	0.9924	0.9919	0.9913	0.9906	0.9898	0.9888	0.9876	0.9860	0.9839	0.9809	0.9764	0.9690	0.9542	0.9103
0.4	0.9912	0.9907	0.9901	0.9894	0.9886	0.9876	0.9864	0.9848	0.9827	0.9797	0.9752	0.9678	0.9530	0.9092
0.6	0.9900	0.9895	0.9889	0.9883	0.9875	0.9865	0.9852	0.9836	0.9815	0.9785	0.9741	0.9667	0.9519	0.9081
0.8	0.9888	0.9883	0.9878	0.9871	0.9863	0.9853	0.9840	0.9825	0.9803	0.9774	0.9729	0.9655	0.9502	0.9070
1.0	0.9876	0.9872	0.9866	0.9859	0.9851	0.9841	0.9829	0.9813	0.9792	0.9762	0.9717	0.9644	0.9496	0.9059
1.2	0.9865	0.9860	0.9854	0.9847	0.9839	0.9829	0.9817	0.9801	0.9780	0.9750	0.9706	0.9632	0.9485	0.9048
1.4	0.9853	0.9848	0.9842	0.9835	0.9827	0.9818	0.9805	0.9789	0.9768	0.9739	0.9694	0.9620	0.9474	0.9038
1.6	0.9841	0.9836	0.9831	0.9824	0.9816	0.9806	0.9793	0.9778	0.9757	0.9727	0.9683	0.9609	0.9462	0.9027
1.8	0.9829	0.9824	0.9819	0.9812	0.9804	0.9794	0.9782	0.9766	0.9745	0.9715	0.9671	0.9598	0.9451	0.9016
2.0	0.9818	0.9813	0.9807	0.9800	0.9792	0.9782	0.9770	0.9754	0.9733	0.9704	0.9660	0.9586	0.9440	0.9005
2.2	0.9806	0.9801	0.9795	0.9789	0.9781	0.9771	0.9758	0.9743	0.9722	0.9692	0.9648	0.9575	0.9429	0.8995
2.4	0.9794	0.9787	0.9784	0.9777	0.9769	0.9759	0.9747	0.9731	0.9710	0.9681	0.9637	0.9563	0.9417	0.8984
2.6	0.9783	0.9778	0.9772	0.9765	0.9757	0.9748	0.9735	0.9720	0.9699	0.9669	0.9625	0.9552	0.9406	0.8973
2.8	0.9771	0.9766	0.9761	0.9754	0.9746	0.9736	0.9624	0.9708	0.9687	0.9658	0.9614	0.9541	0.9395	0.8963
3.0	0.9759	0.9755	0.9749	0.9742	0.9734	0.9724	0.9712	0.9696	0.9676	0.9646	0.9602	0.9529	0.9384	0.8952
3.2	0.9748	0.9743	0.9733	0.9731	0.9723	0.9713	0.9701	0.9685	0.9664	0.9635	0.9591	0.9518	0.9373	0.8941
3.4	0.9736	0.9731	0.9726	0.9719	0.9711	0.9701	0.9689	0.9673	0.9653	0.9623	0.9580	0.9507	0.9362	0.8931
3.6	0.9725	0.9720	0.9714	0.9708	0.9700	0.9690	0.9678	0.9662	0.9641	0.9612	0.9568	0.9495	0.9350	0.8920
3.8	0.9713	0.9708	0.9703	0.9696	0.9688	0.9678	0.9666	0.9651	0.9630	0.9601	0.9557	0.9484	0.9339	0.8909
4.0	0.9702	0.9697	0.9691	0.9685	0.9677	0.9667	0.9655	0.9639	0.9618	0.9589	0.9546	0.9473	0.9328	0.8899
4.2	0.9690	0.9685	0.9680	0.9673	0.9665	0.9656	0.9643	0.9628	0.9607	0.9578	0.9534	0.9462	0.9317	0.8888
4.4	0.9679	0.9674	0.9668	0.9662	0.9654	0.9644	0.9632	0.9616	0.9596	0.9566	0.9523	0.9451	0.9306	0.8878
4.6	0.9667	0.9663	0.9657	0.9650	0.9642	0.9633	0.9621	0.9605	0.9584	0.9555	0.9512	0.9439	0.9295	0.8867
4.8	0.9656	0.9651	0.9646	0.9639	0.9631	0.9621	0.9609	0.9594	0.9573	0.9544	0.9500	0.9428	0.9284	0.8857
5.0	0.9645	0.9640	0.9634	0.9628	0.9620	0.9610	0.9593	0.9583	0.9562	0.9533	0.9490	0.9418	0.9274	0.8847

Note: The number of significant figures given in the columns for coefficient of discharge should not be taken to imply a corresponding accuracy but only to assist in interpolation and analysis.

Values of C_d derived from the above equation are given in Table 10.8.
 Practical limitations in the use of equation (10.44) are

 h should not be less than 0.06 m;
 b should not be less than 0.30 m;
 P should not be less than 0.15 m;
 h/P should not be greater than 1.5;
 h/L should not be greater than 0.57.

The location of the head measurement section is at 2–$3H_{max}$ upstream from the weir block.

Example

Calculate the discharge for a round-nose weir for the following conditions

$h = 0.67$ m

$b = 10$ m

$L = 2$ m

$P = 1$ m.

$$C_d = \left(1 - \frac{0.006L}{b}\right)\left(1 - \frac{0.003L}{h}\right)^{3/2}$$

$$= \left(1 - \frac{0.006 \times 2}{10}\right)\left(1 - \frac{0.003 \times 2}{0.67}\right)^{3/2}$$

$$= 0.999 \times (0.99)^{3/2}$$

$$= 0.985.$$

Now for C_v

$$\frac{C_d bh}{A} = \frac{0.985 \times 10 \times 0.67}{10(1 + 0.67)}$$

$$= 0.395$$

and, from Table 10.1, $C_v = 1.039$. Now from equation (10.44)

$$Q = (\tfrac{2}{3})^{3/2}\, C_d\, C_v\, b\, \sqrt{(g)} h^{3/2}$$

$$= 0.544 \times 0.985 \times 1.039 \times 10 \times 3.132 \times 0.548$$

$$= 9.555 \text{ m}^3 \text{ s}^{-1}.$$

Flumes

General

A flume is a flow measurement device which is formed by a constriction in the channel. The constriction can be a narrowing in the channel or a hump, or both. A flume with a hump in the invert has a discharge equation identical to that of the broad-crested weir. An advantage of the flume over a weir is its capacity to transport sediment. The following standard flumes will be covered:

(a) rectangular
(b) trapezoidal
(c) U-shaped
(d) Parshall.

Rectangular throated flume

The rectangular flume consists of a constriction of rectangular cross-section symmetrically disposed with respect to the approach channel. There are three types:

(a) with side contractions only;
(b) with bottom contraction (or hump) only; and
(c) with both side and bottom contractions. A diagrammatic illustration of the rectangular flume is shown in Fig. 10.16.

The discharge equation is

$$Q = (\tfrac{2}{3})^{3/2} C_v C_d b \sqrt{(g)} \, h^{3/2} \qquad (10.46)$$

where, as in the case of the round-nose weir (equation (10.45))

$$C_d = \left(1 - \frac{0.006L}{b}\right)\left(1 - \frac{0.003L}{h}\right)^{3/2}$$

where L is the length of flume throat and b is the width of throat (Table 10.8).
C_v is found from Table 10.1 taking the cross-sectional area of flow, A, as

$$A = B(h + P) \qquad (10.47)$$

where B is the width of the approach channel and P is the height of the hump

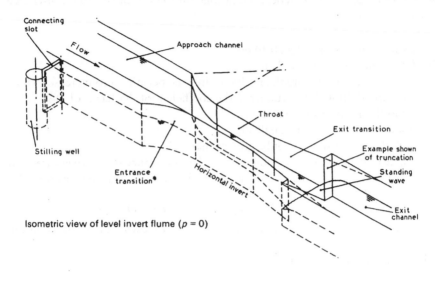

Isometric view of level invert flume ($p = 0$)

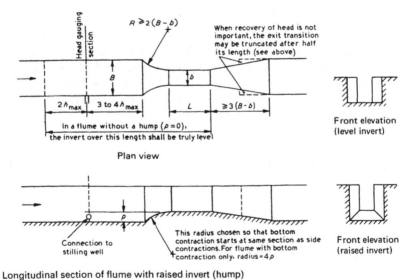

Plan view

Longitudinal section of flume with raised invert (hump)

Figure 10.16 Diagrammatic illustration of the rectangular throated flume.

(with no hump, $P = 0$). Practical limitations on the use of the above discharge equation (10.46) are:

(a) b should not be less than 0.10 m;
(b) h/b should not be more than 3;
(c) h/L should not be more than 0.50 but may be permitted to increase to 0.70 with an additional uncertainty in the coefficient of discharge of 2%;
(d) h should not be more than 2 m;
(e) h should not be less than 0.05 m;
(f) $(bh)/[B(h + P)]$ should not be greater than 0.7;
(g) the head measurement section is located at a distance of between three and four times h_{max} upstream of the leading edge of the entrance transition.

Simplified discharge equation for design or spot measurements

For design purposes or for spot measurements, the following discharge equation may be used.
   Rectangular throated flume

$$Q = 1.8bh^{3/2} \text{ m}^3 \text{ s}^{-1}.$$ (10.48)

Trapezoidal throated flume

The trapezoidal flume has a throat, entrance and exit of trapezoidal cross-section. The construction therefore requires entrance and exit transitions. The flume is normally constructed in the field with the level of the invert of the throat level with the invert of the approach channel. However, the flume may also be constructed with a hump in the throat. A diagrammatic illustration of the trapezoidal flume is shown in Fig. 10.17.
   The discharge equation is

$$Q = (\tfrac{2}{3})^{3/2} C_v C_d C_s b \sqrt{(g)} h^{3/2}$$ (10.49)

where

$$C_d = \left(1 - 0.006x\frac{L}{b}\right)\left(1 - \frac{0.003L}{h}\right)^{3/2};$$ (10.50)

values of x for values of m are obtained from Table 10.9.
   Note: m is the slope of the flume sides (m horizontal: 1 vertical).
   Before C_v can be calculated, C_s requires to be found. C_s is a function of mH/b, where H is the total head allowing for velocity of approach.
   This leads to a slightly more complicated calculation since only gauged head is known and whilst gauged head is satisfactory for design purposes for a

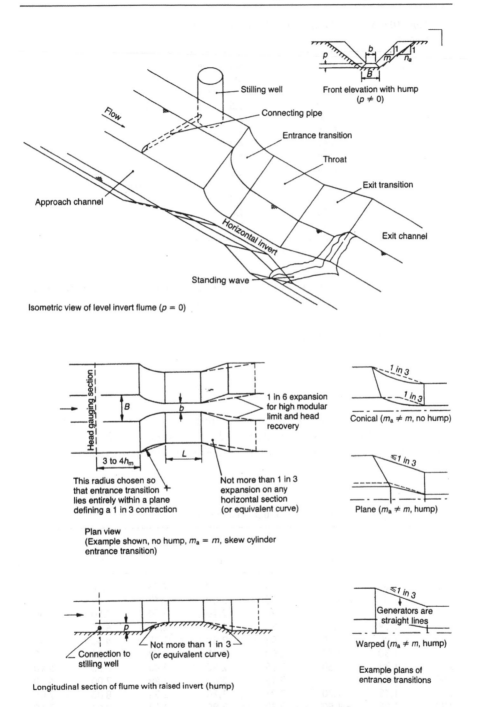

Front elevation with hump
$(p \neq 0)$

Stilling well

Connecting pipe

Entrance transition

Throat

Exit transition

Flow

Approach channel

Horizontal invert

Exit channel

Standing wave

Isometric view of level invert flume ($p = 0$)

Head gauging section

B

b

1 in 6 expansion
for high modular
limit and head
recovery

$\frac{1 \, in \, 3}{}$

$\frac{1 \, in \, 3}{}$

Conical ($m_a \neq m$, no hump)

3 to $4h_m$

L

This radius chosen so
that entrance transition
lies entirely within a plane
defining a 1 in 3 contraction

Not more than 1 in 3
expansion on any
horizontal section
(or equivalent curve)

$\leqslant 1 \, in \, 3$

Plane ($m_a \neq m$, hump)

Plan view
(Example shown, no hump, $m_a = m$, skew cylinder
entrance transition)

$\leqslant 1 \, in \, 3$

Generators are
straight lines

p

Not more than 1 in 3
(or equivalent curve)

Connection to
stilling well

Warped ($m_a \neq m$, hump)

Example plans of
entrance transitions

Longitudinal section of flume with raised invert (hump)

Figure 10.17 Diagrammatic illustration of the trapezoidal throated flume.

Table 10.9 Values of x for use in determining C_d for trapezoidal flumes

Side slope m	x	Side slope m	x
0.1	0.90	1.5	0.30
0.2	0.80	2.0	0.24
0.3	0.75	3.0	0.17
0.4	0.67	4.0	0.13
0.5	0.60	5.0	0.10
0.6	0.57	6.0	0.08
0.7	0.51	7.0	0.07
0.8	0.48	8.0	0.06
0.9	0.45	9.0	0.06
1.0	0.41	10.0	0.05

trapezoidal flume, in order to evaluate discharge successive approximations are required to deduce C_s. This is performed as follows:

as a first approximation assume

$$\frac{mH}{b} = \frac{mh}{b}$$

and obtain C_s from Table 10.10. Find the initial value of C_v from Table 10.1. (Note that in Table 10.1 C_s is used instead of C_d.) Calculate the first approximation of discharge Q from equation (10.49) with the above values of C_d, C_s and C_v.

Table 10.10 Shape coefficient C_s for trapezoidal flumes

$\frac{mh}{b}$	C_s	$\frac{mh}{b}$	C_s	$\frac{mh}{b}$	C_s	$\frac{mh}{b}$	C_s
0.02	1.02	0.90	1.64	1.90	2.36	2.90	3.08
0.05	1.03	1.00	1.70	2.00	2.42	3.00	3.15
0.10	1.07	1.10	1.77	2.10	2.50	3.50	3.50
0.20	1.14	1.20	1.85	2.20	2.58	4.00	3.90
0.30	1.20	1.30	1.93	2.30	2.65	4.50	4.30
0.40	1.30	1.40	2.00	2.40	2.72	5.00	5.00
0.50	1.35	1.50	2.08	2.50	2.80	6.00	5.60
0.60	1.42	1.60	2.15	2.60	2.88	7.00	6.10
0.70	1.48	1.70	2.23	2.70	2.95	8.00	7.00
0.80	1.55	1.80	2.30	2.80	3.02	9.00	7.60

Practical limitations on the use of the discharge equation (10.49) are:

(a) b should not be less than 0.1 m;
(b) h should not be more than 2 m;
(c) h should not be less than 0.05 m;
(d) h/L should not be more than 0.50 but may be permitted to rise to 0.70 with an additional uncertainty in the coefficient of discharge of 2% (L is the length of throat);
(e) at all elevations, the width between the throat walls should be less than the width between the approach channel walls at the same elevation;
(f) the sloping walls of the throat should continue upwards, without change of slope, far enough to contain the maximum discharge to be measured;
(g) the Froude number should not be more than 0.5 and is calculated from

$$\text{Fr} = \frac{\bar{v}}{\sqrt{(gA/W)}}$$

where $\bar{v}$ is the average velocity in the approach channel at the head measurement section, A is the cross-sectional area of flow in the approach channel at the head measurement section, and W is the water surface width in the approach channel at the head measurement section;
(h) the head measurement section is located at a distance of between 3 and $4h_{max}$ upstream from the leading edge of the entrance transition.

Example

Calculate the discharge for a trapezoidal throated flume for the following conditions

$m = 1.00$

$m_a = 1.00$ (side slopes of approach channel – m horizontal: 1 vertical)

$b = 0.50$ m

$B = 2.00$ m (bed width of approach channel)

$P = 0.15$ m

$L = 3.00$ m

$g = 9.81$ m s^{-2}

$h = 1.00$ m.

(a) First approximation

Calculate C_d.
From Table 10.9 for a throat slope m of 1.00

$x = 0.414.$

From equation (10.50)

$$C_d = \left(1 - 0.006 \times 0.414 \frac{L}{b}\right)\left(1 - \frac{0.003L}{h}\right)^{3/2}$$

$$= 0.985 \times 0.973$$

$$= 0.958.$$

Note: C_d retains its value.
Calculate (mH/b).

$$\frac{mh}{b} = 2.00.$$

C_s from Table 10.10 is 2.420;

$$\frac{C_s bh}{A} = \frac{2.420 \times 0.50 \times 1}{(h + P)\,[B + m_a(h + P)]}$$

$$= \frac{1.210}{(1.15)[2.0 + 1(1.15)]} = 0.334.$$

Calculate C_v.
From Table 10.1 $C_v = 1.027$.
Calculate Q from equation (10.49).

$$Q = (\tfrac{2}{3})^{3/2} C_v C_d C_s b \sqrt{(g)} h^{3/2}$$

$$= 0.544 \times 1.027 \times 0.958 \times 2.435 \times 0.5 \times 3.132 \times 1^{3/2}$$

$$= 2.041.$$

(b) Second approximation

From equation (10.14)

$$C_v = (H/h)^{3/2}.$$

Therefore

$$\frac{mH}{b} = \frac{mb}{b}(C_v)^{2/3}$$

$$= \frac{1 \times 1}{0.5} \times 1.027^{2/3}$$

$$= 2.036.$$

C_s from Table 10.10 is 2.445.

$$\frac{C_s bh}{A} = \frac{2.445 \times 0.50 \times 1}{1.15 \times 3.15}$$

$$= 0.337.$$

C_v from Table 10.1 is 1.028.
Therefore

$$Q = 0.544 \times 1.028 \times 0.958 \times 2.445 \times 0.5 \times 3.132 \times 1^{3/2}$$

$$= 2.051 \text{ m}^3 \text{ s}^{-1}.$$

(c) Third approximation

$$\frac{mb}{b} \times (C_v)^{2/3} = \frac{1 \times 1}{0.5} \times 1.028^{2/3}$$

$$= 2.037.$$

C_s from Table 10.10 is 2.450.

$$\frac{C_s bh}{A} = \frac{2.450 \times 0.50 \times 1}{1.15 \times 3.15}$$

$$= 0.338.$$

C_v from Table 10.1 is 1.028.

$$Q = 0.544 \times 1.028 \times 0.958 \times 2.450 \times 0.5 \times 3.132 \times 1^{3/2}$$

$$= 2.055 \text{ m}^3 \text{ s}^{-1}.$$

It will be noted that in this example the difference in discharge between the first approximation and the third approximation is less than 1%.

For spot checks or preliminary design requirements and taking $C_v C_d$ as 1.00, the following relation may be used

$$Q = 1.704bC_s h^{3/2} \text{ m}^3 \text{ s}^{-1}.$$

Taking the above values of b, h and C_s from Table 10.10 gives a value for Q of 2.062 m^3 s^{-1}.

U-throated (round bottomed) flumes

Flumes with cylindrical-shaped inverts are mainly used in sewerage systems where the flow enters from a U-shaped conduit. The sensitivity of a U-shaped flume is greater than that of a rectangular flume especially in the lower range of discharge within the lower semicylinder. However, U-shaped flumes do not have a large variety of situations as do the trapezoidal flumes. A diagrammatic illustration of the U-shaped flume is shown in Fig. 10.18.

The discharge equation is

$$Q = (\tfrac{2}{3})^{3/2} C_v C_d C_u D \sqrt{(g)} h^{3/2} \tag{10.51}$$

where C_u is the shape coefficient and D is the diameter of the throat

$$C_d = \left(1 - \frac{0.006L}{D}\right)\left(1 - \frac{0.003L}{D}\right)^{3/2}. \tag{10.52}$$

Values of C_d may be obtained from Table 10.8.

The practical limitations on equation (10.51) are:

(a) h should not be less than 0.05 m;
(b) h should not be more than 2 m;
(c) D should not be less than 0.1 m;
(d) h/L should not be more than 0.50 but may be permitted to increase to 0.70 with an additional uncertainty in the coefficient of discharge of 2%;
(e) the Froude number should not exceed 0.5;
(f) at all elevations the width between the throat walls should be less than the width between the approach channel walls at the same elevation;
(g) the head measurement section is located at $3h_{max}$ to $4h_{max}$ upstream of the leading edge of the entrance transition.

Computation of discharge

The procedure is similar to that for computing discharge for a trapezoidal flume and is essentially one of calculating C_d, C_v and C_u:

(a) List the values of D, L, P, D_a, where D_a is the width of the approach channel and P is the height of the hump (where installed).
(b) Compute the approach channel cross-sectional area of flow: if the water level lies within the semicircular base portion, $(h + P)$ less than $D_a/2$,

$$A = \tfrac{1}{4} D_a^2(\theta - \sin\theta \cos\theta) \tag{10.53}$$

where

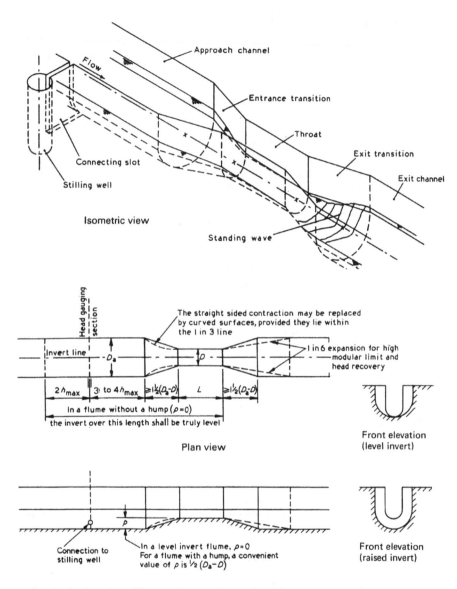

Isometric view

Plan view

Front elevation
(level invert)

Front elevation
(raised invert)

Longitudinal section of flume with raised invert (hump)

Figure 10.18 Diagrammatic illustration of the U-throated flume.

$$\theta = \cos^{-1}\left[\frac{1 - 2(h + P)}{D_a}\right].$$

(10.54)

If the water level is above the semicircular base portion and $(h + P)$ is greater than $D_a/2$,

$$A = \frac{\pi}{8}D_a^2 + [(h + P) - \tfrac{1}{2}D_a]D_a.$$

(10.55)

(c) As a first approximation assume $H/D = h/D$ and obtain C_u from Table 10.11.
(d) Calculate $(C_uDh)/A$ and obtain a first approximation for C_v from Table 10.1 (replacing C_d by C_u and b by D).
(e) Compute C_d from Table 10.8 (C_d retains its value).
(f) Calculate H from $H/h = (C_v)^{2/3}$ (equation (10.14)). C_v and C_u now require refining.
(g) Evaluate H/D and obtain a new value of C_u from Table 10.11.
(h) Evaluate $(C_uDh)/A$ and obtain a new value of C_v from Table 10.1.
(i) Calculate the new value of H from $H/h = (C_v)^{2/3}$.

The above procedure is repeated until sufficient precision has been obtained. Generally three approximations are sufficient.

Example

Calculate the discharge from a U-throated flume for the following conditions

$D = 0.40$ m

$L = 1.00$ m

$D_a = 0.60$ m

Table 10.11 Shape coefficient C_u for U-throated flumes

$\dfrac{h}{D}$	C_u	$\dfrac{h}{D}$	C_u	$\dfrac{h}{D}$	C_u
0.1	0.350	1.1	0.859	2.1	0.924
0.2	0.480	1.2	0.870	2.2	0.928
0.3	0.576	1.3	0.880	2.3	0.931
0.4	0.645	1.4	0.889	2.4	0.934
0.5	0.700	1.5	0.896	2.5	0.939
0.6	0.745	1.6	0.901	2.6	0.940
0.7	0.780	1.7	0.908	2.7	0.943
0.8	0.807	1.8	0.912	2.8	0.945
0.9	0.828	1.9	0.917	2.9	0.948
1.0	0.842	2.0	0.920	3.0	0.950

$P = 0$ m (no hump)

$g = 9.81$ m s^{-2}

$h = 0.25$ m

(a) Cross-sectional area of flow

$$= \frac{\pi}{8} D_a^2 + [(h + P) - \tfrac{1}{2}D_a]D_a$$

$$= \frac{\pi}{8} \times 0.60^2 + (0.25 - 0.30)0.60$$

$$= 0.1115 \text{ m}^2.$$

(b) $\dfrac{h}{D} = \dfrac{0.25}{0.40} = 0.625.$

(c) $C_u = 0.754$ (Table 10.11).

(d) $\dfrac{C_u Dh}{A} = \dfrac{0.754 \times 0.40 \times 0.25}{0.1115}$

$$= 0.676.$$

(e) $C_v = 1.133$ (Table 10.1).

(f) $C_d = \left(1 - \dfrac{0.006 \times 1}{0.4}\right)\left(1 - \dfrac{0.003 \times 1}{0.25}\right)^{3/2}$ (or from Table 10.8)

$$= 0.985 \times 0.982$$

$$= 0.967.$$

(g) $H = h(C_v)^{2/3}$

$$= 0.25 \times 1.133^{2/3}$$

$$= 0.2717.$$

Refine C_v and C_u (second approximation).

(h) $\dfrac{H}{D} = \dfrac{0.2717}{0.4} = 0.6793.$

(i) $C_u = 0.776$ (Table 10.11).

(j) $\dfrac{C_u Dh}{A} = \dfrac{0.776 \times 0.40 \times 0.25}{0.1115}$

$$= 0.696.$$

(k) $C_v = 1.144$ (Table 10.1).

(l) $H = h(C_v)^{2/3}$

$\qquad = 0.25 \times 1.144^{2/3}$

$\qquad = 0.2734.$

Refine C_v and C_u (third approximation).

(m) $\dfrac{H}{D} = \dfrac{0.2734}{0.4} = 0.6834.$

(n) $C_u = 0.777$ (Table 10.11) (no change).

(o) $\dfrac{C_u Dh}{A} = 0.697.$

(p) $C_v = 1.144$ (no change).

(q) $H = 0.2734$ (no change).

Therefore $Q = 0.544 \times 1.144 \times 0.967 \times 0.777 \times 0.40 \times \sqrt{(g)} \times 0.25^{3/2}$

$$\text{(eq. (10.51))}$$

$$= 0.073 \text{ m}^3 \text{ s}^{-1}.$$

For spot checks or preliminary design requirements and taking $C_v C_d$ as 1.00, the following relation may be used

$$Q = 1.704 \, DC_u h^{3/2} \text{ m}^3 \text{ s}^{-1}.$$

Taking the values of D, h and C_u from Table 10.11 gives a value of Q of $0.064 \text{ m}^3 \text{ s}^{-1}$.

The Parshall flume

The Parshall flume has a rectangular cross-section and comprises three main parts: a converging inlet section with a level floor, a throat section with a downward sloping floor and a diverging outlet section with an upward sloping floor.

The control section of the flume is not located in the throat as in the previous flumes described, but near the end of the level floor, or crest, in the converging section. Because of this, Parshall flumes are considered as short throated flumes. Laboratory calibration has been carried out in the modular, free flow range but because of their low modular (submergence) ratio the Parshall flume is sometimes used with a tapping for measuring the downstream water level. A diagrammatic illustration of the Parshall flume is shown in Fig. 10.19.

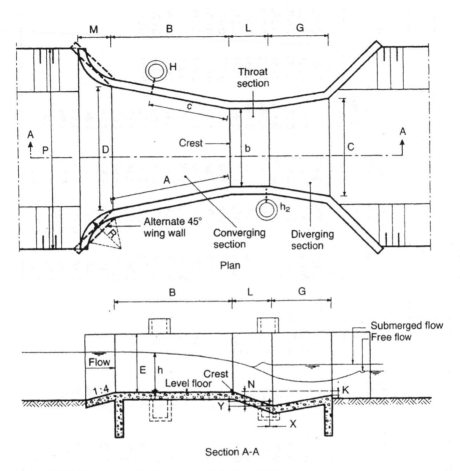

Plan

Section A-A

Figure 10.19 Diagrammatic illustration of the Parshall flume.

The discharge equation is

$$Q = Kh^u \tag{10.56}$$

where K is a dimensional factor which is a function of the throat width b. The power u varies between 1.522 and 1.600.

Table 10.12 summarises the dimensions for the range of Parshall flumes normally used and Table 10.13 gives details of the equations of discharge, discharge range for flumes and the head range.

Practical limitations for the application of equation (10.56) are:

(a) h should not be less than 0.015 m;
(b) h should not be more than 1.83 m;

Table 10.12 Dimensions of standard Parshall flumes. Note flume sizes 0.076 to 2.44 m have approach aprons rising at 1:4 slope and the following entrance roundings: 0.076 to 0.228 m, radius 0.4 m; 0.30 to 0.90 m, radius 0.51 m; 1.2 to 2.4 m, radius 0.60 m

Widths			Axial lengths			Wall depth	Vertical distance below crest		Converging	Gauge points			h_2
Size throat width	Upstream end	Downstream end	Converging section	Throat section	Diverging section	Converging section	Dip at throat	Lower end of flume	Wall length[a]	h dist upstream of crest[b]			
b (m)	D (m)	C (m)	B (m)	L (m)	G (m)	E (m)	N (m)	K (m)	A (m)	c (m)	X (m)	Y (m)	
0.025	0.167	0.093	0.357	0.076	0.204	0.153–0.229	0.029	0.019	0.363	0.241	0.008	0.013	
0.051	0.213	0.135	0.405	0.114	0.253	0.153–0.253	0.043	0.022	0.415	0.277	0.016	0.025	
0.076	0.259	0.178	0.457	0.152	0.30	0.305–0.610	0.057	0.025	0.466	0.311	0.025	0.038	
0.152	0.396	0.393	0.610	0.30	0.61	0.61	0.114	0.076	0.719	0.415	0.051	0.076	
0.229	0.573	0.381	0.862	0.30	0.46	0.76	0.114	0.076	0.878	0.588	0.051	0.076	
0.305	0.844	0.610	1.34	0.61	0.91	0.91	0.228	0.076	1.37	0.914	0.051	0.076	
0.457	1.02	0.762	1.42	0.61	0.91	0.91	0.228	0.076	1.45	0.966	0.051	0.076	
0.610	1.21	0.914	1.50	0.61	0.91	0.91	0.228	0.076	1.52	1.01	0.051	0.076	
0.914	1.57	1.22	1.64	0.61	0.91	0.91	0.228	0.076	1.68	1.12	0.051	0.076	
1.22	1.93	1.52	1.79	0.61	0.91	0.91	0.228	0.076	1.83	1.22	0.051	0.076	
1.52	2.30	1.83	1.94	0.61	0.91	0.91	0.228	0.076	1.98	1.32	0.051	0.076	
1.83	2.67	2.13	2.09	0.61	0.91	0.91	0.228	0.076	2.13	1.42	0.051	0.076	
2.13	3.03	2.44	2.24	0.61	0.91	0.91	0.228	0.076	2.29	1.52	0.051	0.076	
2.44	3.40	2.74	2.39	0.61	0.91	0.91	0.228	0.076	2.44	1.62	0.051	0.076	
3.05	4.75	3.66	4.27	0.91	1.83	1.22	0.34	0.152	2.74	1.83			
3.66	5.61	4.47	4.88	0.91	2.44	1.52	0.34	0.152	3.05	2.03			
4.57	7.62	5.59	7.62	1.22	3.05	1.83	0.46	0.229	3.50	2.34			
6.10	9.14	7.31	7.62	1.83	3.66	2.13	0.68	0.31	4.27	2.84			
7.62	10.67	8.94	7.62	1.83	3.96	2.13	0.68	0.31	5.03	3.35			
9.14	12.31	10.57	7.92	1.83	4.27	2.13	0.68	0.31	5.79	3.86			
12.19	15.48	13.82	8.23	1.83	4.88	2.13	0.68	0.31	7.31	4.88			
15.24	18.53	17.27	8.23	1.83	6.10	2.13	0.68	0.31	8.84	5.89			

[a] For sizes 0.3 to 2.4 m, A = b/2 + 1.2 m.
[b] h is located 2/3A distance from crest for all sizes, distance is wall length, not axial. For symbols see Fig. 10.19.

Table 10.13 Discharge characteristics of Parshall flumes

Throat width b (m)	Discharge range ($m^3 s^{-1} \times 10^{-3}$)		Equation $Q = Kh^u$ ($m^3 s^{-1}$)	Head range (m)		Modular limit h_2/h
	Minimum	Maximum		Minimum	Maximum	
0.025	0.09	5.4	$0.0604h^{1.55}$	0.015	0.21	0.50
0.051	0.18	13.2	$0.1207h^{1.55}$	0.015	0.24	0.50
0.076	0.77	32.1	$0.1771h^{1.55}$	0.03	0.33	0.50
0.152	1.50	111	$0.3812h^{1.58}$	0.03	0.45	0.60
0.229	2.50	251	$0.5354h^{1.53}$	0.03	0.61	0.60
0.305	3.32	457	$0.6909h^{1.522}$	0.03	0.76	0.70
0.457	4.80	695	$1.056h^{1.538}$	0.03	0.76	0.70
0.610	12.1	937	$1.428h^{1.550}$	0.046	0.76	0.70
0.914	17.6	1427	$2.184h^{1.566}$	0.046	0.76	0.70
1.219	35.8	1923	$2.953h^{1.578}$	0.06	0.76	0.70
1.524	44.1	2424	$3.732h^{1.587}$	0.06	0.76	0.70
1.829	74.1	2929	$4.519h^{1.595}$	0.076	0.76	0.70
2.134	85.8	3438	$5.312h^{1.601}$	0.076	0.76	0.70
2.438	97.2	3949	$6.112h^{1.607}$	0.076	0.76	0.70
		in $m^3 s^{-1}$				
3.048	0.16	8.28	$7.463h^{1.60}$	0.09	1.07	0.80
3.658	0.19	14.68	$8.859h^{1.60}$	0.09	1.37	0.80
4.572	0.23	25.04	$10.96h^{1.60}$	0.09	1.67	0.80
6.096	0.31	37.97	$14.45h^{1.60}$	0.09	1.83	0.80
7.620	0.38	47.14	$17.94h^{1.60}$	0.09	1.83	0.80
9.144	0.46	56.33	$21.44h^{1.60}$	0.09	1.83	0.80
12.192	0.60	74.70	$28.43h^{1.60}$	0.09	1.83	0.80
15.240	0.75	93.04	$35.41h^{1.60}$	0.09	1.83	0.80

Note: Each of the discharges in the upper half of the table requires to be divided by 10^3.

(c) the head measurement section (c) for free flow is located at $(b/3) + 0.813$ m (or $2/3A$) upstream from the downstream end of the horizontal crest. Note that this distance is measured along the wall and not axially (see Fig. 10.19).

Dimensions to be strictly followed are given in Fig. 10.19 and Table 10.12.

Compound measuring structures

A compound gauging structure consists of a series of individual weirs or flumes disposed across the width of an open channel. The individual sections of the compound structure are separated by divide walls so that each separate structure is treated as a simple weir or flume and three-dimensional flow conditions are minimised. The computation of discharge for individual sections is based on the established discharge equations.

Compound measuring structures can be tailored to the needs of particular sites and the measurement of low flows can be confined to narrow crest sections therefore increasing their accuracy. High afflux (difference between the upstream and downstream water levels) at peak flows can be avoided by providing wide flanking crests.

Compound structures can be designed to operate in the drowned flow range but the difficulties already stated in Section 10.1 are further exacerbated with compounding.

A diagrammatic illustration of a typical compound triangular profile weir is shown in Fig. 10.20.

The following design guidelines apply to compound measuring structures:

(a) The head measurement section should be based on the recommendations already given for individual structures. It is advisable to have the intake pipe in the vertical side wall with its soffit at or just below the lowest crest level, in the case of a weir, or lowest throat level, in the case of a flume.

(b) For this condition, the divide walls should not extend beyond the head measurement section. Downstream they should extend to the limit of the individual weir or flume.

(c) Divide walls should be at least 0.3 m thick and preferably have semi-circular or semi-elliptical noses.

(d) In order to minimise cross-flows at the cutwaters of the divide walls, the difference in levels between adjacent crests or flume inverts should not exceed 0.5 m.

(e) The heights of the divide walls should normally be the same as the vertical side walls. If they are less, an error will arise in the computation of discharge when the flow reaches the top of the divide walls.

(f) Flow conditions at or near the divide wall cutwaters may be improved if the top level of the foundation slab upstream from a low-crest weir section (or flume) is set below that of the adjacent high crest sections. The increase in approach depths to the lower-crest sections achieved in this way will result in approach velocities to all sections being more nearly equal to one another.

(g) The vertical side walls should be straight and parallel, and continued upstream from the tapping point for a length equal to the maximum head to be gauged before beginning the side-wall curves.

The example given in Fig. 10.20 shows a compound triangular profile weir having three crests, a lower crest of 3 m length and two flanking crests each of 6 m length. The lower central crest is 0.3 m lower than the flanking crests and the approach floor to the central crest has been set 0.3 m lower than the approach floor to the flanking crests. The P value of all three crests is 0.6 m. The weir has been designed to have a modular limit of $h = 1$ m and a crest tapping provided for discharge measurements beyond that head.

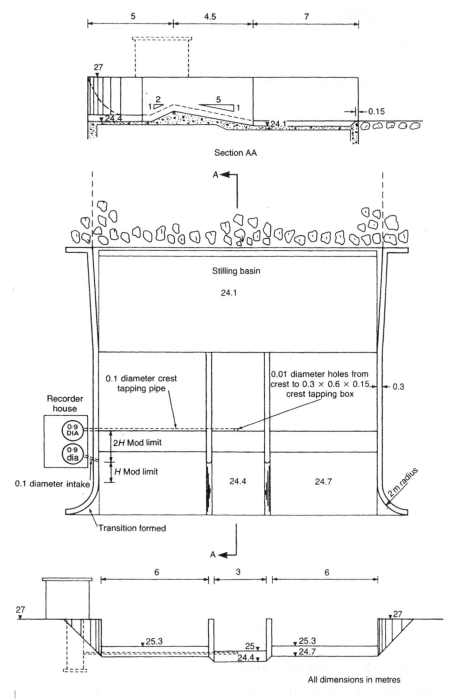

Figure 10.20 Diagrammatic illustration of a typical compound triangular profile (Crump) weir showing crest tapping, arrangement for measuring non-modular (drowned) flow.

Example

Calculate the discharge of a compound triangular profile weir in the free flow condition for the following conditions

flank weir breadth, 2 × 6.10 m = 12.20 m

low central weir breadth   = 3.05 m

difference in crest levels   = 0.305 m

height of low weir   = 0.50 m

gauged head over low weir   = 1.219 m.

(a) For C_v

$$\frac{C_d bh}{A} = \frac{C_d bh}{b(h + P)} = C_d\left(\frac{h}{h + P}\right) = 1.16\frac{h}{(h + P)}.$$

(See under the triangular profile weir section and note that C_d is taken as 1.16.) Then

$$1.16\frac{h}{(h + P)} = 1.16\frac{0.914}{(0.914 + 0.805)} = 0.617$$

(note: the flanking weirs are used in the calculation of C_v), and C_v from Table 10.1 is 1.104.

(b) For the low crest

$$Q = 0.633 \times 1.104 \times \sqrt{(g)} \times b \times 1.219^{3/2} \quad \text{(from equation (10.33))}$$

$$= 0.633 \times 1.104 \times 3.132 \times 3.05 \times 1.346$$

$$= 8.985 \text{ m}^3 \text{ s}^{-1}.$$

(c) For the flanking crests

$$Q = 0.633 \times 1.104 \times 3.132 \times 12.20 \times 0.914^{3/2}$$

$$= 23.333 \text{ m}^3 \text{ s}^{-1}.$$

(d) The total discharge is therefore

$$8.985 + 23.333 = 32.318 \text{ m}^3 \text{ s}^{-1}.$$

10.6 Non-modular (drowned) flow

In the drowned flow range, the discharge is dependent upon both upstream and downstream water levels. Because of the difficulty of measuring the downstream head to an acceptable accuracy, due to turbulence, a crest tapping arrangement is often incorporated in the design of Crump weirs. A typical arrangement is shown in Fig. 10.20 where it can be seen that a separate stilling well is required to measure the crest tapping pressure head (h_p). The ratio h_p/h is related to the drowned flow reduction factor (f) to be applied to the Crump weir modular discharge equation (10.34). Equation (10.34) becomes

$$Q = fC_vC_db\sqrt{(g)}h^{3/2} \tag{10.57}$$

and fC_v can be found from Table 10.14.

It should be noted that if total head (H) is used in the computation (equation (10.34)), f may be found from Table 10.15.

Table 10.14 Values of fC_v in terms of h_p/h and $C_d\dfrac{h}{h+p}\dfrac{b}{B}$ where $C_d = 0.633$ for the Crump weir (drowned flow)

h_p/h	$C_d\dfrac{h}{h+p}\dfrac{b}{B}$					
	0.1	0.2	0.3	0.4	0.5	0.6
0.20	1.00	1.03	1.08	1.17	1.35	
0.24	1.00	1.02	1.08	1.17	1.35	
0.30	0.98	1.01	1.05	1.13	1.32	
0.40	0.95	0.98	1.02	1.09	1.27	
0.50	0.92	0.94	0.98	1.07	1.20	
0.60	0.88	0.90	0.93	0.99	1.13	1.40
0.70	0.82	0.83	0.87	0.92	1.03	1.35
0.80	0.71	0.73	0.76	0.82	0.91	1.20
0.90	0.55	0.58	0.60	0.65	0.70	0.90
0.95	0.40	0.42	0.45	0.46	0.52	0.65

Table 10.15 Values of f in terms of h_p/H (Crump weir drowned flow)

$\dfrac{h_p}{H}$	f
0.20	1.00
0.24	1.00
0.30	0.98
0.40	0.95
0.50	0.92
0.60	0.88
0.70	0.81
0.80	0.71
0.90	0.55
0.95	0.40

Under modular flow conditions h_p/h is constant at 0.20 and the value of f is 1.00. The modular limit (downstream head, h_2 divided by upstream head, h) is 0.75.

A crest tapping may also be incorporated in the design of flat V weirs in order to extend the modular range into the drowned flow range (see Figure 10.12). The discharge equation for drowned flow becomes

$$Q = \frac{4}{5} f C_v C_d \sqrt{(g)} m Z h^{5/2}. \tag{10.58}$$

Values of f may be found from Table 10.16 in terms of h_p/h and $2C_d Z m h^2 / b(h + P)$. In the drowned flow range $h_p/H > 0.4$ and the modular limit is 70%.

Example

Calculate the drowned flow discharge for a flat V weir having the dimensions given in the example on page 341 but with the additional information in the drowned flow range

head = 0.69 m

h_p = 0.52 m.

Therefore

$$\frac{h_p}{h} = \frac{0.52}{0.69} = 0.75,$$

Table 10.16 Values of f in terms of h_p/h and $2C_d z m h^2 / b(h + p)$ for the flat V weir (drowned flow)

h_p/h	$2C_d z m h^2 / b(h + p)$								
	0.1	0.2	0.3	0.4	0.5	0.6	0.7	0.8	0.88
0.40	0.995	0.996	0.996	0.997	0.998	0.999	1.000	1.000	1.000
0.45	0.984	0.984	0.985	0.986	0.987	0.988	0.990	0.992	0.994
0.50	0.968	0.969	0.969	0.970	0.971	0.973	0.975	0.978	0.981
0.55	0.950	0.951	0.951	0.953	0.954	0.956	0.958	0.962	0.965
0.60	0.929	0.930	0.931	0.932	0.934	0.937	0.940	0.943	0.947
0.65	0.905	0.906	0.907	0.909	0.912	0.914	0.918	0.922	0.927
0.70	0.877	0.878	0.879	0.881	0.884	0.887	0.892	0.897	0.902
0.75	0.843	0.844	0.845	0.848	0.851	0.855	0.860	0.867	0.873
0.80	0.789	0.800	0.802	0.805	0.809	0.814	0.820	0.828	0.836
0.85	0.738	0.739	0.742	0.746	0.751	0.758	0.766	0.776	0.786
0.90	0.635	0.638	0.642	0.649	0.657	0.667	0.679	0.694	0.708
0.93	0.483	0.488	0.497	0.510	0.525	0.545	0.567	0.593	0.615

$$2C_d Zmh^2/b(h + p) = \frac{2 \times 0.620 \times 1 \times 20.3 \times 0.49}{36(0.69 + 0.82)}$$

$$= 0.23$$

and from Table 10.16 this value with $h_p/h = 0.75$ gives a value of $f = 0.844$. Therefore from equation (10.58)

$$Q = 0.8 \times 0.84 \times 1.008 \times 0.620 \times \sqrt{(9.81)} \times 1 \times 20.3 \times 0.69^{5/2}$$

$$= 10.56 \text{ m}^3 \text{ s}^{-1}.$$

10.7 Summary of the range of discharge for standard weirs and flumes

Table 10.17 provides a comparison guide to the range of maximum and minimum discharges for similar dimensions for the measuring structures discussed in this chapter. The table is based on the limitations given under each device and is for conditions of free (modular) flow. It is usual in practice, however, that the measurement of discharge will unavoidably go beyond these limitations and records will be required outside the imposed conditions. When this occurs a degradation in the accuracy of the measuring structure has to be accepted.

A column has been added to the table to give the approximate modular (submergence) limit of free flow for each device where relevant. This is given as a percentage of the downstream gauged head divided by the upstream gauged head.

10.8 Summary of uncertainties in the coefficients of discharge

Table 10.18 provides a guide to the uncertainties attainable in the coefficients of discharge of standard weirs and flumes if reasonable skill is taken in their design, construction and operation.

10.9 Non-standard weirs

This chapter has been devoted to measuring structures which have been standardised internationally as a result of considerable research and field application over many years. Provided these structures are designed, maintained and operated according to the relevant international standards, high accuracy can be expected. There are, however, many cases where the need arises to use non-standard weirs for flow measurement. Such weirs are often built for other purposes but may be rated by a current meter if a suitable measuring cross-section is available. Over the years, however, equations of discharge have been developed for many non-standard weirs and these equations may provide

Table 10.17 Comparison of the range of discharge of measuring structures for the specified dimensions

Measuring structure	Weirs					Flumes				Modular limit $h_2/h \times 100$
	P (m)	Min	Max	Min	Max	Throat width b(m)	Throat length L(m)	Min	Max	
		($m^3 s^{-1}$ per metre of crest)		($m^3 s^{-1}$)				($m^3 s^{-1}$)		
Thin plate – full width	0.2	0.005	0.667							
	1.0	0.005	7.70							
Thin plate – contracted	0.2	0.009	0.450							
	1.0	0.009	4.90							
Thin plate – 90° V-notch				0.001	1.80					
Triangular profile (Crump)	0.2	0.01	1.17							75
	1.0	0.01	13.00							
Rectangular weir	0.2	0.02	0.20							66
	1.0	0.02	2.68							
Round-nose weir	0.15	0.02	0.18							80
	1.0	0.02	3.10							70
Flat V weir										
1:10 b=4 m	0.2			0.002	2.80					
1:40 b=80 m	1.0			0.055	630.0					
Rectangular flume						0.5	1.0	0.009	0.30	75–90
						1.0	2.0	0.033	1.70	
Trapezoidal flume m = 5						1.0	4.0	0.27	41.00	75–90
U-shaped flume						D = 0.3	0.6	0.002	0.07	75–90
						D = 1.0	2.0	0.019	1.40	
Parshall flume						0.152	0.3	0.002	0.11	50–80
						1.22	0.6	0.036	1.92	

Table 10.18 Uncertainties attainable in the coefficients of discharge (C_d) of weirs and flumes

Measuring structure	Uncertainty attainable in the coefficient of discharge (68% level)
Rectangular thin plate weirs	±1
Thin plate V-notch	±1
Triangular profile (Crump) weir	±1
Rectangular broad-crested weir	±2
Round-nose broad-crested weir	±2
Flat V weir	±2
Rectangular flume	±2
Trapezoidal flume	±2
U-throated flume	±2
Parshall flume	±2

acceptable results probably within an accuracy of about 10% at high values of head. The weirs of this type considered below are compound thin plate weirs, round crested weirs and the Cipoletti weir.

Compound thin plate weirs

Where, in addition to measuring compensation water or prescribed flows in a reservoired catchment, it is required to measure flood flows, compound thin plate weirs without divide walls are often used. These weirs may have several crest levels, the lowest crest or V-notch being designed to measure very low flows. Coefficients of discharge obtained from standard thin plate weirs are used in the computation of flow but a problem arises in estimating the flows over different crests as the head increases. Also, the permitted limitations for standard weirs may be exceeded – for example the h/p ratio or the maximum permitted head – upon which the standard equations of discharge have been based. Figure 10.21 shows a typical compound thin plate weir having a low flow V-notch, two crest levels and a circular orifice. When the flow is within the V, an accurate estimate of the discharge is obtained, but as the head increases, the accuracy decreases for the above reasons and because of the disturbances caused at the corners of the crests and V-notch.

Round crested weirs

The overflow spillways of dams and barrages may have crest sections of circular or parabolic form and can often be adapted for flow measurement (Figure 10.22). The advantage of the parabolic crest is that, for a given discharge, the minimum possible head is required over a given length of crest.

The general equation of discharge for these weirs is

$$Q = Cbh^{3/2} \text{ m}^3 \text{ s}^{-1} \tag{10.59}$$

Figure 10.21 Compound thin plate weir having a low flow V-notch, two crest levels and a circular orifice. The circular orifice ensures that compensation water from the reservoir flows downstream when the water level falls below the V-notch.

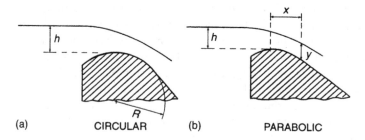

Figure 10.22 Round crested weirs, (a) circular and (b) parabolic, being sections of spillways of dams and often adapted for flow measurement.

where

$$C \text{ for circular weirs} = 2.03(h/R)^{0.07} \qquad (10.60)$$

and

$$C \text{ for parabolic weirs} = 1.86h^{0.1}, \qquad (10.61)$$

where R is the radius of the circular crest. The equations of discharge are then

circular weirs: $Q = 2.03(h/R)^{0.07}bh^{3/2}$ m^3 s^{-1} $\qquad$ (10.62)

parabolic weirs: $Q = 1.86bh^{1.6}$ m^3 s^{-1}. $\qquad$ (10.63)

The limits of application for the above equations are (approximately):

(a) h should not be less than 0.05 m;
(b) h/p should not be less than 3;
(c) h/P_2 should be less than 1.5 (where P_2 is the height of the weir above the downstream channel);
(d) $\dfrac{b}{h}$ should not be less than 2;
(e) the modular limit $h/_2h_1 = 0.3$ (where h_2 is the downstream head above crest level);
(f) the upstream head, h, should be measured at a distance two to three times h_{max} upstream from the weir face.

It should be noted that the discharge coefficient C in equations (10.59), (10.60) and (10.61) has dimensions of the square root of gravity acceleration; that is, it includes g. Many discharge equations for non-standard weirs have in the past been presented in this form and in imperial units. To convert the equation

$$Q = Cbh^{3/2}$$

where b and h are in feet, Q is in ft^3 s^{-1} and C is in dimensions of the square root of gravity acceleration, to the metric equation

$$Q = Cbh^{3/2}$$

where b and h are in metres, Q is in m^3 s^{-1} and C is in dimensions of the square root of gravity acceleration, the imperial equation is multiplied by 0.552.

Note: the above equations of discharge are valid for weirs with vertical upstream faces. For sloping upstream faces (three vertical to three horizontal; two vertical to three horizontal; one vertical to three horizontal) correction factors for the coefficient of discharge are given in Table 10.19. Table 10.20 provides drowned flow reduction factors as a function of h_2/h_1 and h_2/P_2.

Cipoletti weir

The Cipoletti thin plate weir has a trapezoidal section with side-slopes 1:4 (horizontal to vertical) (Fig. 10.23). Cipoletti, who designed the weir in 1887, argued that the inclination of the sides compensated for the decrease in

Table 10.19 Correction factors for the coefficient of discharge for circular and parabolic weirs for various upstream slopes as a function of the h/p ratio (round crested weirs)

h/p	Slope of upstream face		
	3:3	2:3	1:3
3	1.02	1.02	1.01
2	1.01	1.02	1.00
1	1.00	1.01	1.00
0.5	0.99	1.00	1.00

Table 10.20 Drowned flow reduction factors as a function of h_2/h_1 and h_2/P_2 where h_2 is the downstream head, h_1 is the upstream head and P_2 is the height of the weir above the downstream channel (round crested weirs)

$\dfrac{h_2}{P_2}$	h_2/h_1				
	0.1	0.3	0.5	0.7	0.9
0.25	1.00	0.99	0.97	0.93	0.70
0.50	1.00	0.99	0.98	0.95	0.75
1.00	1.00	0.99	0.99	0.97	0.80
2.00	0.98	0.98	0.98	0.96	0.80

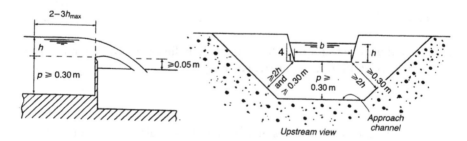

Figure 10.23 Cipoletti thin plate weir showing dimensions.

discharge over a contracted rectangular thin plate weir as the head increased. The equation of discharge is

$$Q = \tfrac{2}{3}\sqrt{(2g)}\,C_v C_d b h^{3/2} \ \text{m}^3 \ \text{s}^{-1} \tag{10.64}$$

where

$$C_d = 0.63$$

and C_v can be obtained from Table 10.1.

The limits of application of equation (10.64) are:

(a) P should be less than $2h_{max}$ with a minimum value of 0.3 m;
(b) the distance from the sides of the weir to the sides of the approach channel should not be less than $2h_{max}$ with a minimum value of 0.3 m;
(c) h should be not less than 0.06 m;
(d) h should not exceed 0.6 m;
(e) h/b should not exceed 0.5 m;
(f) the tailwater level should not be less than 0.05 m below crest level to ensure aeration of the nappe.

The head measuring section should be at $4\text{--}5h_{max}$ upstream from the weir. If necessary, the weir may be placed in a non-rectangular channel.

10.10 Orifices and sluices

Orifices (Fig. 10.24) are normally used in waterworks or in irrigation works to measure small discharges but can also be designed to provide constant discharge under constant head. They are generally circular, rectangular or U-shaped. The general equation for flow through orifices is

$$Q = CAv \text{ m}^3 \text{ s}^{-1} \tag{10.65}$$

where C is the coefficient of discharge;
$\quad$ A is the cross-sectional area of the orifice (m^2);
$\quad$ v is the average velocity through the orifice (m s^{-1}).

Then

$$Q = CA \sqrt{(2gh)} \text{ m}^3 \text{ s}^{-1} \tag{10.66}$$

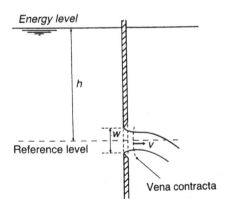

Figure 10.24 Detail of a freely discharging orifice.

where h is the head above the centre of the orifice (m). For large rectangular orifices under low heads

$$Q = \int dq = Cb\sqrt{(2g)} \int_{h_2}^{h_1} h^{1/2} \, dh \quad \text{m}^3 \text{ s}^{-1} \tag{10.67}$$

$$= \tfrac{2}{3}Cb\sqrt{(2g)}(h_1^{3/2} - h_1^{3/2}) \tag{10.68}$$

where h_1 is the distance from the water surface to the lower edge of the orifice and h_2 is the distance from the water surface to the top edge of the orifice.

In practice, however, equation (10.66) is generally used for all orifices under free-flow conditions.

Sluice gates

The free discharge below a sluice gate (Fig. 10.25) is given by the following equation

$$Q = CA\sqrt{[2g(h - a)]} \quad \text{m}^3 \text{ s}^{-1} \tag{10.69}$$

where a is the water level immediately below the sluice gate or taken as equal to nw where h is the upstream water level, w is the gate opening and n is the coefficient of contraction which may be taken as 0.61.

Submerged orifices

The basic discharge equation for submerged orifices and sluice gates (Fig. 10.26) is

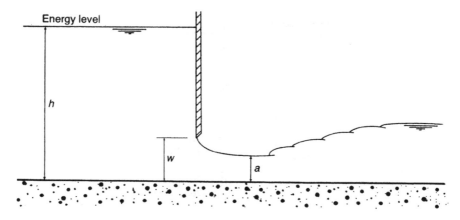

Figure 10.25 Detail of an undershot (sluice gate).

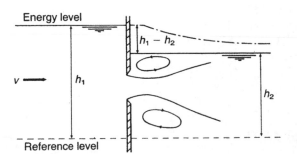

Figure 10.26 Detail of a submerged orifice.

$$Q = CA\sqrt{[2g(h_1 - h_2)]} \text{ m}^3 \text{ s}^{-1} \tag{10.70}$$

where $(h_1 - h_2)$ is the differential head across the orifice or sluice gate, or the difference between the upstream and downstream water levels.

Discharge coefficients

Little recent research has been done on discharge coefficients of orifices and sluice gates but past work indicates the following values with uncertainties of about 5%:

(a) orifices and sluice gates under free-flow conditions 0.62;
(b) orifices and sluice gates under submerged flow conditions 0.60.

It is advisable, however, that until further research is carried out, these values should be checked by current meter in individual cases wherever possible. A photograph of a typical orifice to measure compensation water is shown in Fig. 10.27 (see also Fig. 10.21).

Limits of application for orifices

(a) The upstream edge of the orifice should be sharp and smooth in accordance with the profile shown in Fig. 10.8.
(b) The upstream face of the orifice should be vertical.
(c) For rectangular orifices the top and bottom edges should be horizontal and the sides should be vertical.
(d) The distance from the edge of the orifice to the bed and sides of the approach and tailwater channels should be greater than twice the least dimension of the orifice or in the case of circular orifices not less than the radius of the orifice.

Figure 10.27 U-shaped orifice (right foreground under ladder) to measure compensation water. The head over the rectangular thin plate weir in the centre of the picture is maintained sensibly constant so that the head over the orifice is also constant maintaining a constant designed compensation flow into a side chamber for return to the river. The rectangular thin plate weir measures the flow to supply.

Figure 10.28 Triangular profile (Crump) weir having a single crest.

Figure 10.29 Compound triangular profile (Crump) weir having a low flow central crest and two flanking crests.

Figure 10.30 Flat V weir.

Figure 10.31 View of 120° V notch in foreground and two 90° V notches in background.

Figure 10.32 Rectangular thin plate weir.

Figure 10.33 Compound thin plate weir having five crests.

Figure 10.34 Small 100 mm rectangular standing wave flume under test.

Figure 10.35 Small rectangular standing wave flume under calibration. Note electromagnetic current meter, 'Diptone' electric–tape gauge and 'look-down'
ultrasonic water level gauge.

(e) In order to neglect velocity of approach, the wetted cross-sectional area of
the approach channel where head is measured should be not less than ten
times the area of the orifice.

(f) In the case of submerged orifices the differential head across the orifice
should be not less than 0.03 m.

Example

A submerged U-shaped orifice of cross-sectional area 0.038 m² is under a differential head of 0.235 m. Calculate the discharge taking a coefficient of discharge
of 0.60.

From equation (10.70)

$$Q = CA\sqrt{[2g(h_1 - h_2)]}$$
$$= 0.60 \times 0.038\sqrt{[2 \times 9.81(0.235)]}$$
$$= 0.049 \text{ m}^3 \text{ s}^{-1}.$$

Views of some typical measuring structures are shown in Figs 10.27–10.35.

Further reading

Ackers, P., White, W. R., Perkins, J. A. and Harrison, A. J. M. *Weirs and Flumes for Flow Measurement.* John Wiley and Sons, Chichester 1978.

Addison, H. *Hydraulic Measurements.* Chapman and Hall, London 1940.

Bos, M. G. *Discharge Measurement Structures.* Publication No. 161, Delft Hydraulics Laboratory, Delft 1976.

Clemmens, A. J., Bos, M. J. and Replogle, J. A. *Flume-design and Calibration of Long Throated Measuring Flumes.* ILRI Pub. 54, Wageningen, The Netherlands 1993.

Herschy, R. W., White, W. R. and Whitehead, E. *The Design of Crump Weirs.* Technical Note No. 8, Department of the Environment (Water Data Unit), London 1997.

Herschy, R.W. General purpose flow measurement equations for flumes and thin plate weirs. *Flow Measurement and Instrumentation, Vol. 6 No. 4, Elsevier Science, Oxford pp. 283–293* 1995.

Herschy, R.W. and Fairbridge, R.W. *The Encyclopedia of Hydrology and Water Resources*, Chapman and Hall, London 1998.

Herschy, R.W. Hydrometry Principles and Practices (Ed.) John Wiley and Sons, Chichester 1999.

Horton, R. E. *Weir Experiments, Coefficients and Formulas.* US Geological Survey Water Supply and Irrigation Paper 200 1907.

ISO 14139. Methods of Liquid Flow in Open Channels – Compound Gauging Structures. *HMSO* London, 2000.

ISO 1438/1 *Liquid Flow Measurement in Open Channels: Thin Plate Weirs.* ISO, Geneva, Switzerland, 2008.

ISO 3846, *Liquid Flow Measurement in Open Channels: Free Overfall Weirs of Finite Crest Width (Rectangular Broad Crested Weirs).* ISO, Geneva, Switzerland, 2007.

ISO 4359. *Liquid Flow Measurement in Open Channels: Flumes.* ISO, Geneva, Switzerland, 1983.

ISO 4360. *Liquid Flow Measurement in Open Channels: Triangular Profile Weirs (Crump).* ISO, Geneva, Switzerland, 2007.

ISO 4374. *Liquid Flow Measurement in Open Channels: Round Nose Horizontal Weirs.* ISO, Geneva, Switzerland, 1989.

ISO 4377. *Liquid Flow Measurement in Open Channels: Flat V Weirs.* ISO, Geneva, Switzerland, 2002.

ISO 8368. *Liquid Flow Measurement in Open Channels: Guidelines for the Selection of Flow Gauging Structures.* ISO, Geneva, Switzerland, 1999.

Linford, A. *Flow Measurement and Meters.* E & FN Spon, London, 1949.

Parker, P. A. M. *The Control of Water.* Routledge and Kegan Paul, London, 1949.

White, W.R. Thin Plate Weirs. An investigation into performance characteristics, INT 152 HRS Wallingford 1975.

White, W.R. Standard Specification for flow measurement structures. Proc. of The Institution of Civil Engineers, Water and Maritime Engineering 148 Issue 2 pp. 143–153 paper 12304, 2001.

White, W.R., Whitehead, E. and Forty, E.J. Extending the scope of standard specifications for open channel flow gauging structures. H.R. Wallingford Report SR 564 2001.

Chapter 11

Dilution gauging

11.1 General

The dilution method is generally used for purposes of calibration or for spot gaugings mainly because of the costs of performing a gauging and a chemical analysis of the tracer samples. Nevertheless the method can often provide very accurate results given a suitable reach of river. The outstanding advantage of the dilution technique is that it is an absolute method, because discharge is computed from volume and time only. Tracer concentrations need be determined only in dimensionless relative readings. In rock-strewn shallow streams, the dilution method may provide the only effective means of estimating flow. The main disadvantages of the method are the difficulties in obtaining complete mixing of the tracer without loss of tracer and the problem of obtaining permission in some countries to inject tracers into rivers. There are two basic injection techniques, several sampling techniques and a large number of possible tracers of three main types – chemical, fluorescent and radioactive. The technique is normally carried out by specially trained personnel and although the method is mostly used for smaller rivers, discharges of up to 2000 m^3 s^{-1} have been measured with confidence.

11.2 Principle

The basic principle of the dilution method is the addition of a suitably selected tracer to the flow. Downstream of the injection point, when dispersion throughout the flow is effected, the discharge may be calculated from the determination of the dilution of the tracer.

If this tracer were present in the flow before the injection, the increase in concentration of tracer due to the injection is known as the **concentration of added tracer**. The following methods permit the calculation of the discharge without regard to the prior concentration of tracer, analogous with background noise, as long as this can be assumed constant during the measurement.

Constant rate injection method

A solution of concentration C_1 of a suitably chosen tracer is injected at a constant rate, q, into a cross-section located at the beginning of the measuring reach of the channel in which the discharge, Q, remains constant for the duration of the gauging. At a second cross-section downstream from this reach, at a sufficient distance for the injected solution to be uniformly diluted, the concentration is measured for a sufficient period of time and at a sufficient number of points to enable a check to be made that good mixing has been obtained and that the concentration of added tracer, C_2, has attained a constant value. Under these conditions, if all of the tracer injected passes through the sampling cross-section, the mass rate of tracer at the injection point is equal to that passing through the sampling cross-section.

The mass rate at which the tracer enters the test reach is

$$qC_1. \tag{11.1}$$

Similarly the rate at which the tracer leaves the test reach is

$$(Q + q)C_2. \tag{11.2}$$

Equating these two rates

$$qC_1 = (Q + q)C_2 \tag{11.3}$$

from which

$$Q = \left(\frac{C_1 - C_2}{C_2}\right)q \tag{11.4}$$

where q and Q are in litre s^{-1} or m^3 s^{-1}, and C_1 and C_2 are in mg litre^{-1} or kg m^{-3}. Equation (10.4) can usually be simplified because, in general, C_1 is much larger than C_2. Then

$$Q = \frac{C_1}{C_2}q. \tag{11.5}$$

The discharge can therefore be determined by comparing the concentration of the injected solution with that measured at the sampling cross-section. C_1/C_2 is termed the dilution ratio, N.

Figure 11.1 shows a constant rate injection gauging by the multipoint injection method.

Figure 11.1 Constant rate injection dilution gauging by the multipoint injection method.

Integration ('gulp' or sudden injection) method

A volume V of a solution of concentration C_1 of a suitably chosen tracer is injected over a short period into a cross-section located at the beginning of the measuring reach, in which the discharge Q remains constant for the duration of the gauging. The injection is often performed by a simple steady emptying of a flask of tracer solution. At a second cross-section downstream from this reach, beyond the mixing length (Section 11.4), the concentration of added tracer, C_2, is determined over a period of time sufficiently long to ensure that all the tracer has passed through the second (sampling) cross-section.

If all the tracer injected passes through the sampling cross-section, the discharge is calculated from the following equation

$$M = VC_1 = Q \int_{t_0}^{\infty} C_2(t) \, dt \qquad (11.6)$$

where

> M is the mass of tracer injected;
> V is the volume of injected solution;
> C_1 is the concentration of tracer in the injected solution;
> Q is the river discharge;
> $C_2(t)$ is the concentration of added tracer at the fixed sampling point over the time interval dt;
> t is the elapsed time, taking as the origin the instant at which the injection started;
> t_0 is the time interval of the first molecule of tracer at the sampling cross-section.

Equation (11.6) requires that the value of the integral

$$\int_{t_0}^{\infty} C_2(t)\, dt \qquad (11.7)$$

be the same at every point of the sampling cross-section. This condition is satisfied if the injected solution is well mixed with the flow in the river. In practice, the presence of the tracer is no longer detectable at any point in the sampling cross-section at a certain time $t_0 + T$. The value of T is known as the time of passage of the tracer cloud through the sampling cross-section. Let

$$\bar{C}_2 = \frac{1}{T}\int_{t_0}^{T} C_2(t)\, dt. \qquad (11.8)$$

The practical condition of good mixing is that $\bar{C}_2$ is identical at all points in the cross-section, hence from equations (10.6) and (10.8)

$$Q = \frac{V_1}{T}\frac{C_1}{\bar{C}_2}. \qquad (11.9)$$

Figure 11.2 shows a radioactive 'gulp' injection dilution gauging.

General equations for the calculation of discharge

Equations (11.5) and (11.9) can be written in the general form

$$Q = KN. \qquad (11.10)$$

Putting $N = C_1/C_2$, the dilution ratio for measurement by the constant rate injection method

$N = C_1/C_2$, the dilution ratio for measurement by the integration method;

Figure 11.2 Dilution gauging by the radioactive gulp injection method.

$K = q$ in measurement by the constant rate injection method;
$K = V/T$ in measurement by the integration method.

The dilution ratio (or factor), N, which is the fundamental concept of the dilution method, is determined by comparative analysis of samples of the injection solution and samples taken from the sampling cross-section. Figure 11.3 shows typical curves for constant rate and integration methods. The former may be considered as a large number of small sudden injections made at equal intervals of time. If the pulse from a sudden injection is drawn for each small sudden injection at the appropriate time, and the resulting curves summed vertically, the shape of the curve for the constant rate injection is obtained.

In Fig. 11.3 the plateau indicates where the chemical tracer has attained a steady volume of C_2 which is substituted in equation (11.5) to compute discharge. In the case of the integration method, the area under the curve is determined in order to evaluate the integral

$$\int_t^T C_2(t)\ \mathrm{d}t. \tag{11.11}$$

The application of the discharge equations for both methods may be demonstrated by examples.

Example 1: constant rate injection method

Results of a constant rate injection are as set out. Calculate the discharge Q.

$$q = 7.5\ \mathrm{ml\ s^{-1}}$$
$$C_1 = 5.67\ \mathrm{g\ l^{-1}}$$

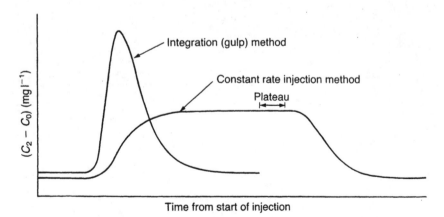

Figure 11.3 Typical time–concentration comparison curves for the constant rate injection method and the integration method.

$C_2 = 0.05 \text{ mg l}^{-1}$.

From equation (11.4)

$$Q = \frac{(C_1 - C_2)}{C_2} q \text{ m}^3 \text{ s}^{-1}$$

$$q = \frac{7.5}{1000} \text{l s}^{-1}$$

$$C_2 = \frac{0.05}{1000} \text{g l}^{-1}.$$

Therefore

$$Q = \frac{5.67 - 0.05/1000}{0.05/1000} \times 7.5/1000 \text{ l s}^{-1}$$

$$= 850.5 \text{ l s}^{-1}$$

or

$$Q = 0.851 \text{ m}^3 \text{ s}^{-1}.$$

Note: equation (11.5), $Q = (C_1/C_2) \, q$, also gives $Q = 0.851 \text{ m}^3 \text{ s}^{-1}$.

Example 2: integration (gulp) method

Two gaugings by the integration method are shown graphically in Fig. 11.4. The background concentration of tracer in both cases was zero; the ordinate therefore represents the sampling concentrations C_2. The samples for the first gauging are shown by circles and for the second gauging by crosses.

The details for each gauging are as follows. Calculate the discharge Q.

First gauging

$$V = 2.91 \times 10^{-3} \text{ m}^3$$

$$\left(\frac{C_2}{C_1}\right)_{\text{mean}} = 3.10 \times 10^{-6}$$

$$t = 15 \text{ min}$$

$$= 900 \text{ s}.$$

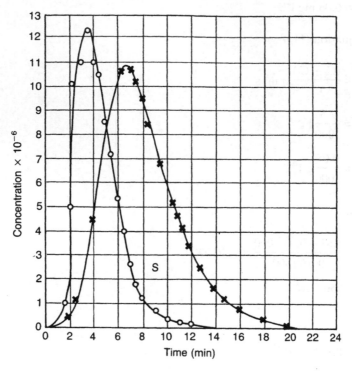

Figure 11.4 Examples of time–concentration curves for the integration method using multiple samples.

Second gauging

$$V = 10.61 \times 10^{-3} \, \mathrm{m}^3$$

$$\left(\frac{C_2}{C_1}\right)_{\mathrm{mean}} = 3.29 \times 10^{-6}$$

$$t = 24 \, \mathrm{min}$$

$$= 1440 \, \mathrm{s}.$$

First gauging

$$Q = \frac{V}{T}\frac{C_1}{C_2} = \frac{2.91}{900} \times \frac{10^6}{3.10} \times \frac{1}{10^3} \ (\text{equation (11.9)})$$

$$= 1.04 \, \mathrm{m}^3 \, \mathrm{s}^{-1}.$$

Second gauging

$$Q = \frac{V}{T}\frac{C_1}{C_2} = \frac{10.61}{1440} \times \frac{10^6}{3.29} \times \frac{1}{10^3}$$

$$= 2.24 \, \mathrm{m}^3 \, \mathrm{s}^{-1}.$$

Example 3: integration (gulp) method

Results of an integration method are as follows. Calculate the discharge Q.

Weight of sodium dichromate = 1500 g
Sampling interval $t^1 = 15$ s
Number of samples $n = 39$
$\bar{C}_2$ (average of 39 samples) = 5.24 mg 1^{-1}.

Equation (11.9) may be written

$$Q = \frac{M}{nt^1(\bar{C}_2)}.$$

Then

$$Q = \frac{1500}{39 \times 15 \times 5.24} \text{ m}^3 \text{ s}^{-1}$$

$$= 0.5 \text{ m}^3 \text{ s}^{-1}.$$

Note: 5.24 mg 1^{-1} = 5.24 g m^{-3}.

11.3 Tracers

The various substances chosen as tracers are selected for their properties which provide ease of detection at low concentrations. The ideal properties of the tracer are:

(a) It is soluble in water;
(b) It is not adsorbed on suspended solids sediments, bed and banks, sample containers, etc.:
(c) It does not react chemically with any of the surfaces in (b);
(d) It is stable in dilute solution in natural waters under the influence of light and temperature. This permits a delay in chemical analysis which may arise from the remoteness of some gauging sites;
(e) It does not have any harmful effects on human health or adverse effects on flora and fauna, particularly fish;
(f) It is readily detectable above the background level at concentrations which are compatible with the accuracy of the measurement desired and the quantity of the tracer it is convenient to inject;
(g) Its use for streamflow measurement is recognised by the various national regulations applying to this procedure.

Types of tracer

There are three main types of tracer used in dilution gauging: (a) chemical, (b) fluorescent and (c) radioactive.

Chemical tracers

The usual tracers are sodium chloride, in the form of common salt, sodium dichromate, lithium chloride, sodium nitrate and manganese sulphate.

The cheapest and most convenient tracer to use in many countries is common salt (NaCl), preferably fine-grained table salt which dissolves quickly. The amount of salt required per m^3s^{-1} of stream discharge depends on the mixing length. A long reach requires more than a short reach for the same discharge. The background conductivity (natural conductivity measured when no salt solution is present) of natural water also affects the minimum amount of salt that can be used. As a rule of thumb, 0.2 kg of salt per m^3s^{-1} of discharge is considered sufficient for natural waters with low background conductivity. Under good conditions, however, discharges of up to 140 m^3 s^{-1} have been measured by the use of not more than 12 kg of salt (86 g of salt per m^3 s^{-1} of discharge).

The minimum concentrations of common salt which can be measured with an accuracy of the order of ±1% by the conductivity meter method of analysis are shown in Table 11.1 against the conductivity of natural water. When the stream has a very low conductivity, the minimum concentrations can be reduced to about 10% of those indicated in the table.

Common salt has a solubility of 3.6 kg to 10 litres at 15°C but under field conditions not more than 2.5 kg is generally used in 10 litres of water. The final minimum concentrations at which common salt should be present at the down-stream sampling cross-section is 2.0 mg litre^{-1} (0.002 kg m^{-3}).

Colorimetric analysis permits the measurement of very low concentrations of sodium dichromate. With final concentrations between 0.2 and 2 mg litre^{-1}, the accuracy of the analysis depends upon the concentration used, and the sensitivity and accuracy of the colorimetric apparatus. The solubility of sodium dichromate in water is relatively high at 600 g litre^{-1} (600 kg m^{-3}).

In the UK, lithium chloride has also been used as a tracer and has a solubility of 637 g litre^{-1} at 0°C; the element lithium can be detected in concentrations down to about 10^{-4} mg litre^{-1} in the laboratory with specialised flame photometers.

Table 11.1 Minimum concentration of NaCl that can be accurately detected

	Conductivity of natural water ($\mu S \times 10^{-3}$)			
	100	20	10	2
Minimum concentration of NaCl measurable with ±1% accuracy (kg m^{-3} × 10^{-3})	10	2	1	1

Fluorescent tracers

Fluorescent dyes, particularly the green dye fluorescein, are used both as mixing indicators and as tracers. However, these dyes are not generally suitable for dilution gauging although they are used for time of travel, retention or dispersion studies. Their main disadvantages are photochemical decay and a tendency to adsorb on suspended solids or surfaces at very low concentrations. Their main use, however, is in checking the passage of a colour cloud in the river which can be checked visually, particularly in preliminary tests to estimate the delay period and mixing distance before sampling. Dyes such as Rhodamine-WT and Sulpho-Rhodamine B Extra can be injected at concentrations of tens of grams per litre and detected at tens of nanograms per litre, depending on the extent to which background substances fluoresce in water. They can therefore be used to assess tracer loss at a second sampling station provided high accuracy is not required. Although fluorescent dyes can be detected in samples that are colourless to the human eye, they can cause taste problems and may colour fish flesh. It is unlikely therefore that permission to use dyes in unpolluted rivers will be readily granted.

Radioactive tracers

Radioactive tracers have considerable advantages when high discharges require to be measured. The injection solution may range in concentration up to tens of curies per litre and most isotopes are accurately detectable down to tens of picocuries per litre, so that significant dilutions are possible as well as flexibility with regard to injection techniques. When chemicals are used, large systems may require such large volumes of saturated solution that difficulties in handling the injection may occur and layering of the dense solution in the river may restrict the degree of mixing.

The two most successful radioisotopes for dilution gauging are probably bromine-82, which may be obtained as irradiated potassium bromide tablets, and tritium in the form of tritiated water (HTO).

Background levels of bromine-82 are usually zero but surface waters at the present time contain levels of tritium of several tens of picocuries per litre (about one order of magnitude above natural levels) caused by the contamination of the hydrosphere from nuclear bomb tests in the 1960s.

Bromine-82 is short-lived with a half-life of only 35.4 h (time taken for the activity to decrease to one half of its initial value) and has been used in a large range of streamflow measurements. It disappears at a predictable rate by radioactive decay and cannot cause prolonged radioactive pollution or contamination.

However, tritium, although it can be considered the ideal water tracer since it is in the form of a water molecule, is long-lived with a half-life of 12.3 years. It may therefore be considered as a pollutant and should be used only for special studies. The concentrations of both isotopes at the sampling station can be kept less than an order of magnitude below drinking water tolerance levels set for these isotopes.

Other isotopes for dilution gauging include sodium-24 (half-life 15 h) and iodine-131 (half-life 8 days), but preliminary tests of both require to be made to assess adsorption losses.

The use of radioactive tracers for dilution gauging requires personnel specially trained in this technique and the method should not be considered without this training.

11.4 Selection of measuring reach

The length of measuring reach is chosen, if possible, to be as short as possible but may be as much as several kilometres in some large rivers. In any event the reach must be of a length at least equal to the mixing length. It is necessary to choose a reach where the river is as narrow and as turbulent as possible, where there are no dead water zones, and where there are numerous transverse currents to promote good lateral mixing. Vegetated zones and zones where the river separates (bifurcates) into branches are avoided. In all cases it is advisable to inject into a zone of high velocity, if possible towards the opposite bank if the injection is performed from one bank (Fig. 11.5).

The fundamental condition for the application of the dilution method is that the tracer should be perfectly mixed with river water at the sampling cross-section. This condition can be expressed by

$$\int_0^\infty C_2(t)\,\mathrm{d}t = \text{constant at all points of the cross-section for the method of integration.}$$

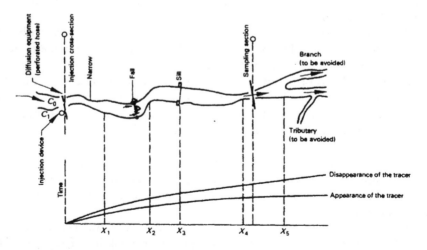

Figure 11.5 Selection of measuring reach.

For the constant rate injection method, this expression reduces to the simpler condition that C_2 is constant over the cross-section at the steady-state plateau condition.

The mixing length is not a fixed value but varies according to the admissible variation in concentration; the smaller the admitted variation, the greater the mixing length. A first trial can be carried out using fluorescein as stated previously.

A concentrated solution of this dye can be injected over a relatively short time at a point situated at the beginning of the measuring reach. Visual examination of the dispersion of the solution makes it possible to determine whether there are any dead water zones or other zones of loss of tracer, and is a first indication of the smallest distance which must separate the point of injection and a convenient sampling cross-section.

This procedure can be improved by making a continuous injection of the fluorescent tracer, whose transverse distribution of concentration is studied at different sections downstream with the aid of a portable fluorimeter, including flow close to the banks.

Mixing length

There are a number of empirical equations for estimating the mixing distance L. All of them are approximate and are normally used for preliminary surveys of the reaches available or the reach selected. The following empirical equations however, have been used successfully in many instances

(a) $L = bQ^{1/3}$ (m)   (11.12)

  where L = distance between the injection site and the sampling station (mix-
     ing length) (m);
    $b = 14$ for mid-stream injection;
    $b = 60$ for injection from one bank;
    Q = stream discharge (m^3 s^{-1}).

(b) $L = \dfrac{0.13\bar{b}^2 C(0.7C + 2\sqrt{g})}{g\bar{d}}$ (11.13)

  where $\bar{b}$ = average width of the channel (m);
    $\bar{d}$ = average depth of the channel (m);
    C = Chezy coefficient (see Chapter 8 for values).

Completeness of mixing

The following equations may be used to assess the completeness of mixing x, say, which is directly related to uncertainty in discharge:

(a) Integration method

$$x = \frac{100 \sum_{1}^{n} (U_i - \bar{U})}{2n\bar{U}} \tag{11.14}$$

where

$$U_i = \int_{t_0}^{\infty} C_{2i}(d) \, dt \tag{11.15}$$

$$\bar{U} = \frac{1}{n} \sum_{1}^{n} U_i \tag{11.16}$$

n = number of sampling points.

(b) Constant rate injection method

$$x = \frac{100}{2n\bar{U}} \sum_{1}^{n} (C_{2i} - \bar{C}_2). \tag{11.17}$$

The value of x in equations (11.14) and (11.17) is determined after field measurements have been carried out, and the equations are valid for values of x less than ten. The expected uncertainty in discharge is about $2x$ so it is important to have the smallest possible value of x for the dilution measurement.

The equations for mixing length ((11.12) and (11.13)) are associated with a value of $x = 1$.

Particular cases

Dead water zones

If dead water zones are present, they can detain the tracer and release it only very slowly. In the case of the constant rate injection method, the result is that the time for the establishment of the plateau concentration is prolonged, and as a consequence so is the time taken for the gauging. In the case of the integration method the time of gauging is also prolonged as it is necessary to continue measurements until all the tracer retained in the dead water zones has passed through the sampling cross-section. This can cause an error in the measurement, because a non-negligible quantity of tracer can pass the downstream section at the end of the test, when concentrations are too low to be measured.

Inflow of water into the measuring reach

Measurements can be made when there is an inflow of water (tributaries or springs) in the measuring reach only if good mixing is achieved in the sampling cross-section. The measured discharge then includes the intermediate inflows of water.

Abstraction or leakage from the measuring reach

If there is a leakage or abstraction between the injection cross-section and the sampling cross-section, the result will be questionable except when the location of this leakage or abstraction is known and is situated at a point where good mixing has already been achieved; in this case, analysis of samples will lead to a value of the discharge of the river upstream of this leakage or abstraction and not to the value of the discharge of the channel in the sampling cross-section.

11.5 Procedure

Constant rate injection method

Duration of injection

The duration of the injection is fixed so that a steady concentration exists for an adequate period, generally 10–15 min, at the sampling station. It will, of course, vary directly as the length of measuring reach, and the extent of the dead water zones and inversely as the mean velocity of the river, which is generally related to the degree of turbulence.

For a given discharge, observations of the moment of arrival and disappearance of tracer in each section may be plotted as curves 1 and 2 in Fig. 11.6. On the other hand, this experiment can be used to determine the minimum mixing length. The selected sampling section corresponding to the mixing length consistent with the precision sought is represented by the straight line S.

If it is desired to obtain a steady regime for a time, Δt, at the selected sampling cross-section, it is necessary to add the time Δt to the time t corresponding to the disappearance of the tracer in the section (Fig. 11.6) and to draw, through the point obtained in this way, a curve 1' parallel to curve 1. The intercept at zero of this curve gives the minimum duration of injection θ. Curve 2' parallel to curve 2 determines the end of the passage of the tracer.

Preparation

The solution to be injected is preferably prepared in a tank other than the supply tank (Figs. 11.7 and 11.8). The supply tank should be of sufficient capacity to avoid the need for the addition of solution during an injection. The solution is drawn from a level above the bottom of the tank to prevent sediment being carried out with the injected solution. No attempt is made to obtain a

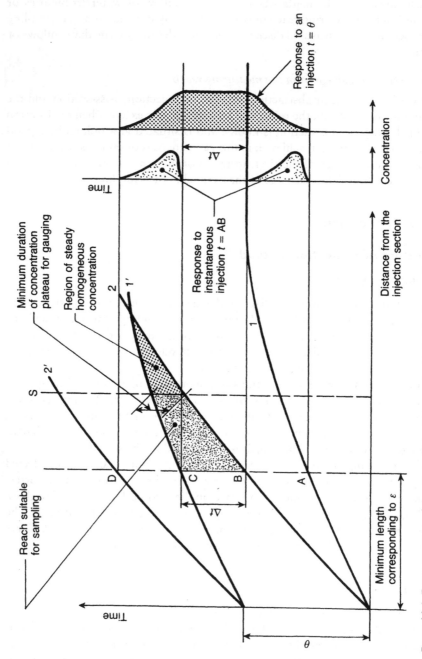

Figure 11.6 Determination of duration of injection: constant rate injection method.

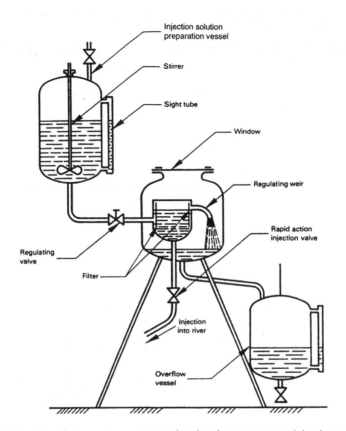

Figure 11.7 Injection device using a constant head tank: constant rate injection method.

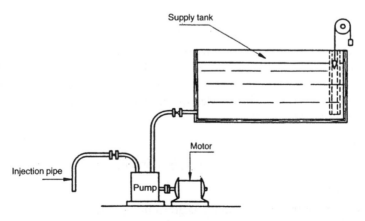

Figure 11.8 Injection arrangement using a rotary volumetric pump: constant rate injection method.

fully saturated solution and no particle of undissolved chemical substance should be present in the supply tank when the solution is drawn off.

Water taken from the open channel is preferably used for preparing the concentrated solution.

The concentration of the injected solution should be homogeneous. Homogeneity of the solution is achieved by vigorous mixing performed with a mechanical stirrer or with a pump.

Injection of concentrated solution

The solution to be injected is introduced into the stream at the chosen injection cross-section. The concentrated solution is injected at a constant rate of flow and is controlled by one of the following devices:

(a) constant head tank
(b) volumetric pump driven by a constant speed motor
(c) Mariotte vessel
(d) floating siphon.

Constant head tank (Fig. 11.7)

The constant head tank is made of such dimensions and volume as permits suitable arrangements to be made internally to avoid any short-circuiting of flow between the inlet and outlet pipes. The crest of the weir controlling the water level in the tank should be sharp, horizontal and of sufficient length to control the head on the outlet pipe. The weir should extend completely across one side of the tank to avoid contractions and preferably be situated immediately above the outlet pipe. Arrangements are included for the collection of any solution which overflows from the tank during a test. A range of injection flow rates may be obtained by using a calibrated set of terminal orifice plates or nozzles of appropriate dimensions.

Volumetric pump driven by a constant speed motor (Fig. 11.8)

A volumetric pump is used of the rotary type capable of delivering a constant discharge irrespective of pressures at inlet and outlet. Care should be taken to ensure that the speed of the pump is constant, the speed being determined either directly or from a knowledge of the supply frequency if driven by a synchronous electric motor.

Mariotte vessel

An illustration of the Mariotte vessel is given in Fig. 11.9.

Floating siphon

An illustration of the floating siphon is given in Fig. 11.10.

When the injection rate is calculated from the volumetric capacity of the pump or from the geometric details of one of the other devices referred to

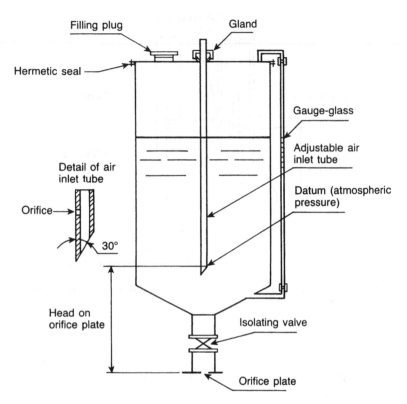

Figure 11.9 Mariotte vessel for injecting concentrated solution: constant rate injection method.

above, a second means of measurement is incorporated in the injection apparatus to serve as a check on the accuracy and consistency of the measurements.

Rate of injection

The rate of injection of the concentrated solution is normally measured by direct observation if the regulating device has been previously calibrated. Where the rate of injection is controlled so that it is constant but not exactly determined, it is ascertained indirectly by measuring the total volume of solution discharged over a measured period of time.

The degree of accuracy with which the rate of injection can be measured depends on the particular measuring device used. The rate of injection is determined from readings taken during the injection period of one of the following characteristics depending on the device used:

(a) level of liquid in the constant head tank in relation to the terminal orifice;

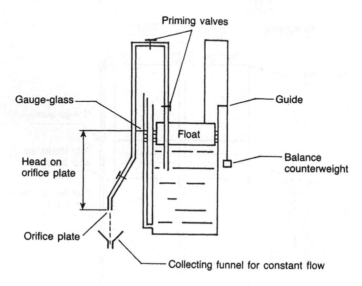

Figure 11.10 Floating siphon for injecting concentrated solution: constant rate injection method.

(b) speed of the pump;
(c) level of the orifice in relation to the atmospheric pressure datum in the Mariotte vessel;
(d) height between the solution surface and the orifice of the floating siphon arrangement;
(e) reading of the flow meter.

For (a), (c) and (d) the rate of injection of the concentrated solution can then be determined from the calibration curves for these devices. The devices are calibrated both before and after the tests. If a volumetric pump is used as in (b) it is advisable that a calibration check be carried out before and after the tests to check that there is no leakage, or alternatively to determine the leakage if this exists, i.e. to calibrate the pump under the same conditions as those experienced during the test.

A flow meter (e) may be installed in the injection apparatus for the specific purpose of measuring the rate of injection. It is calibrated before and after the test with a meter installed in the apparatus. The average of the two calibrations of the device carried out before and after the test is used in the calibration of the flow meter.

Indirect measurement of the rate of injection involves the determination of two factors, the duration of the discharge and the total volume of solution delivered.

The volume may be measured by observing the drop in level in a previously calibrated feed tank or by the use of a positive displacement meter.

Calibrated feed tank

Where the volume of concentrated solution is to be measured in a feed tank, the tank is calibrated and erected so that the vertical axis of the tank is vertical.

Alternatively, the height–volume relationship may be determined by the use of a smaller calibrated vessel, the contents of which are added to the main tank and the height measured in successive steps.

Where the rate of injection is controlled by a constant head tank, the spillage therefrom is collected and measured in a vessel, calibrated to an accuracy depending on the ratio of the amount of spillage to the amount injected. Alternatively, the spillage may be collected and returned to the main tank before the final reading is taken.

Positive displacement meter

Where the volume of concentrated solution is measured by means of a volumetric meter, of a positive displacement type, the meter is connected in the pipeline between the constant head tank and the injection point, and is of such size and so placed in relation to the constant head that sufficient operating head is available to sustain flow through the meter to the injection point at the greatest injection rate required.

Precautions need to be taken to remove all suspended impurities from the concentrated solution before it is delivered to the constant head tank.

The volume of solution delivered is ascertained from the difference between the readings on the meter index before and after the injection.

Sampling

Samples are taken as follows:

(a) *Upstream from the injection station.* At least two samples are taken before and after the injection. If any variation of the background concentration is anticipated along the measuring reach, the samples of the natural water are taken at the sampling station before and after the passage of the injected solution. Care should be taken to ensure that the whole of the injected chemical substance has passed before the second set of samples is collected.

(b) *At the outlet of the injection apparatus.* Three to five samples are taken either just before and just after the injection period or, alternatively, during the injection period. Sampling during the period of injection will affect the total amount and constancy of the injection rate but generally this source of error is negligible.

(c) *At two or three points in the sampling cross-section.* Five to ten samples at each point are taken at regular intervals during the period of steady concentration. These samples are taken by immersing bottles or by pumping. With a view to avoiding the influence of more or less rapid chance variations of the concentration in the sampling section, the time interval between taking successive samples should be as short as possible.

Figure 11.11 shows injection equipment inside a truck for the constant rate injection method; the equipment consists of a small petrol-driven pump for a radioisotope injection and a hosepipe to transport the isotope to the channel.

Integration method

A solution of concentration C_1 is prepared with the necessary mass of tracer, M, with the objective of obtaining a volume suitable for the condition of the injection. It may be possible to select an appropriate solvent other than water for this operation.

Figure 11.11 Complete injection equipment for a dilution gauging, housed in a truck.

For a single measurement of discharge it is possible to use a single device for:

(a) the exact determination of the injected volume V;
(b) the mixing conditions for sampling for the determination of C_1;
(c) the complete and rapid injection of all the measured volume.

For a series of comparable measurements, it is possible to separate the operations of mixing and injection by using several vessels. It is practical, for example, to carry known volumes of the same injection solution in conveniently chosen sealed vessels.

Determination of the injected volume

Three possibilities are available:

(a) use of a calibrated reservoir;
(b) geometrical determination of the volume of the injection solution contained in the reservoir;
(c) measurement by weighing.

Method (c) is the most precise. To determine the volume in this case requires the use of the density of the injected solution at the given temperature, which is obtained by laboratory methods, using a sample of the injected solution.

Injection of the solution

The duration of the injection should be as short as possible, so as to reduce the duration of sampling and the mass of tracer required for a given precision.

However, by extending the duration of injection of the tracer, possibly to several minutes, the maximum concentrations around the point of injection are reduced, and this reduces the effects of density segregation.

Sampling

For the purposes of the measurement, it is necessary to provide the following samples:

(a) two or three samples taken in the measuring cross-section before the injection and upstream from the injection cross-section before and after the measurement to verify that the initial concentration C_0 in the channel remains constant during the measurement;
(b) one or two samples of concentration C_1 of the injection solution.

In the measuring section, it is possible to proceed either by continuous recording, or by sampling. In all cases it is advisable that sampling is performed in at least three points (left bank, centre and right bank) to verify that good mixing has been achieved. This procedure is important for a measurement in an unknown reach or in a known reach under new conditions. For routine measurements where experience can guarantee the quality of mixing, a single sample may be sufficient.

Methods of sampling during the passage of tracer

1 *Continuous recording.* In the conductivity method, and where variations in temperature and natural conductivity of the water are negligible, it is possible to record continuously the conductivity of the water during the passage of the tracer. Continuous recording is also possible when fluorescent or radioactive tracers are used.

Two methods are normally used. Either the sensing element of the measuring apparatus may be directly immersed in the flow or it may be placed in a diversion channel fed by sampling the flow at a reasonably constant rate from a fixed point.

2 *Collection of discrete samples.* Two procedures are used:

(a) Collection of quasi-instantaneous samples at known and recorded times. A minimum of 25 consecutive samples is taken at each point of the cross-section. After analysis, the curve of variation of C_2 with time (or of C_1/C_2 according to the method of analysis) may be traced and planimetered.
(b) Collection of quasi-instantaneous samples at equal and known intervals of time. This procedure leads to the tracing of the curve but does not require the recording of the time of each sample.

3 *Collection of mean sample.* Two procedures are used:

(a) Sampling at constant rate. Several devices can be used to obtain a sample at constant rate, during the passage of the tracer. The simplest method is to siphon through a flexible pipe (inside diameter about 5 mm), provided with a strainer, into a vessel of sufficient volume to accommodate the complete sample. If the site does not lend itself to siphoning operations, a volumetric pump may be used.
(b) Collection of quasi-instantaneous samples at equal intervals of time and mixing of these samples. To approach the conditions in (a) above, the interval of time is reduced; the rate of sampling can reach 20 per minute in some cases.

Experience shows that departures of 5% in the volumes of the various samples do not introduce a significant error in the calculation of the flow if the number of samples is large.

The collection of quasi-instantaneous samples at equal intervals of time can be replaced by the collection of mean samples over the same time intervals. This procedure is less sensitive to variations in the rhythm of sampling and to irregularities in the rate of sampling or in the curve $C_2 = f(t)$.

4 *Obtaining several mean samples beginning at the same instant.* When there is reason to believe that sampling will not continue sufficiently long for all the solution to pass through the sampling cross-section, it is advisable to proceed as follows.

If t is the estimated time of passage of tracer to the sampling cross-section, n mean samples can be taken at constant rate from the same fixed point in the cross-section; these n samples have the same origin of time before the arrival of the tracer and terminate at times $t_1, t_2, \ldots, t_n$.

The most reliable solution, which will, however, necessitate the use of larger quantities of tracer of lower concentrations for the samples finishing at times t_2, $t_3, \ldots, t_n$, is to adopt the value $t_1 = t$. In this case, if t has not been underestimated, all the samples will lead to the true value of discharge. If $\bar{D}_i$, is the dilution determined by analysis for the sample which terminates at time t_i

$$Q = \frac{V}{t_1} \bar{D}_1 = \frac{V}{t_2} \bar{D}_2 = \ldots = \frac{V}{t_n} \bar{D}_n. \qquad (11.18)$$

If t has been underestimated, the apparent discharge calculated from the first samples will be too large, but certain samples such as $t_2, \ldots, t_n$ will give the true value. In the worst case, an extrapolation of the curve $Q(t)$ to its asymptotic value will lead to the estimated true value of the discharge.
A more economical solution is to adopt, for example

$t_3 = t$ with $t_1 < t$ and $t_2 < t$

$t_4 > t$ $t_5 > t$ $\ldots,$ $t_n > t.$

If t is correctly evaluated, the analysis of samples $t_3, t_4, \ldots, t_n$ will lead to the true value of the discharge; t_1 and t_2 will lead to too large a value (t_3 will be the concentration considered as optimal for analysis; t_4, t_5 and t_n will have lower concentrations).

If t has been underestimated, t_5 and t_6 or an extrapolation of the curve $Q(t)$ will lead to the estimated true value of discharge.

If t has been overestimated, all the samples will lead to the true value of discharge.

5 *Collection of several successive mean samples.* An alternative procedure, which requires simpler equipment and has the advantage of distinguishing between the systematic decline in $Q(t)$ and random errors in the analysis, is to

obtain average samples over the intervals zero to t_1, t_1 to t_2, etc., and to analyse these samples individually.

If t_1 is chosen to be equal to t and if t is an underestimate of the real value, then the concentration of the sample taken between t_1 and t_2 will be greater than zero, but later samples will have concentrations tending to zero.

If the sample taken between t_n and t_{n+1} is the last that is significantly different from zero, and the samples represent dilutions $D_1, D_2, \ldots, D_n$, then

$$Q = V/[t_1/D_1 + (t_2 - t_1)/D_2 + \ldots + (t_{n+1} - t_n)/D_{n+1}]. \qquad (11.19)$$

The advantage of this method is that the time t_{n+1} at which the passage of tracer is complete is more easily and reliably estimated. The disadvantage is that much reliance is placed on the first samples.

Whichever method is chosen, it is essential that the sampling point remains the same for all the samples or throughout the duration of sampling.

It may be desirable, particularly in the case of (3) above, to have a detector placed *in situ* upstream from the measuring cross-section to indicate the time at which the first sample must be taken. The distance at which the detector is placed should be sufficient for sampling to start before the appearance of the tracer at the sampling cross-section.

The methods of sampling in (3) above are particularly recommended as they involve the least cumbersome analysis. On the other hand, the methods described in (2) above can involve the use of lighter equipment in the field, and offer the possibility of analysing the effectiveness of discrete sampling methods.

In all cases it is important that the duration of the sampling operation is made at least equal to and synchronous with the duration of passage of the tracer cloud through the measuring cross-section.

11.6 Comparison between the two dilution methods

The differences between the constant rate injection method and the integration method are concerned principally with the necessary equipment and field operations. The advantages and disadvantages can be grouped into the three categories below.

General

The constant rate injection method provides more information on the quality of mixing and the possible effects of a variation in discharge; it is generally possible to obtain as many samples on the plateau of concentration and distributed in time and space as are necessary for verification.

For a fixed duration of injection, Fig. 11.6 shows that the duration, T, say, of the plateau, will not be sufficient for all cross-sections close to the intersection of curves 1′ and 2 and will be zero for all cross-sections downstream from the intersection.

The possible reach is therefore limited for the constant rate injection method. The method of integration, in the absence of loss of tracer by adsorption, is less constrained. For this reason, and for large streams, the integration method will often be preferred.

An advantage of the integration method is the use of a smaller quantity of tracer, and this can be very important in terms of cost or public acceptability, in particular when abstraction of water, downstream from the injection point, requires a certain stream water quality. However, the peak values of concentration of tracer in the river will be higher if, for analytical reasons, the mean value for the integration method is chosen to be of similar magnitude to the plateau value for the constant rate method.

Figure 11.12 Gulp injection of BR-82 in River Thames by the extended gulp technique from truck on bridge. Note hose pipe transporting isotope into river.

The total duration of an integration measurement will always be shorter than that of a constant rate measurement. However, the difference will only be significant for small streams.

Injection of tracer

The constant rate method requires more sophisticated equipment for the injection of tracer to obtain a constant and measurable discharge, but conversely there is no need for precise measurement of times or volumes in the field.

The integration method, on the other hand, requires no compiex injection equipment, and this is an important advantage, particularly when highly active radiotracer solutions are to be injected. Nevertheless, the volume injected must be known precisely. Moreover, the higher concentrations of tracer encountered

Figure 11.13 A mountain stream (Schwarzach River) in the Austrian Alps where the dilution method would be most suitable for streamflow measurement.

Figure 11.14 Measurement of discharge of small stream with conductivity probe.

near the injection point may cause problems of toxicity and density segregation. These are diminished by increasing the duration of the injection. In this case the rate of injection of tracer for a non-instantaneous injection need not be constant, provided that a precise measurement can be made of the volume of solution injected. Fig. 11.12 shows a gulp injection in the River Thames.

Sampling and analysis of results

In general, for the integration method, sampling is more complex than for the constant rate method. This is due to the necessity to measure mean values of concentration instead of the quasi-instantaneous values obtained on a concentration plateau. Sampling for the integration method requires either the taking of a large number of samples on a rigid time schedule or the use of sophisticated apparatus. The analysis of results by statistical methods is identical for the two methods but, in general, the constant rate method will supply a larger amount of data, permitting greater precision in the final determination of the discharge.

The dilution method is seldom used for continuous records of discharge at network stations. Nevertheless in certain circumstances, for example the gauging of mountain torrents, it may be the only available method (Fig. 11.13).

For the measurement of small discharges of the order of 1 m^3 s^{-1}, a conductivity probe with added software is now available (Fig. 11.14) for use with sodium chloride which produces good results.

Further reading

ISO 9555/1. *Liquid Flow Measurement in Open Channels: Dilution Methods – General.* ISO, Geneva, Switzerland 1994.

ISO 9555/2. *Liquid Flow Measurement in Open Channels: Dilution Methods – Radioactive Tracers.* ISO, Geneva, Switzerland 1992.

ISO 9555/3. *Liquid Flow Measurement in Open Channels: Dilution Methods – Chemical Tracers.* ISO, Geneva, Switzerland 1993.

ISO 9555/4. *Liquid Flow Measurement in Open Channels: Dilution Methods – Fluorescent Tracers.* ISO, Geneva, Switzerland, 1992.

ISO 11656. *Dilution Methods: Mixing Length in Open Channels.* ISO, Geneva, Switzerland 1993.

Tilrem, O. A. *Manual on Procedures in Operational Hydrology.* Ministry of Water, Energy and Minerals of Tanzania 1979.

White, K. E. Dilution methods In *Hydrometry: Principles and Practices* (ed. R. W. Herschy). John Wiley and Sons, Chichester 1978.

Chapter 12

The ultrasonic method of streamflow measurement

12.1 General

The need for new methods of stream gauging arose basically from the requirements to measure the flow in rivers at locations where conventional methods proved unsuitable. The moving boat method is one such method (Chapter 2), the electromagnetic method another (Chapter 2), and the ultrasonic method a third.

The ultrasonic method is particularly applicable to rivers up to about 300 m or more in width, where (a) there is no stable stage–discharge relation, and (b) a measuring structure is unsuitable or not feasible. The method is therefore appropriate under conditions of backwater from dams, tides, or other causes, and where the installation of a measuring structure would either prove too expensive or sufficient afflux, or head, is not available.

Further important advantages of the ultrasonic method are that there is no constriction or obstruction in the river, the output is an average velocity (or discharge), the system is capable of measuring reversed flow and the measurement is continuous.

Figure 12.1 shows an analogue display taken from an ultrasonic gauge on a river influenced by tides which illustrates the versatility of the system in the measurement of reversed flow. Because of the effect of the tide, Fig. 12.1 shows that downstream flow is gradually reduced until the flow passes through zero and into negative, or reversed, flow. At the same time the stage increases as storage increases. When the tidal cycle reaches its peak the negative flow decreases until it passes through zero again and ultimately both flow and stage return to their former state.

12.2 Principle

Ultrasonic river gauging is based on the continuous measurement of stream velocity at chosen depths by recording the difference in time for sound pulses sent obliquely across the river in opposite directions. Sound waves travelling downstream propagate at a higher velocity than those travelling upstream due

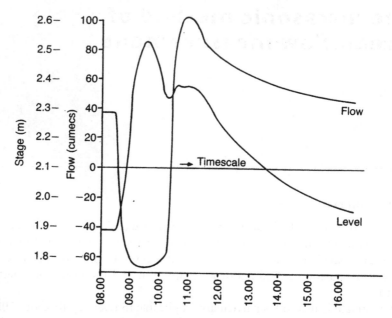

Figure 12.1 Autographic record, produced from paper tape output, showing operation of ultrasonic gauge on river under tidal influence – River Severn at Ashleworth, UK, 15.3.76.

to the component of stream velocity parallel to the acoustic path. Since the stream velocity is much lower than the velocity of sound in water, the difference between upstream and downstream travel time is very small compared to the individual travel times. This requires that an accurate time measurement is made in order to achieve the required accuracy. Transducers are mounted on each bank of the river to transmit and receive these sound pulses. The angle between the transducer line, or flight path, and the direction of river flow is normally made between 45° and 60°. The stream velocity component across the acoustic path can be measured to within ± 0.1%. Simultaneously the average depth of flow is measured using a float recorder or equivalent instrument. The discharge is then calculated by a velocity–area method from the mean velocity component along the flight path, the average depth of flow and the channel width. To smooth out fluctuations in the flow, velocity measurements are taken several times a second and averaged over several minutes.

Single path system

In the single path system, measurement of velocity is made at one depth only with the pair of transducers normally set at 0.6 of the most frequently occurring depth. If the transducers are movable, they are used to calibrate the system by taking line velocity measurements throughout the depth as in the vertical velocity distribution method (Chapter 2). Instead of verticals being used, however, as in the velocity–area method, horizontal paths are scanned at an infinite number of 'verticals'. If the transducers are fixed, a relation is established between flow, as measured by current meter, and the ultrasonic line velocity.

The single path system is normally used when the variation in stage is small for a larger percentage of the time.

Multipath system

The more common multipath system incorporates several pairs of transducers to provide velocity measurements at various water depths and is particularly suitable for rivers having a wide range in stage or with irregular velocity distributions. The transducer assembly is normally installed on a sloping steel section, or beam, set flush with the river banks.

A microprocessor controls the system, selecting which transducer pairs are submerged. Each pair of transducers is energised in turn, starting with the lowest. This sequence is repeated continuously for the prescribed period, usually 15 min, and a mean velocity for each path calculated. A weighted summation of these velocities, allowing for the spacing between paths and the length of each, provides the total discharge. As with the single path system, the velocities, the depth and the discharge are displayed and recorded.

The multipath system, with fixed transducers, is now favoured, by both manufacturer and user. Also, the cost of extra transducers is small in proportion to the total cost of the installation. Normally four to six pairs are satisfactory, although as many as twelve pairs have been used. Figure 12.2 illustrates diagrammatically a multipath transducer assembly on a small channel.

12.3 Theory

Referring to Fig. 12.3, which illustrates diagrammatically the stream velocity components, let

L = path length (m);
$\bar{v}_L$ = line velocity, the velocity of axial flow at the transducer height (m s^{-1});
C = velocity of sound in water (about 1500 m s^{-1});
θ = angle which the acoustic path makes with the direction of flow (degrees);
$\bar{d}$ = average depth of flow along AB (m);
t_{AB} = travel time from A to B;
t_{BA} = travel time from B to A;
$\bar{v}_p$ = component of stream velocity along the acoustic path (m s^{-1}).

There are two common methods of determining V_L:

(a) *Travel time difference method*

$$t_{AB} - t_{BA} = \Delta t = \frac{L}{C - v_p} - \frac{L}{C + v_p}. \tag{12.1}$$

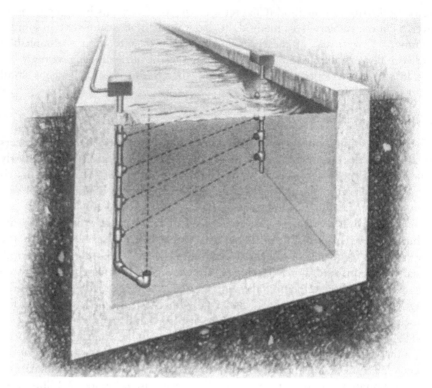

Figure 12.2 Diagrammatic sketch of a 4-path ultrasonic measuring system in a small rectangular channel: also shown is an ultrasonic 'look-up' water-level gauge.

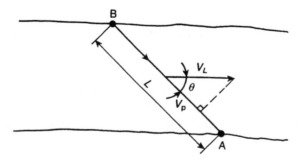

Figure 12.3 Ultrasonic method: velocity components.

Then

$$\Delta t = \frac{2Lv_\mathrm{p}}{C^2 - v_\mathrm{p}^2} \tag{12.2}$$

$$= \frac{2Lv_\mathrm{p}}{C^2} \tag{12.3}$$

where v_p^2 is small compared with C^2 and since

$$C = \frac{L}{t} \tag{12.4}$$

then

$$v_p = \frac{\Delta t L}{2\bar{t}^2} \tag{12.5}$$

where

$$\bar{t} = \frac{t_{BA} + t_{AB}}{2} \tag{12.6}$$

and

$$v_L = \frac{v_p}{\cos \theta}. \tag{12.7}$$

Then

$$v_L = \frac{\Delta t L}{2\bar{t}^2 \cos \theta} \tag{12.8}$$

and

$$Q = A\bar{v}_L = \bar{v}_L \, dL \sin \theta \tag{12.9}$$

where $\bar{v}_L$ is the average line velocity or

$$Q = \frac{L^2 \bar{d} \, \Delta t \tan \theta}{2\bar{t}^2}. \tag{12.10}$$

It will be noted that the solution is independent of the velocity of sound in water.

(b) *Frequency difference method*

From equation (12.1)

$$\frac{1}{t_{BA}} - \frac{1}{t_{AB}} = \frac{2v_p}{L} \tag{12.11}$$

or

$$v_p = \frac{L}{2}\left(\frac{1}{t_{BA}} - \frac{1}{t_{AB}}\right)$$ (12.12)

and from equation (12.7)

$$v_L = \frac{L}{2\cos\theta}\left(\frac{1}{t_{BA}} - \frac{1}{t_{AB}}\right)$$ (12.13)

and if $\bar{v}_L$ is the average velocity (equation (12.9))

$$Q = \bar{v}_L \, \bar{d} L \sin\theta$$

then

$$Q = \frac{L^2}{2}\left(\frac{1}{t_{BA}} - \frac{1}{t_{AB}}\right)\bar{d}\tan\theta.$$ (12.14)

In a single path system, $\bar{v}_L$ is the path velocity when the transducers are located at the depth of flow giving the average velocity in the cross-section. In a multi-path system, $\bar{v}_L$ is the average velocity of the paths. In practice the discharge is computed in various slices by multiplying the velocity by the width (equation (12.9)) and the thickness of the slice and summing these separate measurements (Fig. 12.4).

12.4 Site selection

When making a decision as to whether a site is suitable for an ultrasonic streamflow gauge the following basic factors should be noted:

(a) A reliable source of power is available; a portable system is now marketed, however, which is battery operated.
(b) The channel width normally considered economically feasible is in the range 20–300 m although the system operates outside this range.
(c) The channel is free of weed growth at all times as weeds disperse (attenuate) the beam.
(d) Suspended solids may have a significant effect on the signal attenuation caused by both reflection and scatter from sediment particles suspended in the stream. Generally, as a guide, sections having concentrations of over 1000 mg litre^{-1} for significant periods of time are avoided. If it is necessary to use an ultrasonic gauge under such conditions, the gauge may cease to function temporarily at a certain concentration level but return to normal when that level has fallen.

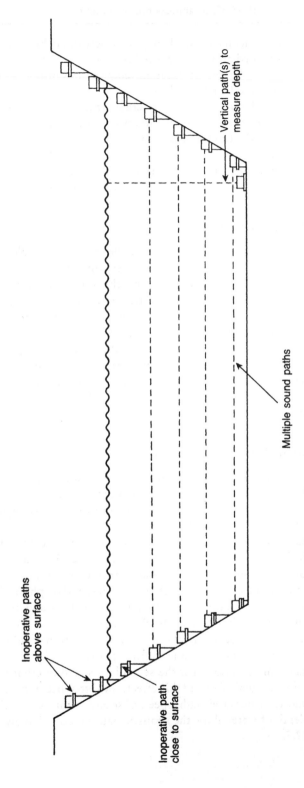

Inoperative paths
above surface

Inoperative path
close to surface

Vertical path(s) to
measure depth

Multiple sound paths

Figure 12.4 Diagrammatic illustration of multipath ultrasonic system. Also shown is an ultrasonic depth gauge.

(e) Attenuation of the acoustic signal may also be caused by reflection and scatter from entrapped air usually caused by rapids or waterfalls.

(f) Upstream inflows to the river at different temperatures or high salt content may cause the beam to be refracted.

(g) The measuring reach is one which is preferably straight and uniform, and at cross-sections in the reach between the upstream and downstream transducers, the velocity distribution is similar.

(h) The bed is stable.

(i) Wide shallow rivers are avoided, otherwise the beam may reflect from the surface or bed and cause gross uncertainties (Section 12.6).

12.5 Operating frequency

If there were no sound absorption in water, the highest possible operating frequency could be used as this increases the system timing accuracy and allows closer spacing between the acoustic path and the surface or bottom of the channel. However, sound absorption by water and scattering by particulate matter and entrapped air increases with increasing frequency. While increasing transmitter power helps somewhat, there is an upper limit achievable and a compromise is therefore reached in determining the operating frequency. Generally the frequency range used in ultrasonic systems varies between 200 and 500 kHz, the lower frequencies being used in wide rivers. Manufacturers, however, are able to modify their systems in order to use the appropriate frequency for the given conditions of the site.

12.6 Minimum depth requirement

As already stated in Section 12.4, special problems arise for the paths nearest to the surface and bed. If the transducers are too near these boundaries, reflected signals from the boundary can interfere with the direct transmission and cause inaccuracies in the time measurement.

Sound is reflected from the water surface and, to a lesser extent, from the channel bed. The bed may even be a net absorber of sound. As an acoustic wave propagates across a river (generally as a cone of around $5°$ width) it will intersect with the water surface and be reflected, suffering a $180°$ phase change in the process. The secondary wave will proceed across the river and arrive at the opposite bank. Its arrival will be sensed by the target transducer later than the direct wave, and the difference in arrival time will be a function of the difference in the respective lengths of the direct and indirect paths.

Errors in signal timing will occur if the secondary signal interferes with the first cycle of the direct signal. To avoid this effect, the difference in the two paths should exceed one acoustic wavelength (speed of sound/frequency). This will be achieved if the depth of water above the acoustic path exceeds that given by the equation (Fig. 12.5)

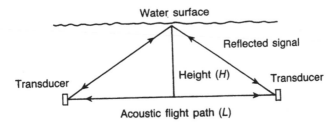

Figure 12.5 Schematic illustration to show signal interference from water surface or bed to determine minimum depth requirement.

$$H = 27 \sqrt{\left(\frac{L}{f}\right)} \tag{12.15}$$

where H is the minimum depth, in metres;

   L is the path length, in metres;

   f is the transducer frequency, in hertz.

A similar restriction may apply to the channel bed, particularly if it is smooth and hence reflects, rather than absorbs, an acoustic signal.

12.7 Crossed paths

A problem which arises in ultrasonic streamflow gauging is the measurement of path angle θ. In the design of an ultrasonic station this is normally a surveying problem and can usually be resolved with acceptable precision. This is particularly the case when the channel is straight, with no bends either upstream or downstream. However, if oblique flow is suspected, usually due to a bend, or bends, upstream the estimation of the path angle often presents a problem. Oblique flow is difficult to diagnose and measure as the oblique angle may vary with width, depth and stage. Figure 12.6 illustrates oblique flow velocity components v'_L in relation to path angle θ and, from the discussion of oblique flow in Chapter 5, if v'_L is the oblique flow at angle ϕ to v_L then the axial velocity is $v'_L \cos \phi$.

Now, if it is assumed that velocities are uniform throughout the reach and normal to the cross-section then (equation (12.7))

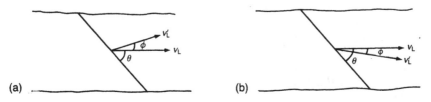

(a)   (b)

Figure 12.6 Schematic diagram to demonstrate oblique flow v'_L: (a) oblique flow φ outside path angle θ; (b) φ within path angle θ.

$$v'_{L} = v_{L} = \frac{v_{p}}{\cos \theta}$$

and

$$Q_{u} = \frac{Av_{p}}{\cos \theta} \qquad (12.16)$$

where Q_{u} is the discharge computed by the ultrasonic gauge.

If the magnitude of the velocity does not change, but ϕ becomes significant by an unknown amount due to a shift in the proportionate flows entering the reach, then $\bar{v}$ will no longer equal v_{L} which is the velocity normal to the cross-section. Thus v_{L} is reduced and equal to $v'_{L} \cos \phi$ (Fig. 12.6(a)) and the estimated true discharge is $Q = Av'_{L} \cos \phi$. The value of v_{p} recorded by the ultrasonic gauge is also changed and becomes equal to

$$v'_{p} = v'_{L} \cos (\theta + \phi). \qquad (12.17)$$

However, Q'_{u} does not account for ϕ and will become

$$Q'_{u} = \frac{Av'_{p}}{\cos \theta} = \frac{Av'_{L} \cos (\theta + \phi)}{\cos \theta} \qquad (12.18)$$

and the ratio between the computed and estimated true discharges is

$$\frac{Q'_{u}}{Q} = \frac{Av'_{L} \cos (\theta + \phi)/\cos \theta}{Av'_{L} \cos \theta} \qquad (12.19)$$

which reduces to

$$\frac{Q'_{u}}{Q} = 1 - \tan \theta \tan \phi \qquad (12.20)$$

for the condition shown in Fig. 12.6(a).

Similarly when ϕ is within θ the relation between the ultrasonic recorded discharge and the estimated true discharge is

$$\frac{Q'_{u}}{Q} = 1 + \tan \theta \tan \phi. \qquad (12.21)$$

Equations (12.20) and (12.21) are evaluated in Table 12.1.

Table 12.1 Multipliers to adjust the ultrasonic discharges when the flow is oblique for various values of φ, the oblique angle, and θ, the path angle

θ (degrees)	φ (degrees)								
	−4	−3	−2	−1	0	+1	+2	+3	+4
30	1.040	1.030	1.020	1.010	1.000	0.990	0.980	0.970	0.960
45	1.070	1.052	1.035	1.017	1.000	0.983	0.965	0.948	0.930
50	1.083	1.062	1.042	1.021	1.000	0.979	0.958	0.938	0.917
60	1.121	1.091	1.060	1.030	1.000	0.970	0.940	0.909	0.879

To overcome the complex problem of oblique flow, crossed paths are generally used, as shown in Fig. 12.7. By averaging the velocities from each path, the estimated true discharge is obtained since the second path gives corresponding errors due to oblique flow in the opposite direction. In a multipath system, a crossed path is normally required for each transducer path in the opposite direction, so that two pairs are installed at each transducer designed depth.

Example

Details of a crossed-path system are as follows (Fig. 12.7)

$$\text{Path length AB(1)} = 62.95 \text{ m}$$
$$\text{Path length CD(2)} = 65.92 \text{ m}$$
$$\text{Average depth of flow across AB, } d_1 = 2.1 \text{ m}$$
$$\text{Average depth of flow across CD, } d_2 = 2.1 \text{ m}$$
$$\theta_1 = 59.3°$$
$$\theta_2 = 59.9°$$
$$v_L(\text{AB}) = 0.595 \text{ m s}^{-1}$$
$$v_L(\text{CD}) = 0.508 \text{ m s}^{-1}.$$

Then the area of flow related to AB is

$$\text{AB} \sin \theta_1 d_1 = 62.95 \sin 59.3° \times 2.1$$
$$= 113.668 \text{ m}^2$$

and the area of flow related to CD is

$$\text{CD} \sin \theta_2 d_2 = 65.92 \sin 59.9° \times 2.1$$
$$= 119.765 \text{ m}^2.$$

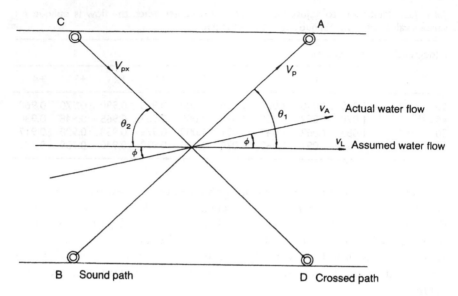

Figure 12.7 Schematic illustration showing crossed-path system: AB and CD are the crossed paths, θ_1 and θ_2 are the path angles, respectively, and φ is the angle of oblique flow velocity v_A.

The discharge related to path AB, Q_{AB}, is

$$Q_{AB} = 113.668v_L(AB)$$
$$= 113.668 \times 0.595$$
$$= 67.632 \text{ m}^3 \text{ s}^{-1}.$$

The discharge related to path CD, Q_{CD}, is

$$Q_{CD} = 119.765 \times v_L (CD)$$
$$= 119.765 \times 0.508$$
$$= 60.840 \text{ m}^3 \text{ s}^{-1}.$$

Therefore, since the discharges are not equal, oblique flow is present. Then the estimated true discharge is

$$Q = \frac{67.632 + 60.840}{2},$$

therefore

$$Q = 64.236 \text{ m}^3 \text{ s}^{-1}.$$

The value of the oblique angle ϕ may be found from equations (12.20) and (12.21) by substituting the appropriate values of discharges Q, Q'_u and angle θ. For AB, using equation (12.21)

$$64.236 = \frac{67.632}{1 + \tan 59.3 \tan \phi};$$

then

$$\phi = 1.799°.$$

Similarly, using equation (12.20) for CD

$$64.236 = \frac{60.840}{1 - \tan 59.9 \tan \phi};$$

then

$$\phi = 1.756°.$$

Therefore, averaging

$$\phi = 1.78° \text{ (approximately)}.$$

Substituting this value in equations (12.20) and (12.21) or from Table 12.1, equation (12.20) gives

$$\frac{Q'_u}{Q} = 0.948$$

and equation (12.21) gives

$$\frac{Q'_u}{Q} = 1.052.$$

Therefore if path AB alone is installed the recorded velocities or discharges require multiplication by 0.948 (5%) and conversely if path CD alone is installed the recorded values require multiplication by 1.052 (5%). Alternatively the angle θ_1 or θ_2 may be adjusted in the processor. It can be seen from the above example that the crossed-paths technique conveniently overcomes the problem of oblique flow.

12.8 Depth measurement

Since the average depth of flow is estimated across the line of the transducers, a detailed survey of the bed is necessary. This survey is extended upstream by about one channel width and downstream by a similar amount. Depending on the result of the survey, consideration is given to any improvement of the bed configuration by cleaning or dredging. Since the bed level is entered into the electronic processor for the calculation of flow area, this survey is repeated during the operation of the station. The frequency of such a survey depends on the stability of the bed.

The average bed level is related to stage which is normally measured by a float-operated digital recorder or a recorder coupled to a shaft encoder. The stage may also be measured by a single transducer mounted under the water with its beam pointing vertically towards the surface (Fig. 12.4). The ultrasonic pulse transit time to the surface and its reflection to the same transducer gives a measure of river depth. For this device, compensation is made for variation in the speed of sound in water (due to changes in either temperature or density) by measuring the speed of sound at each level using the horizontal paths.

Other ultrasonic water level gauges which have the transducer positioned above the water surface are also available (Fig. 3.18).

12.9 Reflector system

The installation of a typical ultrasonic flow gauge requires a cable to be laid in the bed to carry the signals across the river since active elements on the far bank require power. A reflector system incorporates the transducers on one bank only with a reflector on the far bank installed in the form of a length of angled steel (Fig. 12.8). The reflector may also be in the form of a third transducer.

In a reflector system, the main and reflected beams make different angles in the direction of flow and give a first-order cancellation of errors caused by oblique flow.

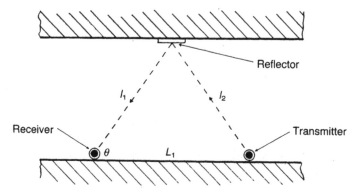

Figure 12.8 Schematic illustration of a typical reflector system.

Referring to Fig. 12.8 and from equation (12.13) and letting l_1 and l_2 be the lengths of each path where $l_1 + l_2 = L$,

$$\frac{2\bar{v}_L \cos \theta}{L} = \frac{1}{t_1} - \frac{1}{t_2} \tag{12.22}$$

and

$$\bar{v}_L = \frac{L(1/t_1 - 1/t_2)}{2 \cos \theta}. \tag{12.23}$$

Then

$$\bar{v}_L = \frac{(l_1 + l_2)^2}{2L_1} \left(\frac{1}{t_1} - \frac{1}{t_2} \right) \tag{12.24}$$

where L_1 is the distance between the receiver and transmitter. It will be noted that equation (12.24) does not include θ and therefore avoids the necessity of measuring it.

12.10 Responder system

Where a cable link either above or below the river is impractical a responder system may be considered. A responder is an electronic device that transmits an acoustic signal upon receipt of one and essentially echoes and strengthens the signal (Fig. 12.9).

A signal is transmitted downstream from transducer A along a diagonal path to transducer B. The received signal is then routed to the responder where it

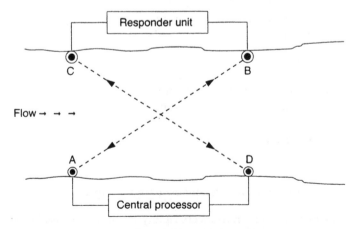

Figure 12.9 Schematic illustration of a typical responder system.

triggers a transmit signal after a minimal delay. This signal again travels downstream along another diagonal path from transducer C to transducer D and is received back at the central processor. For the upstream travel time, the above procedure is reversed. This procedure produces longer signal paths. The downstream path is ABCD and the upstream path is DCBA. The path angle does not need to be taken into account in the equation for computing line velocity.

Responder systems have more electronic components than reflector systems or conventional configurations that use cable links to all transducers. A responder system requires the use of a central processor, a responder unit, and a minimum of four transducers to produce single path velocity data. A failure of any component of the system will render the station incapable of producing valid data.

Computations

The equation for deriving line velocity with a responder system is

$$\bar{v} = \frac{L_R}{2}\left(\frac{1}{t_d} - \frac{1}{t_u}\right) \tag{12.25}$$

where L_R is the distance from A to B plus the distance from C to D;
$\qquad t_d$ is the time for the signal to travel in the downstream direction and is

$$t_d = t_{AB} + t_{te} + t_{tr} + t_{CD} + t_{CP}$$

where t_{AB} is the signal travel time from point A to point B;
$\qquad t_{te}$ is the time delay in the cables between the transducers and the electronics;
$\qquad t_{tr}$ is the time delay in the responder electronics;
$\qquad t_{CD}$ is the signal travel-time from point C to point D;
$\qquad t_{CP}$ is the time delay in the central processor;
$\qquad t_u$ is the time for the signal to travel in the upstream direction and is

$$t_u = t_{DC} + t_{te} + t_{tr} + t_{BA} + t_{CP}$$

where t_{DC} is the signal travel time from point D to point C;
$\qquad t_{BA}$ is the signal travel time from point B to point A.

12.11 System design

Hardware

The basic multipath ultrasonic streamflow gauge comprises:

(a) An array of ultrasonic transducers rigidly mounted on special beams and brackets on the river banks (Fig. 12.2). Normally these beams are set in concrete and follow the contour of the bank. The paths are therefore of different lengths but in order to retain a constant path angle, θ, the beams are aligned to face each other. Transducer alignment is generally performed by swinging the transducers to maximise the received signals. Provision is made for simple replacement of the transducer or cable, or both, in the event of failure or damage.

(b) Multiplexers and head amplifiers mounted in waterproof boxes adjacent to the transducer arrays.

(c) An electronic control unit containing power supplies; a timing unit to measure path transit times; a microprocessor to control the multiplexers, compute discharge and control the various output peripheral options; a station data module to store various station parameters such as transducer levels, path lengths and number of paths, path angles, station code, bed level and other relevant data; programmable memory module; real-time clock and calendar; and a visual numerical LED display under control of keyboard switches.

(d) In the reflector system, a reflector is required at each transducer level and in a responder system responders are necessary at each path level.

Processing instrumentation

The instrumentation is normally based on a microprocessor bus-bar system which gives a high degree of flexibility and reliability in providing output options.

Output options generally include numerical displays as standard. These displays may include, in addition to discharge at preset intervals (usually every 15 min), average depth, each line velocity measurement, v_L, path lengths and path angle θ as well as station information and real time.

Output options are available for chart recorder, paper tape punch, magnetic cassette recorder, solid-state logger, interface to radio, telephone or satellite, and interface to a central computer.

Transducer electronics

The transducers are driven from the head amplifiers and multiplexers mounted near the transducer assemblies ((b) above). The multiplexers route the control signals to the appropriate amplifiers under the control of the microprocessor. Electronic counters measure from the time the transducers are driven to the times the two receive pulses are detected for each pair of transducers. The counters thereupon transfer to the next path using the processor to control the multiplexer and the process is repeated for each pair of transducers under water.

Signal processing

The degree of complexity of measuring times of multiple paths and combining the measurements is made economical by use of the microprocessor to carry out the necessary computations.

Further use of the processor is made to ensure that data fit a sensible pattern, that velocity at any level does not change significantly compared with other levels and that the pulse arrivals fit into defined windows in time to ensure that the raw data are valid.

If a path fails it is ignored and, by interpolation or extrapolation, the discharge is calculated using the remaining paths.

Many time measurements are made over a period of several minutes. Velocity pulsations are averaged out and any temporary loss of transmission, caused by shipping or other temporary obstructions, is taken into consideration in the calculation by the processor.

Software

The use of distributed software and hardware subroutines is such that where software extensions are required this can be achieved by plugging in additional modules. This facility permits the format for an output peripheral, including its operating sequence, such as data recording intervals and actual parameters recorded, to be entered into a programmable memory on the interface to the peripheral. A wide variety of data recording options is available to the user with the PROM programmed to the user's specification. The processor may also be programmed to enable a number of alarm options to be available. These are normally in the form of indicator lamps backed up by quantitative data displayed on the numerical LED display under control of a keyboard and, if necessary, recorded on the recording medium. Alarms may also include flagging faulty transducers and transmission failure.

12.12 Advantages limitations and accuracy

The ultrasonic method of stream gauging is now established as an operational system to measure discharge at sites which hitherto could not have been gauged satisfactorily, if at all, by traditional methods. Many installations are to be found, notably in the USA, UK, Canada, Japan, Switzerland, France, Germany and The Netherlands. The largest path length operating in North America is probably on the Niagara River – one of some 500 m. The largest depth measured is on the Columbia River where the measuring section has one of some 40 m. The largest site in Europe is on the River Rhine at Arnhem (Fig. 12.10) where the path length is about 140 m. Figure 12.11 shows a view of the ultrasonic station on the River Thames at Reading, UK, with a path length of 55 m, installed in 1992.

(a)

(b)

Figure 12.10 (a) View of ultrasonic streamflow station on the River Rhine near Arnhem; (b) close-up view of one of the two transducer mountings.

(a)

(b)

Figure 12.11 Ultrasonic streamflow station on the River Thames at Reading Bridge showing (a) the four transducer mountings and (b) the panel on the outside wall of the instrument house showing a continuous digital display of the flow. The depth range of flow is 2.45 m–2.80 m with a flow range of 4.00 m^3 s^{-1}–200 m^3 s^{-1}. The path lengths are 55 m and three crossed paths are used, making six paths in all.

Some of the advantages, as well as the limitations, of the ultrasonic method have already been discussed but other features may be summarised as follows.

Accuracy

The uncertainty obtainable in measurement in a well-designed system is of the order of 5%.

Calibration

The multipath system is self-calibrating and it is customary to check the system by other means, normally by current meter. The current meter, however, is a fairly crude instrument compared to the ultrasonic gauge. The minimum speed of response of a current meter is of the order of 0.03 m s^{-1} and at these velocities the uncertainty in current meter measurements is about 20%. In addition, current metering relies on point measurements whilst the ultrasonic measurement is an integrating method. Nevertheless a good current meter measurement, taken when stage is constant and at velocities above 0.5 m s^{-1}, normally gives an indication of the operation of the ultrasonic gauge.

Range of velocities

The ultrasonic gauge is capable of measuring velocities as low as a few millimetres per second. There is no practical upper limit.

Cost

The cost of an ultrasonic system is relatively independent of width of river. The capital cost of a system is more expensive than a velocity–area station (but cheaper to operate) and less expensive than a measuring structure (but possibly dearer to operate). For small installations, however, the cost is relatively high.

Electronics

There are no moving parts, and good serviceability and automatic signal and data testing ensure only accurate readings are accepted for output.

Channel section

The system is not suitable for wide, very shallow rivers (Section 12.6) nor for streams containing weeds, continuous high sediment load or entrained air.

The preferred bed configuration is a flat section free from obstructions and deep holes or depressions. Regular bed surveys are important to ensure that no significant change has occurred or, if there has been change, to enter the new average bed level into the system programme.

Further reading

Cobb, E. D. Broad band acoustic Doppler current profiler. *Flow Measurement and Instrumentation* (ed. R. W. Herschy), 4(1), Butterworth-Heinemann, Oxford, pp. 35–7 1993.

Gibbard, D. Design and installation of ultrasonic and electromagnetic flow gauging stations. *Flow Measurement and Instrumentation* (ed. R. W. Herschy), 4(1), Butterworth-Heinemann, Oxford, pp. 25–33 1993.

Green, M. J. and Herschy, R. W. New methods. In *Hydrometry: Principles and Practices* (ed. R. W. Herschy). John Wiley and Sons, Chichester, pp. 247–76 1978.

Herschy, R. W. New methods of river gauging. In *Facets of Hydrology* (ed. J. C. Rodda). John Wiley and Sons, Chichester, pp. 119–61 1976.

ISO 6416 (2004). *Liquid Flow Measurement in Open Channels: Measurement of Discharge by the Ultrasonic Method.* ISO, Geneva, Switzerland 2004.

ISO 6418. *Liquid Flow Measurement in Open Channels: Ultrasonic Velocity Meters.* ISO, Geneva, Switzerland 1984.

Kinosita, T. Improvement of ultrasonic flowmeter in rivers in Japan. IAHS Symposium, Exeter, Publication No. 134, 1982.

Chapter 13

Accuracy

13.1 Introduction

The accuracy or, more correctly, the error of a measurement of discharge may be defined as the difference between the measured flow and the true value. The true value of the flow is unknown and can only be ascertained (within close limits) by weighing or by volumetric measurement. An estimate of the true value has therefore to be made by calculating the uncertainty in the measurement, this being defined as the range in which the true value is expected to lie expressed at the 95% confidence level (Fig. 13.1). Although the error in a result is therefore, by definition, unknown, the uncertainty may be estimated if the distribution of the measured values about the true mean is known.

The uncertainty in the measurement of an independent variable is normally estimated by taking N observations and calculating the standard deviation, where N is preferably at least 30. Using this procedure to calculate the uncertainty in a gauging, however, would require N consecutive measurements of discharge with different current meters at constant stage which is clearly impractical. An estimate of the true value has therefore to be made by examining all the various sources of error in the measurement.

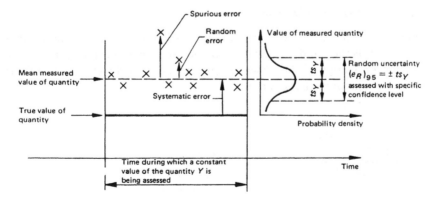

Figure 13.1 Basic statistical terms.

In applying the theory of statistics to streamflow data it is assumed that the observations are independent random variables from a statistically uniform distribution. This ideal condition is seldom met in hydrometry. For example, the measurement of velocity, by current meter, cannot be independent since the velocity itself is fluctuating with time due mainly to pulsations in flow. The measurement of stage at any particular moment in time is not strictly independent since the flow passing the reference gauge is continuous. River flow is by nature, therefore, non-random, each hourly, daily, monthly and annual discharge being dependent on the previous hourly, daily, monthly and annual discharge. The cause of this is mainly the lag in the rainfall runoff relation, although in most catchments this may be an oversimplification. However, it is generally accepted by statisticians that the departure of river gauging data from the theoretical concept of errors is not serious provided care and attention are exercised in sampling techniques.

It should be stressed that the statistical analysis of river flow data is only applicable if the field data have been obtained by acceptable hydrometric principles and practices, i.e. to the relevant national or international standards. In this connection, a large responsibility is imposed on the hydrometric observer who has to ensure that the current meters, water level recorders, reference gauges and other equipment are all in good order, that the measurements are as precise as possible and that they are meticulously recorded. If the raw data are questionable for any reason it is the observer's duty to record this suspicion in the records. No amount of computer processing or statistical analysis will correct a wrong measurement. In addition, the observer should avoid getting into the position, for example, of producing a very precise measurement of the wrong water level, as may well happen under certain circumstances in the field. Statistical analysis is an aid to improving the presentation of the hydrometric data for the user's benefit but the final quality of the data depends on the hydrologist.

13.2 Standard deviation

The most used statistical term in streamflow to estimate the uncertainty in a measurement is the standard deviation. Standard deviation is a measure of the dispersion or scatter of the observations about the arithmetic mean of the sample and is given by the equation

$$s_Y = \left[\frac{\sum_{i=1}^{n} (Y_i - \bar{Y})^2}{N - 1} \right]^{1/2} \tag{13.1}$$

where s_Y = standard deviation of the observations;

$\bar{Y}$ = arithmetic mean of the observations, Y_i;

Y_i = independent random observation of the variable Y;

N = number of observations.

In a linear regression, for example the stage–discharge curve, the term used is **standard error of estimate** which is numerically similar to standard deviation except that the linear regression replaces the arithmetic mean and $(N - 1)$ is replaced by $(N - 2)$. The standard error of estimate is given by the equation

$$S_e = \left(\frac{\Sigma d^2}{N - 2} \right)^{1/2} \qquad (13.2)$$

where S_e = standard error of estimate;

d = deviation of an observation from the computed value taken from the regression.

If a sample of streamflow data fits a bell-shaped symmetrical distribution, known as the normal distribution, then by statistical inference the dispersion of the observations about the mean is measured in standard deviations. Then, on average, 68% of the observations will lie within one standard deviation of the mean, 95% will lie within two standard deviations of the mean, and almost all (actually over 99%) will lie within three standard deviations of the mean (Fig. 13.2). Generally the 95% level of confidence is used in streamflow and provided the sample contains at least 30 observations the standard deviation is multiplied

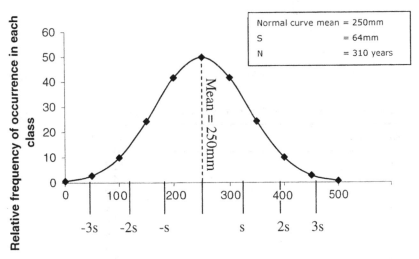

Figure 13.2 Normal distribution curve for annual runoff in the River Thames at London for period 1697–2006 (310 years) (values from 1697 to 1882 computed by rainfall-runoff model by P.C. Saxena, K.S. Rajagopalan and A. Choudhury CWPRS Pune, India). Values from 1883 to 2006 provided by the UK National River Flow Archive, Centre for Ecology and Hydrology, Wallingford.

by 2. When the sample size is small, it is necessary to correct the statistical results that are based on a normal distribution by means of Student's t values. Student's t is a factor which compensates for the fact that the uncertainty in the standard deviation is large with small sample sizes. These values at the 95% level are given in Table 13.1.

Degrees of freedom for small samples

In a sample, the number of degrees of freedom is the sample size, N. When a statistic is calculated from the sample, the degrees of freedom associated with it are reduced by one for every estimated parameter used in calculating the statistic. For example, from a sample of size, N, $\bar{Y}$ is calculated and has N degrees of freedom, and the standard deviation is calculated from equation (13.1) and has $(N - 1)$ degrees of freedom because $\bar{Y}$ is used to calculate standard deviation. In calculating the standard error of estimate from a curve fit, the number of degrees of freedom which are lost is equal to the number of estimated coefficients for the curve; in the case of a stage–discharge relation C and n are the coefficients and two degrees of freedom are therefore lost and $(N - 1)$ is replaced by $(N - 2)$.

Percentage standard deviation

The standard deviation is normally expressed as a percentage in streamflow by the following equation

$$s_Y = \frac{\left[\dfrac{\displaystyle\sum_{i=1}^{n} (Y_i - \bar{Y})^2}{N - 1} \right]^{1/2}}{\bar{Y}} \times 100. \tag{13.3}$$

Table 13.1 Values of Student's t at the 95% confidence level

Degrees of freedom	t
1	12.7
2	4.3
3	3.2
4	2.8
5	2.6
6	2.4
7	2.4
10	2.2
15	2.1
20	2.1
30	2.0
60	2.0
∞	1.96

Expressed in this form, the standard deviation is also known as the coefficient of variation (CV).

Standard deviation of the mean

The standard deviation of the mean, $s_{\bar{Y}}$, is an estimate of the uncertainty of the computed mean and expressed by the equation

$$s_{\bar{Y}} = \frac{s_Y}{\sqrt{N}}.$$

(13.4)

13.3 Standard error of the mean relation

The standard error of the mean relation, S_{mr}, is similar to $s_{\bar{Y}}$ but refers to the uncertainty of the regression (for example, the stage–discharge curve). In this case, however, equation (13.4) gives the value of S_{mr} at the centroid of the regression only and the value varies curvilinearly along the regression, being $s_Y \sqrt{N}$ at the centre and a maximum at the extremes.

This is illustrated in Fig. 13.3 where the S_e limits are shown as parallel to the regression and the S_{mr} limits are shown as curves. Also shown in Fig. 13.3 are the bell-shaped normal distributions for both S_e and S_{mr}. In theory the mean and standard deviation of the discharge observations are computed from samples of N discharges at M stages throughout the range (M and N both being at least 30) producing M normal distribution bell-shaped curves from MN discharge measurements. The standard deviation of the mean is computed from each of the M distributions and the points defining the mean, standard deviation and standard deviation of the mean for each of the M normal distributions joined to form the regression and the confidence limit curves of standard error of estimate and standard error of the mean. At the 95% level the standard error of estimate confidence limits will contain 95% of the discharges, and the standard error of the mean confidence limits defines the uncertainty in the regression and the stage–discharge relation which is the uncertainty of estimates of discharge from stage from the rating equation.

The application of standard deviation to streamflow data can be demonstrated by the following example. Table 13.2 gives the annual runoff values in millimetres from the gauging station on the River Thames at London for the years 1697 to 2006, a total of 310 years (N) (Fig. 13.2). Calculation gives the following values

mean   = 250 mm

standard deviation s_Y = ± 64 mm;

standard deviation of the mean

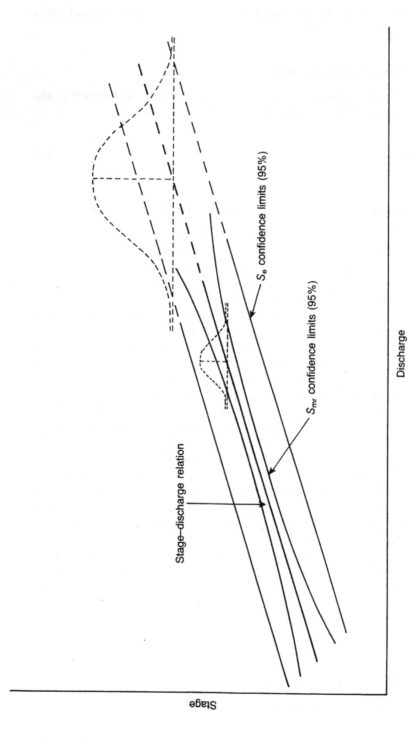

Stage

Discharge

Figure 13.3 Illustration of confidence limits of a stage–discharge curve. The standard error of estimate confidence limits are straight and parallel on either side of regression and contain 95% of the current meter observations; the standard errors of the mean confidence limits are curved on either side of the regression, being narrowest at the centre and maximum at the extremes; the stage–discharge relation will, on average, be within these limits for 95% of the time (19 years out of 20), assuming that there is no change in the hydraulic characteristics of the flow (e.g. change in control, change of datum).

Stage–discharge relation

S_e confidence limits (95%)

S_{mr} confidence limits (95%)

Table 13.2 Annual runoff (mm) for the River Thames at London for the period 1697–2006

Year	Run-off (mm)	Year	Run-off (mm)	Year	Run-off (mm)	Year	Run-off (mm)	Year	Run-off (mm)	Year	Run-off (mm)
1697	131	1743	144	1789	306	1835	287	1881	265	1927	371
1698	285	1744	269	1790	195	1836	305	1882	258	1928	331
1699	165	1745	234	1791	235	1837	215	1883	314	1929	218
1700	191	1746	250	1792	252	1838	220	1884	165	1930	281
1701	224	1747	236	1793	237	1839	305	1885	196	1931	287
1702	220	1748	177	1794	242	1840	162	1886	263	1932	270
1703	269	1749	262	1795	199	1841	351	1887	192	1933	220
1704	180	1750	265	1796	189	1842	237	1888	209	1934	98
1705	148	1751	349	1797	251	1843	254	1889	218	1935	244
1706	284	1752	186	1798	239	1844	283	1890	135	1936	326
1707	144	1753	218	1799	265	1845	228	1891	232	1937	391
1708	215	1754	240	1800	258	1846	279	1892	193	1938	166
1709	317	1755	290	1801	227	1847	167	1893	166	1939	332
1710	199	1756	293	1802	139	1848	324	1894	264	1940	287
1711	256	1757	234	1803	193	1849	245	1895	210	1941	282
1712	259	1758	252	1804	294	1850	180	1896	182	1942	216
1713	303	1759	191	1805	235	1851	273	1897	248	1943	189
1714	104	1760	244	1806	251	1852	350	1898	127	1944	124
1715	272	1761	198	1807	162	1853	307	1899	168	1945	157
1716	152	1762	246	1808	228	1854	174	1900	204	1946	262
1717	246	1763	272	1809	290	1855	239	1901	149	1947	302
1718	233	1764	286	1810	267	1856	245	1902	110	1948	159
1719	169	1765	278	1811	206	1857	207	1903	369	1949	170
1720	262	1766	219	1812	302	1858	149	1904	259	1950	231
1721	234	1767	282	1813	198	1859	252	1905	118	1951	418
1722	165	1768	307	1814	231	1860	303	1906	183	1952	248
1723	121	1769	237	1815	223	1861	215	1907	207	1953	163
1724	219	1770	253	1816	310	1862	286	1908	221	1954	259
1725	146	1771	188	1817	307	1863	183	1909	198	1955	217
1726	273	1772	304	1818	329	1864	176	1910	293	1956	171
1727	232	1773	256	1819	351	1865	314	1911	206	1957	226
1728	305	1774	307	1820	268	1866	370	1912	355	1958	305
1729	188	1775	301	1821	372	1867	281	1913	281	1959	247
1730	216	1776	231	1822	270	1868	230	1914	221	1960	378
1731	76	1777	225	1823	292	1869	280	1915	385	1961	264
1732	190	1778	224	1824	374	1870	175	1916	400	1962	218
1733	162	1779	224	1825	249	1871	198	1917	283	1963	248
1734	231	1780	195	1826	233	1872	303	1918	258	1964	198
1735	253	1781	173	1827	269	1873	221	1919	339	1965	186
1736	285	1782	326	1828	322	1874	176	1920	258	1966	326
1737	297	1783	238	1829	229	1875	269	1921	119	1967	309
1738	285	1784	289	1830	281	1876	248	1922	166	1968	342
1739	375	1785	178	1831	339	1877	355	1923	265	1969	278
1740	165	1786	229	1832	178	1878	284	1924	360	1970	240
1741	181	1787	181	1833	224	1879	361	1925	330	1971	279
1742	194	1788	150	1834	178	1880	271	1926	292	1972	237

Year	Run-off (mm)	Year	Run-off (mm)	Year	Run-off (mm)	Year	Run-off (mm)	Year	Run-off (mm)	Year	Run-off (mm)
1973	134	1979	327	1985	262	1991	154	1997	134	2003	254
1974	310	1980	262	1986	289	1992	233	1998	277	2004	238
1975	275	1981	294	1987	283	1993	284	1999	284	2005	146
1976	133	1982	316	1988	262	1994	304	2000	410	2006	206
1977	333	1983	248	1989	192	1995	287	2001	395		
1978	263	1984	240	1990	204	1996	178	2002	319		

Note: The above values of runoff have been naturalised (adjusted to account for the major public water supply abstractions upstream of the gauging station)

$$s_{\bar{Y}} = \frac{64}{\sqrt{310}} = \pm 4 \text{ mm.}$$

At the 95% level

$$s_Y = \pm 128 \text{ mm}$$

$$s_{\bar{Y}} = \pm 8 \text{ mm.}$$

From these results it can be stated that:

(a)   The best estimate of the long-term annual runoff is the mean = 250 mm with an uncertainty of $\pm$ 8 mm.

(b)   The probability is that the mean in future will lie between 242 and 258 mm in 19 years out of 20.

(c)   The uncertainty, or dispersion, of the annual runoff values about the mean is $\pm$ 128 mm.

(d)   By statistical inference, therefore, 95% of the 310 values in Table 13.2 should be within the range 250–128 mm and 250 + 128 mm, or 122 and 378 mm. That is, at least 295 values should be within this range and no more than 15 outside the range. A study of Table 13.2 will reveal that only 12 values are outside the range, the years 1714, 1723, 1771, 1902, 1905, 1916, 1934, 1937, 1951, 1960, 2000 and 2001. Therefore 298 values are within the range.

(e)   Therefore the probability is that future values of annual runoff will be, in 19 years out of 20, within the range 122 and 378 mm.

All we can conclude with certainty, however, is that the annual values, on average 19 years out of 20, will be within the confidence band 122–378 mm, a 50% scatter on either side of the mean. We have more confidence in predicting the long-term annual mean, however, which on average, in 19 years out of 20, will be contained within the narrow confidence band of 242–258 mm.

By dividing the data into blocks any trend in the annual runoff can be tested by calculating the mean and standard deviation of each block.

	years	Mean (mm)	standard deviation (mm)
1697–1883	187	251	53
1883–2006	124	248	72
1697–2006	310	250	63
1987–2006	20	252	75

It can be seen that the mean has remained substantially constant but the standard deviation shows fluctuations from 53 mm to 75 mm, the latter being the value for the last 20 years, featuring the high annual values of runoff for the years 1994, 2000, 2001 and 2002. Otherwise the data shows no hint of a trend although the dispersion about the mean has increased.

13.4 Nature of errors

In general, the result of a measurement is only an estimate of the true value of the quantity subjected to measurement. The discrepancy between the true value and the measured value is the measurement error. The measurement error, which cannot be known, causes an uncertainty about the correctness of the measurement result.

The measurement error is a combination of component errors which arise during the performance of various elementary operations during the measurement process. For measurements of composite quantities, which depend on several component quantities, the total error of the measurement is a combination of the errors in all component quantities. Determination of measurement uncertainty involves identification and characterisation of all components of error, and the quantification of the corresponding uncertainties.

The Guide to the Expression of Uncertainty in Measurement (the Guide) and the Hydrometry Uncertainty Guidance (HUG) treat measurement uncertainty using concepts and formulae for probability distribution, expected values, standard deviation, and correlations of random variables. The standard deviation of the measurement error is taken as the quantitive measure of uncertainty. The *Guide* does not make use of the traditional categorisation of errors as random and systematic. That categorisation may be difficult to apply in practice. For example, in streamflow it is sometimes difficult to distinguish between random and systematic errors as some errors may be a combination of the two. For instance, where a group rating is used for current meters, each of the meters forming the group may have a plus or minus systematic error which is randomised to obtain the uncertainty in the group rating. A construction error in the length of a weir crest may be randomised by measuring the crest say n times after construction to obtain the mean and standard deviation, each measurement being independent of the other. It recommends that the components of uncertainty are characterised by estimates of standard deviation, which are termed standard uncertainty with recommended symbol u_i where i identifies the component in question, and which are equal to the positive square root of the estimated variance u_i^2. The uncertainty components are combined using formulae for combination of standard deviation of possibly correlated random variables. The resultant uncertainty, which takes all sources and components of uncertainty into account, is called the combined uncertainty and is denoted by u. It also introduces the concept of Type A and Type B methods of evaluation of uncertainty to make a distinction between uncertainty evaluation by statistical analysis of replicate measurements and by other means (subjective or judgement). Type A evaluation of uncertainty is by statistical analysis of repeated observations to obtain statistical estimates of the standard deviation of the observations. Type B evaluation is by calculation of the standard deviation of an assumed probability distribution based on scientific judgement and consideration of all available information, which may include previous measurements and calibration data and experience of general knowledge of the behaviour and properties of relevant instruments. By consideration of correlations, either Type A or Type B method of evaluation can be used for evaluation of either systematic of random uncertainty components (see Section 13.5).

13.5 Theory of errors

If a quantity Q is a function of several measured quantities $x, y, z, \ldots$, the error in Q due to errors $\delta x, \delta y, \delta z, \ldots$, in x, y, z $\ldots$ respectively is given by

$$\delta Q = \frac{\delta Q}{\delta x}\delta x + \frac{\delta Q}{\delta y}\delta y + \frac{\delta Q}{\delta z}\delta z + \ldots . \tag{13.5}$$

The first term in equation (13.5), $(\partial Q/\partial x)\delta x$, is the error in Q due to an error δx in x only (i.e. corresponding to $\delta y, \delta z, \ldots$, all being zero). Similarly the second term $(\partial Q/\partial y)\delta y$ is the error in Q due to an error δy in y only. Squaring gives

$$\delta Q^2 = \left(\frac{\partial Q}{\partial x}\delta x\right)^2 + 2\frac{\partial Q}{\partial x}\frac{\partial Q}{\partial y}\delta x\delta y + \left(\frac{\partial Q}{\partial y}\delta y\right)^2 + \ldots . \tag{13.6}$$

Now the terms $(\delta Q/\delta x)(\delta Q/\delta y)\delta x\delta y$, etc, are covariance terms and, since they contain quantities which are as equally likely to be positive or negative, their algebraic sum may be conveniently taken as being either zero or else negligible as compared with the squared terms. Equation (13.6) then becomes

$$\delta Q^2 = \left(\frac{\delta Q}{\delta x}\delta x\right)^2 + 2\frac{\delta Q}{\delta x}\frac{\delta Q}{\delta y}\delta x \delta y + \left(\frac{\delta Q}{\delta y}\delta y\right)^2 + \ldots \tag{13.7}$$

i.e. the error in Q, δQ, is the sum of the squares of the errors due to an error in each variable. Now

$$\frac{\delta Q}{\delta x} = yz, \frac{\delta Q}{\delta y} = xz, \frac{\delta Q}{\delta z} = xy \tag{13.8}$$

and

$$\delta Q = \left[(yz\delta x)^2 + (xz\delta y)^2 + (xy\delta z)^2 + \ldots\right]^{1/2}. \tag{13.9}$$

Dividing by $Q = xyz$

$$\frac{\delta Q}{Q} = \left[\left(\frac{yz}{xyz}\delta x\right)^2 + \left(\frac{xz}{xyz}\delta y\right)^2 + \left(\frac{xy}{xyz}\delta z\right)^2 + \ldots\right]^{1/2} \tag{13.10}$$

and

$$\frac{\delta Q}{Q} = \left[\left(\frac{\delta x}{x}\right)^2 + \left(\frac{\delta y}{y}\right)^2 + \left(\frac{\delta z}{z}\right)^2 + \ldots\right]^{1/2} \tag{13.11}$$

where $(\delta x)/x$ and $(\delta y)/y$ and $(\delta z)/z$ are fractional values of the errors (standard deviations) in x, y and z, and if they are each multiplied by 100 they become percentage standard deviations. Let u_Q be the percentage standard deviation of Q and

u_x = percentage standard deviation of x
u_y = percentage standard deviation of y
u_z = percentage standard deviation of z

then

$$u_{(Q)} = \pm (u_x^2 + u_y^2 + u_z^2 + \ldots)^{1/2} \tag{13.12}$$

which is generally referred to as the root-sum-square equation for the estimation of uncertainties. It is this equation that is employed to estimate the uncertainty in a current meter gauging.

Similarly, if

(i) $Q = \dfrac{x}{y}$

then

$$u_{(Q)} = \pm (u_x^2 + u_y^2)^{1/2}. \tag{13.13}$$

(ii) $Q = x \pm y$

$$\delta Q^2 = \left(\frac{\delta Q}{\delta x} \delta x\right)^2 \pm \left(\frac{\delta Q}{\delta y} \delta y\right)^2$$

$$\delta Q = (\delta x^2 \pm \delta y^2)^{1/2}$$

$$u_{(Q)} = \frac{100\delta Q}{Q} = \frac{100}{Q} \left[\frac{x^2\delta x^2}{x^2} \pm \frac{y^2\delta y^2}{y^2}\right]^{1/2}$$

therefore

$$U(Q) = \pm \frac{(x^2 u_x^2 + y^2 u_y^2)^{1/2}}{(x \pm y)}. \tag{13.14}$$

(iii) $Q = Cbh^n$ (weir equation)

then

$$U(Q) = \pm (u_c^2 + u_b^2 + n^2 u_h^2)^{1/2} \tag{13.15}$$

which is the error equation for weirs, where C is the coefficient of discharge, b is the length of crest, n the exponent of h (usually 3/2 for a weir and 5/2 for a V-notch), and h is the head over the weir.

13.6 The error equation

Velocity–area method

The general form of the working equation for computing discharge in the cross-section is

$$Q = \sum_{i=1}^{m} (b_i d_i \bar{v}_i) \tag{13.16}$$

where Q is the total discharge in the cross-section, and b_i, d_i and $\bar{v}_i$ are the width, depth, and mean velocity of the water in the ith of the m verticals or segments into which the cross-section is divided.

Equation (13.16) assumes that a sufficient number of verticals have been taken in the cross-section but if this is not the case then equation (13.16) should be multiplied by a factor F so that

$$Q = F \sum_{i=1}^{m} (b_i d_i \bar{v}_i) \tag{13.17}$$

where F may be greater or less than unity. Equation (13.17), therefore, requires to be optimised until sufficient verticals are employed so as to make F unity.

The uncertainties in the individual components of discharge are expressed as relative percentage standard uncertainties, corresponding to percentage coefficients of variation (standard deviation of error divided by expected value of the measured quantity).

The relative (percentage) combined standard uncertainty in the measurement is given by the following equation

$$u(Q)^2 = u_m^2 + u_s^2 + \frac{\sum_{j=1}^{m} ((b_i d_i \bar{v}_i)^2 (u_{b_i}^2 + u_{d_i}^2 + u_{v_i}^2))}{(\Sigma b_i d_i \bar{v}_i)^2} \tag{13.18}$$

where $u(Q)$ is the relative (percentage) combined standard uncertainty in discharge;

ub_i, ud_i and u_{vi} are the relative (percentage) standard uncertainties in the breadth, depth, and mean velocity measured at vertical i;

u_s is the relative uncertainty due to calibration errors in the current meter, breadth measurement instrument, and depth sounding instrument and may be expressed as

$$u_s = (u_{cm}^2 + u_{bm}^2 + u_{ds}^2)^{\frac{1}{2}}.$$

An estimated practical value of 1% may be used for this expression.

u_m is the relative uncertainty due to the limited number of verticals; and m is the number of verticals.

The mean velocity $\bar{v}_i$ at vertical i is the average of point measurements of velocity made at several depths in the vertical. The uncertainty in $\bar{v}_i$ estimated as follows

$$u(\bar{v}i)^2 = u_{pi}^2 + (\tfrac{1}{n_i}) (u_{ci}^2 + u_{ei}^2) \tag{13.19}$$

where

up_i = uncertainty in mean velocity $\bar{v}i$ due to the limited number of depths at which velocity measurements are made at vertical i;

n = number of depths in the vertical at which velocity measurements are made;

u_{ci} = uncertainty in point velocity at a particular depth in vertical i due to variable responsiveness of current meter;

u_{ei} = uncertainty in point velocity at a particular depth in vertical i due to velocity fluctuations (pulsations) in the stream.

Therefore

$$u(Q)^2 = u_m^2 + u_s^2 + \frac{\Sigma((b_i d_i \bar{v}_i)^2 \, (u_{bi}^2 + u_{di}^2 + u_{pi}^2 + (\frac{1}{n_i}) \, (u_{ci}^2 + u_{ei}^2))}{(\Sigma(b_i d_i \bar{v}_i))^2}. \qquad (13.20)$$

If the measurement verticals are placed so that the segment discharges $(b_i d_i \bar{v}_i)$ are approximately equal and if the component uncertainties are equal from vertical to vertical, then equation (13.20) simplifies to

$$u(Q) = [u_m^2 + u_s^2 + (\tfrac{1}{m}) \, (u_b^2 + u_d^2 + u_p^2 + (\tfrac{1}{n}) \, (u_c^2 + u_e^2))]^{\frac{1}{2}}. \qquad (13.21)$$

Although the segment discharge or the contributing uncertainties in a gauging are seldom, if ever, equal, it has been found in practice that equation (13.21) gives results which are not significantly different from those given by equation (13.20). Where the verticals number 20 or more, the value of $u(Q)$ using either equation is generally less than 7%. The discrepancies therefore between the two equations become rather academic and it is clear that for routine gauging $u(Q)$ can be conveniently calculated from equation (13.21). For special studies, however, which require more precise estimation of uncertainty $u(Q)$, it is advisable to use equation (13.20).

It will be evident that equations (13.20) and (13.21) can be employed to obtain a required value of $u(Q)$ by giving special consideration to the individual uncertainties in the equations. Individual uncertainties may usually be reduced by increasing the number of verticals, increasing the number of points in the verticals or, in the case of measuring structures, by reducing the uncertainty in the head measurement.

It is recommended that, whenever possible, the user should determine independently the values of the component uncertainties in the above equations. However, for routine gauging, values are given in section 13.8 that are the result of many investigations carried out since 1965.

It should be noted that since the individual components of uncertainty presented in section 13.8 are based on statistical analyses of the spread of replicate measurements on prior observations, rather than on repeated observations during the actual course of the measurement of discharge, they may be considered as Type B evaluations of uncertainties.

The concepts Type A and Type B methods of evaluation of uncertainty have now been introduced by ISO and other standard organisations to make a distinction between uncertainty evaluation by statistical analysis of replicate measurements and uncertainty evaluation by other subjective or judgement means. Type A evaluation of uncertainty is by statistical analysis of repeated observations to obtain statistical estimates of the standard deviations of the observations. This evaluation may be carried out automatically during the measurement process by the instrumentation. Type B evaluation is by calculation of the standard deviation of an assumed probability distribution based on scientific judgement and consideration of all available knowledge, which may include previous measurements and calibration data and experience of general knowledge of the behaviour and properties of relevant instruments. The examples of uncertainty calculations presented later are of Type B.

Type A or Type B methods of evaluation can be used for either systematic or random uncertainty components. The components of uncertainty are characterised by estimates of standard deviation, which are now termed standard uncertainty, with the symbol u_i where i identifies the component in question, and which are equal to the positive square root of the estimated variances u_i^2. The uncertainty components are combined (including random and systematic components) using equations for combination of standard deviations of possibly correlated random variables. The resultant uncertainty, which takes all sources and components of uncertainty into account, is called the combined uncertainty and described as u. Component uncertainties are expressed numerically as percentages at the 68% level of confidence (this is a departure from previous editions of this book). Expanded uncertainties are explicitly identified with a coverage factor of two corresponding to a level of confidence of approximately 95%.

Example

It is required to calculate the uncertainty in a current meter gauging from the following particulars:

(a) Number of verticals used in the gauging = 20
(b) Number of points taken in the vertical = 2
(c) Average velocity in measuring section = above 0.3 m/s
(d) Exposure time of current meter at each of two points = 3 min
(e) Rating of current meter = individual rating

Component uncertainties are obtained for this example from Section 13.6 (Tables 13.3–13.8). The values of the component uncertainties are expressed as percentage standard uncertainties (level of confidence approximately 68%):

u_m 2.5 (Table 13.8)
u_s 1.0 (see above)
u_b 0.5 (Table 13.3)
u_d 0.5 (Table 13.4)
u_p 3.5 (Table 13.6)
u_c 0.9 (Table 13.7)
u_e 3.0 (at 0.2 depth)(Table 13.5)
   3.0 (at 0.8 depth)(Table 13.5)
   Total $u_\mathrm{e} = (3^2 + 3^2)^{1/2} = 4.2$

Therefore, from equation (15.21)

$$u(Q) = [u_m^2 + u_s^2 + (\tfrac{1}{m}) (u_b^2 + u_d^2 + u_p^2 + (\tfrac{1}{n}) (u_c^2 + u_e^2))]^{1/2}$$

and

$$u(Q) = [2.5^2 + 1^2 + (\tfrac{1}{20}) (0.5^2 + 0.5^2 + 3.5^2 + (\tfrac{1}{2}) (0.9^2 + 4.2^2))]^{1/2}$$

giving

$$u(Q) = 2.89\%, \text{ say } 3\%.$$

Expanded uncertainty, u, coverage factor $k = 2$, approximate level of confidence 95%

$$u_{(k = 2)}(Q) \quad = ku(Q)$$

$$= 2 \times 3$$

$$= 6$$

$$u(Q) \quad = 6\%.$$

Now, if the measured flow is Q m^{3}s^{-1}, the result of the measurement is expressed as

Q m^{3}s^{-1} ± 6% (expanded uncertainty, coverage factor $k = 2$, approximate level of confidence = 95%).

Note: the above uncertainty calculation is a Type B evaluation of uncertainty since the component uncertainties are based on previous measurements and calibration data.

Note: it can be seen from the example that if better accuracy was required this could be achieved by increasing the number of verticals (Section 13.8, Table 13.6).

13.7 Weirs and flumes

The error equation for the estimation of the uncertainty in a single determination of discharge for a weir is as follows
if

$$Q = Cbh^n$$

then

$$u_{(Q)} = \pm(u_c^2 + u_b^2 + n^2 u_h^2)^{1/2} \qquad\qquad (13.22)$$

where c = coefficient of discharge;

  b = length of crest;

  h = gauged head;

  n = exponent of h, usually 3/2 for a weir and 5/2 for a V-notch;

  u_Q = percentage random uncertainty in a single determination of discharge;

  u_c = percentage uncertainty in the value of the coefficient of discharge;

  u_b = percentage uncertainty in the measurement of the length of crest;

  u_h = percentage uncertainty in the measurement of gauged head.

All values of uncertainties are percentage standard deviations at the 95% level.

It is usually convenient to include any small systematic uncertainty components in the terms u_c, u_b and u_h in equation (13.22). If the coefficient C, for example, has been established in the laboratory this is usually performed by a graphical relation of Q versus h. The uncertainty of the relation may therefore be taken as random, any systematic bias being due to systematic errors in the instrumentation. This should be insignificant in calibrations carried out in national laboratories (although it is possible for c to have a systematic error assigned to it when the calibration is applied in the field). It is also usual to investigate, on site, any suspicion of a systematic uncertainty in the gauge zero, or in the recorder, and correct these. If, however, an allowance is made in equation (13.22) for zero error, this allowance should ensure that any systematic bias in the head measurement is included in u_h.

An allowance of 0.1% in u_b should more than compensate for any systematic error in the steel tape or other means for measuring the length of crest.

Examples

Measuring structures: single determination of discharge

(i) Calculate the discharge and the uncertainty in discharge for a triangular profile weir given the following details (uncertainties at 68%)

Gauged head, h	0.67 m
Breadth of weir, b	10 m
Crest height, P	1 m
Coefficient of discharge, C_d	1.163
Coefficient of velocity, C_v	1.054
Uncertainty in $C_v C_d$	± 0.5%

Uncertainty in measurement of breadth   0

Uncertainty in head measurement   ± 3 mm.

Calculations

The discharge equation is equation (10.13)

$$Q = (\tfrac{2}{3})^{3/2} \, C_d C_v b v(g) h^{3/2}$$

and from Chapter 10 (10.35)

$$= (\tfrac{2}{3})^{3/2} \times 1.163 \times 1.054 \times (9.81)^{1/2} \times 10 \times (0.67)^{3/2}$$

giving

$$Q = 11.46 \text{ m}^3\text{s}^{-1}.$$

The uncertainty in this value of Q is found as follows

Let u_Q represent the percentage uncertainties in Q. Then the total percentage uncertainty in Q is from equation 13.22

$$u_{(Q)} = \pm(u_c^2 + u_b^2 + (3/2)^2)u_h^2)^{1/2}$$

where u_c is the uncertainty in the combined coefficients of $C_d C_v$.

Now

$$u_b = 3/670 \times 100 = \pm 0.45\%$$

and

equation (13.22) becomes

$$u_Q = \pm (0.5^2 + 0 + 2.25 \times 0.45^2)^{1/2}$$

$$= \pm 0.84\%.$$

Expanded uncertainty u, coverage factor $k = 2$ approximate level of confidence 95%

$$u_{k=2}(Q) = Ku(Q)\%$$

$$= 2 \times 0.84\%$$

$$= 1.68\%.$$

Now if the measured flow is 11.46 m^{3}s^{-1} the result of the measurement is expressed as $11.46 \text{ m}^3\text{s}^{-1} \pm 1.68\%$ (expanded uncertainty coverage factor $k = 2$ approximate level of confidence 95%) (Type B evaluation of uncertainty).

(ii) Details of a triangular thin plate weir (V-notch) discharge are given as follows; calculate the uncertainty in the discharge u_Q (uncertainties at 68%)

Discharge	$0.007 \text{ m}^3\text{s}^{-1}$
Gauged head, h	0.121 m
Notch angle, a	90°
Height of notch, P	0.30 m
Uncertainty in head measurement	1 mm
Depth of notch	0.440 m
Top width of notch	0.440 m
Uncertainty in depth of notch	± 1.0 mm
Uncertainty in top width of notch	± 0.5 mm
n	2.5

Uncertainty in C_d (Table 10.18) 1%.

$$u_h = \pm \frac{100}{121}\%$$

then

$$u_h = \pm 0.8\%.$$

Now

$$\tan \tfrac{a}{2} = \frac{\tfrac{1}{2} \text{ top width of notch}}{\text{height of notch}} = \frac{b'}{b} \quad \text{say}$$

and u_b in equation (13.22) may be found from

$$u \tan a/2 = \pm (u_b^2 + u_h^2)^{1/2}\%$$

$$= \pm 100 \left[\left(\frac{1.0}{220}\right)^2 + \left(\frac{0.5}{440}\right)^2 \right]^{1/2} \%$$

$$= 0.5\%.$$

Then

$$u \tan \tfrac{d}{2} = \pm 0.5\%$$

and from equation (13.22), putting $n = 2.5$

$$u_{(Q)} = \pm \, (1.0^2 + 0.5^2 + 2.5^2 \times 0.8)^{1/2}.$$
$$= \pm \, (1 + 0.25 + 6.25 \times 0.64)^{1/2}$$
$$= \pm \, 5.25^{1/2}$$
$$= \pm \, 2.25\%.$$

Expanded uncertainty u, coverage factor $k = 2$ approximate level of confidence 95%

$$u_{k=2}(Q) = Ku(Q)\%$$
$$= 2 \times 2.25\%$$
$$= 4.5\%.$$

Now if the measured flow is $0.007 \, \text{m}^3\text{s}^{-1}$ the result of the measurement is expressed as $0.007 \, \text{m}^3\text{s}^{-1} \pm 4.5\%$ (expanded uncertainty coverage factor $k = 2$ approximate level of confidence 95%) (Type B evaluation of uncertainty).

13.8 Values of uncertainties

Suggested values of the uncertainties in the components used in streamflow measurement are given it the following tables (13.3–13.8). It is advisable, however, for users to either confirm these values for a particular gauging site or to establish their own values. The values of the uncertainties given in all the tables are plus or minus percentage standard deviations at the 68% level of confidence.

Uncertainties in width (u_b)

The standard uncertainty in the measurement of width should be no greater than 0.5%. As an example, the uncertainty introduced for a particular range finder having a base distance of 800 mm varies approximately as given in Table 13.3.

Uncertainties in depth (u_d)

For depths up to 0.300 m the standard uncertainty should not exceed 1.5%, and for depths over 0.300m the uncertainty should not exceed 0.5%.

Table 13.3 Example of uncertainties in width measurements, u_b

Range of width m	Absolute uncertainty m	Relative uncertainty %
0 to 100	0–0.15	0.15
101 to 150	0.15–0.25	0.2
151 to 250	0.3–0.6	0.25

As an example, the standard uncertainty in depth in an alluvial river whose depth varied from 2 m to 7 m and where the velocity varied up to 1.5 m/s was, for these conditions, of the order 0.05 m measured using a suspension cable.

As another example, measurements of depth were taken with a sounding rod up to a depth of 6 m, and beyond that value by a log line with standard air-line and wet-line corrections. These observations were made within the velocity range of 0.087 ms to 1.3 ms. Absolute uncertainties (in metres) were determined and relative uncertainties were computed based on the mid-range depth, the results being as given in Table 13.4.

Uncertainties in determination of the mean velocity

Times of exposure (u_e)

The standard uncertainty in point velocity measurement taken at different exposure times and points in the vertical, shown in Table 13.5 are given as a guide and should be verified by the user. The values are given as standard uncertainties in percent, level of confidence approximately 68%.

Number of points in the vertical (u_p)

The uncertainty values shown in Table 13.6 were derived from many samples of irregular vertical velocity curves. They are given as a guide and should be verified by the user.

Table 13.4 Examples of uncertainties in depth measurements, u_d

Range of depth m	Absolute uncertainty m	Relative uncertainty %	Remarks
0.4–6	0.02	0.65	With sounding rod
6–14	0.025	0.25	With log-line and air- and wet-line corrections

Note: Column 3 relative uncertainties computed from column 2 absolute uncertainties using mid-range depths 3.2 m and 10 m.

Table 13.5 Percentage uncertainties in point velocity measurements due to limited exposure time, u_e

Velocity m/s	Point in vertical							
	0.2D, 0.4D or 0.6D				0.8D or 0.9D			
	Exposure time min							
	0,5	1	2	3	0,5	1	2	3
0.050	25	20	15	10	40	30	25	20
0.100	14	11	8	7	17	14	10	8
0.200	8	6	5	4	9	7	5	4
0.300	5	4	3	3	5	4	3	3
0.400	4	3	3	3	4	3	3	3
0.500	4	3	3	2	4	3	3	2
1.000	4	3	3	2	4	3	3	2
over 1.000	4	3	3	2	4	3	3	2

Table 13.6 Percentage uncertainties in the measurement of mean velocity at a vertical, due to a limited number of points in the vertical, u_p

Method of measurement	Uncertainties %
Velocity distribution	0.5
5 points	2.5
2 points (0.2 & 0.8 D)	3.5
1 point (0.6D)	7.5
surface	15

Table 13.7 Percentage uncertainties in current meter rating, u_c

Velocity measured m/s	Uncertainties %	
	Individual rating	Group or standard rating
0.03	10	10
0.10	2.5	5
0.15	1.25	2.5
0.25	1.0	2.0
0.50	0.5	1.5
over 0.50	0.5	1.0

Rotating-element current-meter rating (u_c)

The uncertainty values shown in Table 13.7 are given as a guide and are based on experiments performed in rating tanks.

Number of verticals (u_m)

The uncertainty values shown in Table 13.8 are given as a guide and should be verified by the user.

Table 13.8 Percentage uncertainties in the measurement of mean velocity due to limited number of verticals, u_m

Name of verticals	Uncertainties %
5	7.5
10	4.5
15	3.0
20	2.5
25	2.0
30	1.5
35	1
40	1
45	1

13.9 The uncertainty in the stage–discharge relation

The equation for the stage–discharge relation may be expressed in the general form

$$Q = C(h + a)^n \qquad (13.23)$$

where C is a coefficient, h is the stage, a is a datum correction denoting the value of stage at zero flow in order to linearise the relation and n is an exponent usually in the range 1.5–2.5.

In order to estimate the uncertainty in the relation, equation (13.23) is linearised by a logarithmic transformation of the form

$$\ln Q = \ln C + n \ln (h + a). \qquad (13.24)$$

The procedure is then one of estimating S_e, the standard error of estimate and S_{mr}, the standard error of the mean (Fig. 13.3).

S_e is calculated from

$$S_e = \left[\frac{\Sigma(\ln Q_i - \ln Q_c)^2}{N - 2} \right]^{1/2} \qquad (13.25)$$

where Q_i is the current meter observation and Q_c is the discharge taken from the rating curve corresponding to Q_i and $(h + a)$, where $Q_c = C(h + a)^n$.

Similarly, S_{mr} may be determined from

$$S_{mr} = \pm\, tS_e \left(\frac{1}{N} + \frac{[\ln\,(h + a) - \overline{\ln(h + a)}]^2}{\Sigma\,[\ln\,(h + a) - \overline{\ln\,(h + a)}]^2}\right)^{1/2} \times 100 \qquad (13.26)$$

where t is Student's t correction for the sample size at the 95% confidence level for N gaugings and may be taken as two for 20 or more gaugings.

S_{mr} is calculated for each gauging on the stage–discharge curve at the relevant value of $(h + a)$. The limits will be curved, having a minimum value at the average value of $\ln\,(h + a)$. If the stage–discharge relation comprises one or more break points, S_e and S_{mr} are calculated for each segment and $(N - 2)$ degrees of freedom are allowed for each segment. At least 20 current meter observations should be available in each range before a statistically acceptable estimate can be made of S_e and S_{mr}. A typical stage–discharge relationship is tabulated in Table 13.9.

Substituting in equation (13.25) for S_e from Table 13.9 gives

$$S_e = \left(\frac{0.029}{30}\right)^{1/2} = 0.031.$$

Therefore $tS_e = 2 \times 0.031 \times 100 = 6.2\%$ at the 95% level.
This equation defines two parallel straight lines on either side of the stage–discharge curve and distant $2S_e$ (6.2%) from it, as shown in Fig. 13.3. Therefore 95% of all the current meter observations, on average, will be contained within these limits. Similarly S_{mr} may be found from equation (13.26) and substituting for S_{mr} from Table 13.9 gives

$$2S_{mr} = 6.2\left(0.031\,25 + \frac{[\ln\,(h - 0.115) + 0.4869]^2}{27.9238}\right)^{1/2}.$$

The value of $2S_{mr}$ (95% level) for each current meter observation at stage $(h + a)$ may be so calculated and the results plotted on each side of the stage–discharge curve to give symmetrical limits with a minimum value as before at the average value of $\ln\,(h + a)$.

Therefore substituting for observation number 1 in Table 13.9 gives

$$2S_{mr} = 6.2\left(0.031\,25 + \frac{(-1.8515 + 0.4869)^2}{27.9238}\right)^{1/2} = 1.94\%.$$

Similarly for observation number 18

$$2S_{mr} = 6.2\left(0.031\,25 + \frac{(-0.5009 + 0.4869)^2}{27.9238}\right)^{1/2} = 1.10\%$$

and for observation number 32

$$2S_{mr} = 6.2 \left(0.031\ 25 + \frac{(1.1709 + 0.4869)^2}{27.9238}\right)^{1/2} = 2.23.$$

A summary for all the S_{mr} values so calculated is given in Table 13.9. The equation gives inner symmetrical curved limits on either side of the stage–discharge curve (as shown in Fig. 13.3).

The above procedure should be satisfactory for most stage–discharge relations. If asymmetrical limits are preferred, the procedure is as follows

$$S_{mr} \text{ (upper 95\% confidence limit)} = 100(e^z - 1) \qquad (13.27)$$

$$S_{mr} \text{ (lower 95\% confidence limit)} = 100(1 - e^{-z}) \qquad (13.28)$$

where z is the right-hand side of equation (13.26) excluding the factor of 100.

Using the same example, therefore, for observation number 1 (where $z = 0.0194$), the upper confidence limit for S_{mr} becomes

$$100(e^{0.0194} - 1) = 100\ (1.0196 - 1) = 1.96\%$$

and the lower confidence limit becomes

$$100\left(1 - \frac{1}{1.0196}\right) = 1.92\%.$$

As noted above, S_{mr} will have a minimum value when $\ln(h + a) = \overline{\ln(h + a)}$ and equation (13.26) reduces to

$$S_{mr} = \frac{tS_e}{\sqrt{N}} \qquad (13.29)$$

$$= \frac{6.2}{(32)^{1/2}} = \pm 1.1\%$$

which is the value given from equation (13.26) for observation numbers 18–20 in Table 13.9 where $\ln(h + a)$ is approximately equal to $\overline{\ln(h + a)}$.

Uncertainty in the daily mean discharge

The value of discharge most commonly required for design and planning purposes is the daily mean discharge, which may be calculated by taking the average of the number of observations of discharge during the 24 h period.

The uncertainty in the daily mean discharge for a velocity–area station may be calculated from the following equation

$$U_{dm} = \frac{\Sigma[(S_{mr}^2 + n^2 U^2 (h + a))^{1/2} Q_h]}{\Sigma Q_h}$$

(13.30)

where U_{dm} is the uncertainty in the daily mean discharge (95% confidence level) and Q_h is the discharge corresponding to $(h + a)$.

The procedure is as follows:

1   Calculate $U(h + a)$ for each of the N values of discharge used to compute the daily mean using a ± 3 mm uncertainty.
2   Calculate U_{dm} from equation (13.30) using the appropriate value of S_{mr}.

A typical calculation for hourly values of discharge is given in Table 13.10.

The corresponding equation for a measuring structure is similar and may be expressed as follows

$$U_{dm} = \frac{\Sigma[(U_c^2 + n^2 U^2 (h + a))^{1/2} Q_h]}{\Sigma Q_h}$$

(13.31)

where U_c is the uncertainty in the coefficient of discharge. Note that U_b, the uncertainty in the length of crest (width of throat), has been neglected and note also that in this case the value of a is zero.

The uncertainty in the monthly mean and annual discharge may be estimated from the following equations

$$U_{mm} = \frac{(U_{dm} Q_{dm})}{\Sigma Q_{dm}}$$

(13.32)

and

$$U_{aa} = \frac{(U_{mm} Q_{mm})}{\Sigma Q_{mm}}$$

(13.33)

where U_{mm} is the uncertainty in the monthly mean discharge (95% level), U_{aa} is the uncertainty in the annual discharge, Q_{dm} is the daily mean discharge, and Q_{mm} is the monthly mean discharge.

Table 13.9 Values required for calculation of S_e and S_{mr}

Observation no	$h + a$	$\ln(h + a)(= x)$	$(x_i - \bar{x})^2$	Q_i	$\ln Q_i(= y_i)$	Q_c	$\ln Q_c(= y_c)$	$(y_i - y_c)^2$	$2S_{mr}\%$
1	0.157	−1.8515	1.8621	2.463	0.9014	2.323	0.8428	0.00342	1.97
2	0.158	−1.8452	1.8450	2.325	0.8437	2.345	0.8523	0.00007	1.96
3	0.188	−1.6713	1.4028	2.923	1.0726	3.060	1.1184	0.00209	1.80
4	0.192	−1.6502	1.3533	3.242	1.1762	3.160	1.1506	0.00065	1.78
5	0.219	−1.5187	1.0646	3.841	1.3457	3.865	1.3520	0.00003	1.66
6	0.259	−1.3509	0.7465	4.995	1.6084	4.996	1.6086	0.00000	1.52
7	0.278	−1.2801	0.6292	5.410	1.6882	5.568	1.7170	0.00083	1.46
8	0.279	−1.2765	0.6235	5.422	1.6905	5.598	1.7224	0.00101	1.46
9	0.287	−1.2483	0.5797	5.883	1.7721	5.846	1.7658	0.00004	1.44
10	0.295	−1.2208	0.5386	6.154	1.8171	6.097	1.8078	0.00008	1.42
11	0.348	−1.0556	0.3234	7.376	1.9982	7.851	2.0606	0.00389	1.30
12	0.405	−0.9039	0.1739	9.832	2.2856	9.902	2.2927	0.00005	1.22
13	0.433	−0.8370	0.1226	11.321	2.4266	10.968	2.3950	0.00099	1.19
14	0.461	−0.7744	0.0826	12.372	2.5154	12.072	2.4909	0.00060	1.16
15	0.465	−0.7657	0.0777	11.825	2.4702	12.233	2.5041	0.00115	1.16
16	0.501	−0.6911	0.0417	13.826	2.6266	13.711	2.6182	0.00007	1.14
17	0.511	−0.6714	0.0340	14.102	2.6463	14.132	2.6484	0.00000	1.14
18	0.606	−0.5009	0.0002	19.020	2.9455	18.345	2.9094	0.00130	1.11
19	0.624	−0.4716	0.0002	19.970	2.9852	19.185	2.9541	0.00096	1.11
20	0.632	−0.4589	0.0008	20.280	3.0096	19.563	2.9736	0.00129	1.11
21	0.681	−0.3842	0.0105	21.204	3.0542	21.931	3.0879	0.00113	1.12
22	0.731	−0.3133	0.0301	23.996	3.1779	24.442	3.1963	0.00033	1.13
23	0.926	−0.0769	0.1681	36.242	3.5902	35.098	3.5581	0.00102	1.22
24	1.225	0.2029	0.4758	54.591	3.9999	53.855	3.9863	0.00018	1.38
25	1.411	0.3443	0.6909	67.327	4.2096	66.859	4.2026	0.00004	1.49

Table 13.9 – Continued

Observation no	$h + a$	$\ln(h + a) \; (= x)$	$(x_i - \bar{x})^2$	Q_i	$\ln Q_i \; (= y_i)$	Q_c	$\ln Q_c \; (= y_c)$	$(y_i - y_c)^2$	$2S_{m}\%$
26	1.646	0.4983	0.9706	79.050	4.3701	84.631	4.4383	0.00465	1.62
27	1.895	0.6392	1.2681	110.783	4.7076	104.989	4.6538	0.00288	1.74
28	2.517	0.9231	1.9881	162.814	5.0926	162.095	5.0882	0.00001	2.02
29	3.150	1.1474	2.6709	227.600	5.4276	228.478	5.4314	0.00001	2.24
30	3.165	1.1522	2.6866	228.800	5.4328	230.145	5.4387	0.00003	2.25
31	3.191	1.1603	2.7133	228.500	5.4315	233.044	5.4512	0.00038	2.26
32	3.225	1.1709	2.7483	236.600	5.4664	236.854	5.4674	0.00000	2.27

$$\overline{\ln(h + a)} = -0.4869; \; \Sigma(x_i - \bar{x}) = 27.9238; \; \Sigma(y_i - y_c)^2 = 0.029\,18$$

$(y_i - y_c)^2 = (\ln Q_i - \ln Q_c)^2$

$(x_i - \bar{x})^2 = [\ln(h + a) - \overline{\ln(h + a)}]^2$

Discharge equation: $Q = 39.479(h - 0.115)^{1.53}$.

Note: The above analyses may be conveniently performed by a suitable software programme.

Table 13.10 Typical computation for the uncertainty in the daily mean discharge using hourly values of discharge

Time	h (m)	(h – 0.115) (m)	Q_h (m³s⁻¹)	$U_{(h+a)}$ (%)	$2S_{mr}$ (%)	$[2S_{mr}^2 + n^2 U^2 (h+a)]^{1/2} Q_h$
0 900	1.225	1.110	46.314	0.4	1.3	64.84
1 000	1.565	1.450	69.707	0.3	1.5	115.53
1 100	1.971	1.856	101.699	0.2	1.8	183.06
1 200	2.293	2.178	129.906	0.2	1.9	246.82
1 300	2.520	2.405	151.186	0.2	2.0	302.37
1 400	2.670	2.565	165.850	0.2	2.0	331.70
1 500	2.789	2.674	177.814	0.2	2.0	355.63
1 600	2.872	2.767	186.328	0.2	2.1	391.29
1 700	2.929	2.814	192.255	0.2	2.1	403.74
1 800	2.981	2.876	197.717	0.1	2.1	415.21
1 900	3.034	2.929	203.339	0.1	2.2	447.34
2 000	3.067	2.952	206.867	0.1	2.2	455.11
2 100	3.082	2.967	208.478	0.1	2.2	458.65
2 200	3.065	2.950	206.653	0.1	2.2	454.64
2 300	3.026	2.911	202.487	0.1	2.2	445.47
2 400	2.975	2.860	197.084	0.1	2.1	413.88
0 100	2.915	2.800	190.793	0.2	2.1	400.66
0 200	2.845	2.730	183.543	0.2	2.1	385.44
0 300	2.747	2.632	173.558	0.2	2.0	347.11
0 400	2.628	2.513	161.697	0.2	2.0	323.39
0 500	2.495	2.380	148.788	0.2	2.0	297.57
0 600	2.365	2.250	136.534	0.2	1.9	259.41
0 700	2.257	2.142	126.635	0.2	1.9	240.61
0 800	2.164	2.049	118.320	0.2	1.8	212.98
			Σ 3883.56			
		Daily mean	161.815			Σ 7952.45

$$U_{dm} = \frac{\Sigma \left[(2S_{mr}^2 + n^2 U^2 (h+a))^{1/2} Q_h \right]}{\Sigma Q_h}$$

$$= \frac{7952.45}{3883.56}$$

$$= \pm 2\%$$

Then daily mean discharge = 161.815 m³s⁻¹ ± 2%.

Systematic error in the stage–discharge curve

A systematic error in the stage–discharge curve has not been included in the above example because the curve is usually established using several different current meters, the inference being that any systematic error in the meters is randomised. However, the possible presence of a systematic error in the stage–discharge curve may require to be investigated. In effect the curve may shift along both the stage axis and the discharge axis, the resultant shift being maximum when stage and discharge shifts have opposite signs. The direction of shift, however, will be unknown and plus and minus values require to be assigned to both (say ± 3 mm for stage and ± 1% for discharge). The resultant of these two factors produces a bandwidth within which the curve will be expected to lie due to the systematic uncertainties alone. The analysis for S_e and S_{mr} is unaffected and is carried out independently. The result, however, should then be presented as

$$\text{Discharge} = Q + S_{mr} + E_s$$

where E_s is the estimated systematic uncertainty.

An investigation of the reference gauge zero, however, would determine the sign and value of the stage shift, and if possible the error could be corrected on site or allowed for in plotting $(h + a)$ values. Any systematic error in the current meter rating tank, however, can only be found from a direct comparison of a selection of rating tanks by rating the same current meters in each.

Note that, by tradition, the dependable variable Q in the stage–discharge equation is plotted on the abscissa with stage on the ordinate. This procedure has no effect on the calculations or on the analysis of the uncertainties.

It should be noted that, in an ideal situation, the current meter observations would in fact fall on the stage–discharge curve, the relation being permanent, and S_e and S_{mr} both being zero. The scatter about the curve experienced in practice, however, is due principally to the uncertainty in the current meter observations, the uncertainty in stage measurement, the instability of the station control, changing conditions in the channel due to scour or accretion and seasonal changes in the river regime. In view of these factors it is possible in some cases for S_e to be larger than the uncertainty in the current meter measurements (U'_Q). The current meter measurements may indeed be of high accuracy for the conditions prevailing at the time of measurement but may be influenced by one or more of the above factors when plotted on the stage–discharge curve.

13.10 Uncertainties in individual methods

The uncertainties associated with each method of open channel flow depend on a number of factors, but the most important ones are:

(a) hydraulic conditions of flow;
(b) measurement of head or stage;
(c) number of verticals taken in a current meter measurement;
(d) coefficient of discharge in the weirs and flumes methods;
(e) the stage–discharge relation in the velocity–area methods;
(f) operation and maintenance of the station.

For good measurement practice carried out to ISO standards the attainable uncertainties in a single measurement of discharge may be taken as shown in Table 13.11.

Table 13.11 Attainable uncertainties in a single measurement of discharge

Method	± Percentage uncertainty (95% level)
Current meter measurement	5
Floats	10–20
Slope–area	10–20
Fall–discharge	10–20
Dilution techniques	5
Thin plate weir	5
Thin plate V-notch	2
Triangular profile (Crump) weir	5
Flat V weir	5
Rectangular profile weir	5
Round nosed weir	5
Flumes	5
ADCP	5
Moving boat	5
Ultrasonic	5
Electromagnetic	5–10

The values of uncertainties in Table 13.11 may be considered attainable for average flow conditions but may require to be modified for conditions of extreme low flows or floods in accordance with error equations given in Section 13.5.

It has been shown in Section 13.9 that the uncertainty in the daily mean discharge, monthly mean discharge and annual discharge can be expected to be much better than the uncertainties shown in Table 13.11.

Further reading

CEN/ISO TS 25377 (in press) *Hydrometric Uncertainty Guide* 2007.

Dymond, J. R. and Christian, R. Accuracy of discharge determined from a rating curve. *Hydrological Sciences Journal*, **27**(4), 493–504 1982.

Guide to the expression of uncertainty 1SO, Geneva 1995.

Herschy, R. W. *Accuracy of existing and new methods of river gauging*. PhD thesis. The University of Reading 1975.

Herschy, R. W. Accuracy. In *Hydrometry: Principles and Practices* (ed. R. W. Herschy). John Wiley and Sons, Chichester, pp. 353–97 1978.

Herschy, R. W. The accuracy of current meter measurements. *Proceedings of the Institution of Civil Engineers*, Part 2, 65 TN 187, pp. 431–7 1978.

Herschy, R. The velocity area method. *Flow Measurement and Instrumentation*, **4**(1) pp. 7–10 1993.

Herschy, R. The stage–discharge relation. *Flow Measurement and Instrumentation*, **4**(1); pp. 11–15 1993.

Herschy, R. W. The analysis of uncertainties in the stage–discharge relation. *Flow Measurement and Instrumentation*, 5(2) pp. 188–190 1994.

Herschy, R. W. Uncertainties in hydrometric measurements in Hydrometry Principles and Practices. (Second Edition) John Wiley and Sons Chichester 1998.

ISO 748. *Liquid Flow Measurement in Open Channels: Velocity–Area Methods*. ISO, Geneva, Switzerland 2008.

ISO 1088 *Collection and processing of data for determination of uncertainties in flow measurement* 2007.

ISO 1100/2. *Determination of the Stage–Discharge Relation*. ISO, Geneva, Switzerland 1981.

ISO 1438 *Thin plate weirs* 2007.

ISO 4360 *Triangular profile weirs* 2007.

ISO 5168. *Measurement of Fluid Flow: Estimation of Uncertainty of a Flow-rate Measurement*. ISO, Geneva, Switzerland 2005.

ISO 7066/1. *Uncertainty in Linear Calibration Curves*. ISO, Geneva, Switzerland 1997.

ISO 7066/2. *Uncertainty in Non-linear Calibration Curves*, ISO, Geneva, Switzerland 1988.

Littlewood, I.G. *Stage-flow rating curve programme*, Personal Communication 2008.

Lintrup, M. A new expression for the uncertainty of a current meter discharge measurement. *Nordic Hydrology*, 20, 191–200, 1989.

Pelletier, P. M. Uncertainties in the determination of river discharge – a literature review. *Eighth Canadian Hydrotechnical Conference*, Montreal, 1987.

Chapter 14

Hydrometric data processing

14.1 Introduction

River flow data are the foundations of water management. Hydrometric data are required for resource assessments, regulatory purposes, river management and, in a digested form, to direct policy development and help draft legislation. As with much environmental monitoring, the need for river flow data becomes particularly compelling during periods of actual or anticipated change. As yet, there is limited evidence of flow regime changes attributable to global warming but most scenario-based climate change projections suggest that the broad hydrological stability which characterised the 20th century may not continue through the 21st. However, climate change is but one of a number of driving forces which can affect river flow patterns, others include changes in land use, agricultural practices and patterns of water use. Similarly, the information needs of an increasingly broad community of river flow data users evolve through time in response to changes in, for example, the legislative framework, national policy objectives, and the increasing engagement of a range of stakeholders in water issues and management.

Hydrometric data processing and archiving systems provide the bridge between the monitoring station network and the information user. In the face of change, such systems need to adapt to ensure that data provision matches contemporary user requirements. Correspondingly over much of the world the last 15 years has seen a change from the provision of largely stand-alone hydrometric data processing facilities to their integration into modern hydrological information systems. These exploit new gauging techniques, telemetry, flexible database management capabilities and web technology to provide a more comprehensive and responsive service to users of river flow data. To match the expanding range of applications for river flow data, access to an increasing range of reference, technical and spatial information is also required to help maximise the data's fitness for purpose. An enduring characteristic of any effective system is the need to ensure that river flow data quality and consistency is maintained and improved. This is of particular importance in relation to flows in the extreme ranges – capturing and archiving these are of pivotal importance to the development of improved water management strategies and engineering design procedures.

14.2 Hydrometric data processing

Figure 14.1 illustrates the components in a typical hydrometric data processing flow chart. The processing and validation of river level and flow data is a core function but the system's overall success depends on the effective integration of each of the components. This implies a continuing dialogue between data providers and data users (including policy makers) to ensure that the range and format of the available outputs match the needs of the user community.

As importantly, the number and distribution of gauging stations in national or regional networks needs to be kept under continuing review to maximise both the cost-effectiveness of the network and its overall information output. Network appraisal mechanisms have evolved from simple gauging station density guidelines, often based on physiographic regions or stream ordering conventions, to more sophisticated statistical analyses using both catchment and river regime characteristics to determine how representative individual catchments are of a river basin, region or landscape type. An increasingly important factor

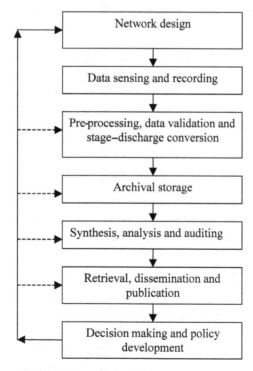

Figure 14.1 Hydrometric data processing flow chart.

is the need to index the contribution that individual gauging stations make (as a consequence of their hydrometric performance and the characteristics of the catchment they monitor) to the development and application of regionalisation techniques, which allow flows to be estimated for ungauged sites.

Modern hydrometric data processing systems typically embrace a number of interlinked components:

(a) A data acquisition module (normally with the capability to handle telemetered data);
(b) A graphical user-interface to provide access to the data for quality assurance purposes, the development and application of stage–discharge relations and other core data-processing activities;
(c) An archiving capability (commonly a relational database, networked in major applications);
(d) A dissemination capability (embracing standard data retrievals, routine reports, web downloads, etc.);
(e) A set of application tools (e.g. for the analysis of extreme flows or the detection of hydrological trends).

In hydrometric terms the critical components are the derivation of the stage–discharge relation and the application of suitable quality assurance mechanisms. The utility of a hydrometric database depends on the ready availability of suitable data-sets of appropriate accuracy. The presence of large volumes of erroneous data can easily undermine the confidence of users in any archiving enterprise. Many factors – including network design, instrument performance, the stability of the stage–discharge relation, staff education, training and motivation – play a part in determining the quality and continuity of archived river flow data. An important additional factor can be the statutory or institutional framework within which the hydrometric data are gathered – diffuse responsibilities shared between a number of organisations can provide little incentive for individual teams/organisations to establish the professional stewardship that the wide community of users requires.

14.3 Quality assurance of river flow data

It will be clear from the above that the quality control of hydrometric data may be expected to involve a wide range of interlinked activities. Considering hydrometric data acquisition as a production line, it is useful to recognise the following reasonably distinct areas where quality control procedures may be applied to good effect.

Hydrometric field practice and the recording of water level

Virtually every part of a river flow archiving system depends for its input, either directly or indirectly, on the original measurement process and hence the quality of the archived data is governed largely by the quality of the recorded stage (or velocity) data. A continuing commitment to station maintenance and good field practice is the only way to ensure that precise and representative river level data are recorded.

The checking of river stage data

Many hydrometric data processing systems now incorporate a facility for the automatic checking of water level data (e.g. range checks and maximum acceptable differences in successive data values). However, the existence of impressive computer software alone is insufficient to guarantee the quality of river level time series. Error recognition is a computer assisted – not computer controlled – procedure and the integrity of the final data will reflect the expertise, enthusiasm and commitment of field and office personnel together with the priority afforded by management to data validation activities.

The stage–discharge relation

In addition to the hydraulic characteristics of the gauging station (or gauging reach) itself, the procedures used to derive a gauging station calibration and the form in which it is expressed may limit the accuracy of the computed discharges. A knowledge of the physical characteristics and behaviour of the river concerned together with an appreciation of the hydraulic and statistical principles underlying the calibration exercise is necessary to achieve the most productive interaction with computer-based calibration programmes (see Chapter 4). To facilitate reprocessing of stage data when rating changes have been detected, and to permit data users to appreciate how historical flow computations have been effected, it is essential that a register of calibrations be maintained.

The validation and flagging of archived flow data

Hydrological data, along with most categories of environmental data, may be most effectively validated in one, or a combination of three, comparative modes: temporal, spatial and with other related variables.

Chart recorders

With the advent of optical shaft encoders and battery-operated solid-state electronic loggers, chart recorders may eventually become superseded. However, where these have been used, they form an important historical archive, especially if the charts contain the observer's notes. Because of the curvature of the rating curve, it should be noted that individual stages are converted to discharge and these discharges averaged for the day to produce daily mean discharges.

Data transmission

Many countries are now using the public telephone network for telemetry links from streamflow stations and this is expected to increase in the future.

Any telemetry system may be regarded as consisting of essentially four elements: the sensor, an encoding device to convert the sensor output to a digital format suitable for transmission, a transmission system linking the sensor to a receiving station and a data reception and distribution facility.

A typical telemetry system may consist of a stilling well and float-driven shaft encoder interfaced to a solid-state logging device linked to a telephone network to a processing centre (Chapter 3). An alarm provision may be provided, but normally 15-minute stage values are stored on site for overnight transmission to microcomputers where the data are initially validated and converted to flow.

Waste water treatment works and waterworks flows

Because of head limitations, discharges in wastewater treatment works are normally measured by rectangular standing wave flumes (or similar devices), while in open channel flow in waterworks, thin plate weirs or V-notches are employed. In either case, a staff gauge is usually installed. These, where used at reservoirs for compensation water or prescribed flows, may have a coloured plastic bar to indicate the required (statutory) head related to the compensation or prescribed flow.

In wastewater treatment works, solid-state recorders may be used and the head measured by ultrasonic look-down gauges. Where stilling wells and float recorders are used in wastewater treatment works, the problem arises of having to clean them continually at regular intervals. In these works, solid-state recorders may be designed to record flows in millions of litres per day (mld), cubic metres per second (cumecs) or litres per second (ls) and, in addition, the total volume for the day in the appropriate units. These recorders normally operate from mains electricity supply and display values of flow and head at very short intervals (seconds).

Any comprehensive quality control system should attempt to provide for the routine screening of all submitted data to identify obviously erroneous figures. No system will ever identify all possible errors; what is required is a practical, efficient set of procedures designed to minimise the volume of significant errors on the river flow archive. Many attempts have been made to incorporate sophisticated data flagging options to complement the validation procedures. Properly implemented these can allow archive users to assess the suitability of particular data-sets better for given applications. Flagging data to allow an audit trail to be established is very valuable but any provision for a comprehensive use of data flags needs to be critically reviewed in light of the resources needed to implement and sustain such a programme. Only if a clear information gain can be demonstrated should more elaborate checking procedures be incorporated into the quality assurance programme.

River flow hydrograph appraisal

Experience gained in many hydrometric monitoring programmes strongly indicates that the most productive means of quality assurance involves the appraisal of hydrographs by personnel familiar with the regime characteristics of the stations under review. By examining a sequence of sub-daily or daily mean flows (typically for a calendar year or water-year) for a target gauging station together with the corresponding hydrograph for a nearby or analogous gauging station, unusual or erroneous flow sequences can be readily identified (see Figure 14.2). This validation approach may be enhanced by the inclusion of the long-term mean hydrograph and shaded envelopes representing the daily highest and lowest flows on record. The envelopes provide a context in which to assess the flow variability through the featured year and help to focus attention on flow sequences or anomalies of most significance – those falling close to, or outside, the existing daily extremes; any which extend the range of recorded variability merit close scrutiny. Appraisal of the flow pattern within the

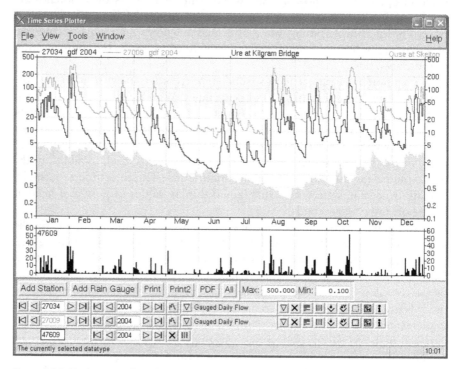

Figure 14.2 Daily river flow hydrographs and a daily rainfall histogram for monitoring stations in the UK.

featured year may be further facilitated by the inclusion of a daily rainfall histogram for a representative rain gauge within the river basin.

Auditing of hydrometric data

The utility of river flow data reflects their accuracy but is also strongly influenced by a number of other factors. Even a small proportion of missing data can greatly reduce the ability to derive meaningful summary statistics (e.g. annual runoff totals or 30-day minima). As importantly, the nature of hydrometric measurement determines that missing data tends to cluster disproportionately in the extreme flow ranges. High levels of accuracy are unlikely to be achievable in such circumstances but the inclusion of an auditable and flagged estimate rather than leaving a gap in the record will produce significant benefits in relation to the overall utility of the time series.

During periods of hydrological stress (floods and droughts) in particular, the overriding operational responsibilities of measuring authorities may necessarily impact on the quality and consistency of the hydrometric data initially available. Also in such circumstances, extreme flows may be computed using stage–discharge relations extrapolated well beyond the range of the gaugings. It is good practice therefore to adopt formal audit procedures, typically activated on an annual basis, to review the quality of the flow data, identify hydrometric issues (e.g. the need for additional high flow gaugings), infill record gaps where practical and, critically, examine any extreme flows (those with an estimated return period of greater than 20 years) to ensure they represent best estimates – with details given of estimation procedures. Such auditing should be undertaken by experienced personnel with the appropriate hydrometric expertise.

14.4 Retrieval and dissemination of river flow data

The value of river flow data is reflected in the volume of its usage and the breadth of its application. Demand for river flow data has never been greater; this demand, together with the need to match the requirements of an ever-wider community of users, can create significant data management challenges. Fortunately information technology developments have revolutionised accessibility to, and dissemination of, hydrometric data – albeit very unevenly across the globe. Electronic data acquisition and telemetry systems allow a rapid linkage between the initial field sensing of water levels and the provision of data to the end-user, and the World Wide Web provides the potential for unprecedented access to river flow data.

Efficient data retrieval systems allow the user of river flow archives to concentrate on analysis and interpretation by minimising the task of locating, collating and processing hydrometric data. Any comprehensive retrieval system should feature a wide range of data selection mechanisms, facilities to compute summary statistics, the ability to generate standard reports and flexibility in the choice of output format.

In practice many requests for data will fall into one of several commonly occurring types. A standard suite of retrieval programmes should be developed to handle such requests. The retrieval suite should anticipate the range of options likely to be required for any given retrieval type and sequence the user-prompts accordingly. For instance, the generation of a flow duration curve may involve decisions concerning the plotting scale, the units to apply and whether single or n-day means should be used as the basis of the percentile computation. By generalising the software a broad spectrum of requests can be catered for; the operator being required to respond to a series of menu choices and screen prompts.

It should be recognised that gauged river flows, on their own, will satisfy the information needs of a minority of users only. In many parts of the world, particularly where artificial influences impact significantly on natural flow regimes, professional stewardship of the basic data requires that the flow data be complemented by sufficient reference, spatial and descriptive information for the user to judge its suitability for a range of applications and to effectively interpret analyses based on the basic data. Figure 14.3 illustrates the dangers of using unmoderated river flow time series. The 10-year running mean plot of annual gauged runoff for the River Thames, for example, displays a compelling long term decline over the 1883–2006 period. However, once the runoff figures are adjusted to account for the increasing upstream abstractions to meet London's water needs, the runoff trend is fundamentally different (see the naturalised time series in Table 13.2). In order to help maximise the utility of hydrometric data, the UK's National River Flow Archive has developed comprehensive Gauging Station Information Sheets for all primary stations in the national network. An example is shown in Figure 14.4.

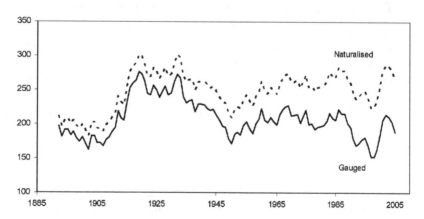

Figure 14.3 Gauged and naturalised runoff (10-yr running averages) for the River Thames at London.

National River Flow Archive
Data Retrieval Service

NRFA Gauging Station Information Sheet

Kennet at Theale

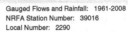

Measuring Authority: Environment Agency
Grid Reference: 41 (SU) 649 708
Station Type: Crump weir

Gauged Flows and Rainfall: 1961-2008
NRFA Station Number: 39016
Local Number: 2290

Daily Flow Hydrograph

Max. and min. daily mean flows from 1961 to 2008 excluding those for the featured year (2006; mean flow: 7.34 m³s⁻¹)

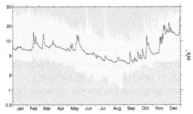

Flow Duration Curve

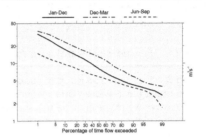

Flow Statistics

(Units: m³s⁻¹ unless otherwise stated)

Mean flow	9.81	
Mean flow (ls⁻¹/km²)	9.49	
Mean flow (10⁶m³/yr)	310.0	
Peak flow / date	70.0	11 Jun 1971
Highest daily mean / date	47.1	6 Feb 2001
Lowest daily mean / date	0.925	21 Aug 1976
10 day minimum / end date	1.102	28 Aug 1976
60 day minimum / end date	1.460	28 Aug 1976
240 day minimum / end date	2.781	24 Sep 1976
10% exceedance (Q10)	17.420	
50% exceedance (Q50)	7.995	
95% exceedance (Q95)	3.762	
Mean annual flood	37.3	
IH Baseflow index	0.87	

Rainfall and Runoff

	Rainfall (1961-2006) mm					Runoff (1961-2008) mm				
	Mean	Max/Yr		Min/Yr		Mean	Max/Yr		Min/Yr	
Jan	77	157	1995	12	1997	38	83	2003	11	1976
Feb	54	139	1990	8	1993	36	74	1995	11	1976
Mar	62	162	1981	15	1997	38	70	2001	11	1976
Apr	57	175	2000	2	1984	32	63	2001	9	1976
May	59	131	1979	8	1990	26	46	2000	7	1976
Jun	57	177	1971	5	1962	21	47	1971	5	1976
Jul	49	100	1988	14	1999	17	37	2007	4	1976
Aug	63	152	1977	6	1995	14	29	2007	4	1976
Sep	66	169	1974	12	1997	13	25	1968	7	1976
Oct	78	159	2000	4	1978	16	36	1966	8	1997
Nov	81	194	2002	31	1990	21	52	2000	10	1978
Dec	83	158	1989	15	1988	29	89	2000	11	1991
Year	786	1059	2002	579	1964	300	464	2000	124	1976

Station and Catchment Characteristics

Station level	(mOD)	43.4
Sensitivity	(%)	6.5
Bankfull flow		
Catchment area	(km²)	1033.
Maximum altitude	(mOD)	297
FSR slope (S1085)	(m/km)	1.46
1961-90 rainfall (SAAR)	(mm)	759
FSR stream frequency (STMFRQ)	(junctions/km²)	
Urban extent	(0-1)	0.0137
Flood Attenuation Index	(0-1)	0.9564

Factors Affecting Runoff

● Runoff influenced by groundwater abstraction and/or recharge.
● Runoff reduced by industrial and/or agricultural abstraction.
● Regulation from surface water and/or ground water.

Station and Catchment Description

Crump weir (15.9m broad) equipped with pressure tapping (not used) & d/s recorder. Cableway installed in 1999; subsequently removed. Some subsidence: fall of 41mm across the weir crest. Modular up to 24 cumecs and all but highest flows contained. Bypassing above 29 cumecs, hence flood flows may be underestimated. Little net impact of abstractions and discharges (minor contribution to Kennet & Avon canal) but augmentation from WBGS during droughts. High baseflow component but responsive contribution from the R. Enbourne. Dmfs 04-24/12/98 estimated by NRFA (using 39103) due to gauging hut refurbishment.

A mainly pervious catchment (Chalk with significant Drift cover), but the lowest quarter is largely impermeable. A primarily rural catchment with scattered settlements (Newbury is the largest town); significant urban growth along the Kennet valley.

Summary of Archived Data

Gauged Flows and Rainfall

	0 1 2 3 4 5 6 7 8 9
1960s	- e A A A A A A A A
1970s	A A A A A A A A A A
1980s	A A A A A A A A A A
1990s	A A A A A A A A A A
2000s	A A A A A A A a e

Key:	All rain-fall	Some or no rain-fall
All daily, all peaks	A	a
All daily, some peaks	B	b
All daily, no peaks	C	c
Some daily, all peaks	D	d
Some daily, some peaks	E	e
Some daily, no peaks	F	f
No gauged flow data	=	-

Naturalised Flows

Key:

All daily, all monthly	A
Some daily, all monthly	B
Some daily, some monthly	C
Some daily, no monthly	D
No daily, all monthly	E
No daily, some monthly	F
No naturalised flow data	=

Centre for Ecology and Hydrology, Wallingford, Oxon OX10 8BB, UK.Tel. (01491) 838800.   3rd June 2008

Figure 14.4 UK National River Flow Archive Gauging Station Summary Sheet.

Publication of river flow data

Guidelines for the format and content of hydrological publications are set out in the World Meteorological Organization's *Guide to Hydrometeorological Practices* (WMO No. 168). Historically, most countries published river flow data in Yearbooks in a range of different formats. This approach has been largely superseded as user needs have become more extensive and complex. However, the release of data via the Internet, hampered in some countries by

Table 14.1 Functions of Hydrological Yearbooks

1 To serve as an introduction to national/regional hydrological information services
2 To catalogue available water resources information (and identify deficiencies)
3 To provide a representative snapshot of the state of water resources through time establishing a series of benchmarks
4 To provide a supplementary archive – a safeguard against the loss or destruction of the original data (or electronic database)
5 To allow evidence-based assessments of hydrological trends to made – and updated
6 To help motivate field and office staff – and familiarise them with the end products of their endeavours
7 To provide tangible evidence of a return on substantial public investment
8 To promote discipline and concern for accuracy within the national regional data processing system
9 To provide a clear incentive to keep databases up to date, allowing contemporary hydrological conditions to be assessed
10 To increase awareness of hydrological and water resources issues and facilitate the widest usage of streamflow data
11 To promote hydrological research through the ready availability of hydrometric and other relevant data

Table 14.2 The UK National River Flow Archive website – main components

Component	Contents
River Flow Data	Time series downloads for 200 index gauging stations throughout the UK. Introduction to the complementary data retrieval facilities of the National River Flow Archive (NRFA).
Water Watch	Outputs from the National Hydrological Monitoring programme; annual, monthly and occasional summaries of hydrological conditions and water resources variability throughout the UK.
UK Gauging Network	Location maps, reference information, hydrographs, flow duration curves, together with data availability and descriptive material relating to all primary gauging stations.
Publications	Details of publications in the Hydrological data UK series and other associated publications and reports.
Catchment Spatial Details	Spatial characterisations (topography, rainfall, land use and hydrogeology) of more than 1200 catchments in the UK monitoring network.

commercial considerations, has yet to fully capitalize on the available technology to provide comprehensive, and timely, access to match the rapidly growing demand for hydrometric information. In exploiting the Internet's potential, the full range of functions previously performed by yearbooks needs to be recognised (see Table 14.1) as well as providing access to a broad range of reference, spatial and other information of interest to the user community. Correspondingly, web-based data download facilities should be complemented by ready access to information relating to the gauging network and capabilities, the hydrometric performance of individual gauging stations, reports based on the archived river flow data (e.g. assessments of water resources status or appraisals of flood magnitude), and research reports relating to issues of hydrometric importance. Table 14.2 gives details of the primary website modules maintained by the UK National River Flow Archive (*http://www.ceh.ac.uk/data/nrfa/index.html*)

Further reading

HYDATA. *Hydrological Database and Analysis System*. The Institute of Hydrology, Wallingford, Oxon., OX10 8BB, UK 1992.

HYDROLOG. Data Management Software Products, GAUGEMAN, Hydrologic Ltd., Leominster, HR6 8DQ, UK 1992.

Laize, C. L. R., Marsh, T. J. and Morris, D. G. Capitalising on regionalisation methodologies to support the appraisal and evolution of hydrometric networks. *ICE Water Management*, (in press) 2007.

Marsh, T. J. The acquisition and processing of river flow data. In *Hydrometry, Principles and Practices* (ed. R. W. Herschy). John Wiley and Sons, Chichester, pp. 399–427, 1978.

Marsh, T. J., Lees, M. L. and Littlewood, I. G. *Surface Water Data Processing: A Guide to the Water Data Unit's Facilities*. DOE Water Data Unit 1991.

Marsh, T. J. *The United Kingdom Surface Water Network*. World Meteorological Organization Casebook on Hydrological Networks in Europe, Geneva, 1991.

Marsh, T. J. Hydrometric Data Acquisition and Archiving Systems. MSc Lecture at the University of Newcastle, UK, 1991.

Marsh, T.J. Maximizing the utility of river flow data in *Hydrometry Principles and Practices. Second edition* (ed. R.W. Herschy). John Wiley and Sons, Chichester 1998.

Marsh, T. J. Capitalising on river flow data to meet changing user needs – a UK perspective. *Flow Measurement and Instrumentation 13 (2002)*, pp. 291–298, 2002.

World Meteorological Organization. *Manual on Stream Gauging, Vol. 11. Computation of Discharge*. WMO No. 519, Geneva, 1980.

World Meteorological Organization. *Guide to Hydrological Practices*, Vol. 1. WMO No. 168, Geneva, 1987.

Chapter 15

Flow in pipes

15.1 Closed conduit flowmeters

Hydrologists are often required to audit closed conduit flowrate in which venturi meters or orifice plate meters, etc., are employed to measure flow in pipes running full. These meters are based on the principle that when water passes through a contraction in a pipe, it accelerates. The resulting increase in kinetic energy is balanced by a decrease in the static pressure at that point in the pipe and the pressure drop caused by the contraction is proportional to the square of the flowrate for a given flowmeter.

The general relation between flowrate and pressure drop for a differential pressure meter is

$$Q = \frac{Ca\sqrt{(2gh)}}{(1 - a^2/A^2)^{1/2}} \text{ m}^3 \text{ s}^{-1} \tag{15.1}$$

or

$$Q = \frac{CE\pi d^2}{4} \sqrt{(2\,gh)} \text{ m}^3 \text{ s}^{-1} \tag{15.2}$$

where C is the coefficient of discharge and

$$E = \frac{D^2}{\sqrt{(D^4 - d^4)}} \tag{15.3}$$

a = cross-sectional area of the contraction (m^2);
A = cross-sectional area of the pipe (m^2);
d = diameter of the contraction (m);
D = diameter of the pipe (m);
h = head difference between contraction and adjacent pipe (differential pressure in metres of water).

It should be noted that the value of C is a function of Reynolds number (equation (15.5)) but normally is constant above a Reynolds number of about 3×10^5 based on the diameter of the contraction.

The most common types of flowmeters for water measurement are the square-edged orifice plate and the venturi tube (Figs 15.1 and 15.2). Typical values of the coefficient of discharge for these are 0.6 and 0.98 respectively.

The orifice plate is simply a plate with a hole in it (usually concentric) and installed transversely in a pipe with differential pressure tappings before and after the plate. There should be at least four tappings in each plane of pressure measurement, distributed evenly around the pipe. This applies to all types of differential pressure meters so that the pressure distribution is constant across each of the two planes of pressure measurement.

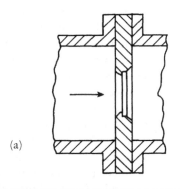

(a)

(b)

Figure 15.1 Orifice plate meter, (a) diagrammatic sketch and (b) view of meter in application.

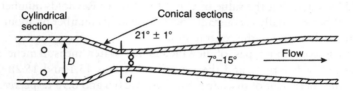

Figure 15.2 Venturi tube meter.

Venturi meters consist of a cylindrical 'throat' section preceded by a short contraction and followed by a longer expansion to allow pressure recovery. They are more expensive to make and install than an orifice plate but pressure recovery ensures a much lower head loss than with an orifice plate.

The Dall tube meter (Fig. 15.3) was invented to ensure a low head loss. It consists of two cones each with a substantial included angle between which is a circumferential slot. The abrupt change of boundary contour results in a flow curvature which increases the differential head produced and the sudden reduction in cross-sectional area at the upstream pressure tapping gives a local pressure increase which also augments the pressure differences.

Manometers are most commonly used to measure differential pressure. Care should be exercised to ensure the leads to the manometer are free from air bubbles and frequent 'bleeding' of the leads to eliminate bubbles is essential. If a U-tube manometer is used to determine the differential pressure, the effective density of the manometer liquid is mercury density minus water density.

Differential meters have a small range (turndown ratio) and because of the square-law relation between flowrate and differential pressure, a 3:1 flow range is about the most which can be measured without changing the manometers or pressure transducers.

Other types of flowmeter include the electromagnetic flowmeter (Fig. 15.4), commonly referred to as the 'Magflow' meter, which is becoming more popular because of its minimal head loss and large range of flow (typically 30:1) and is less sensitive to pipework installation effects. These meters become increasingly

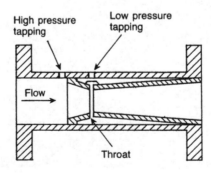

Figure 15.3 Dall tube meter.

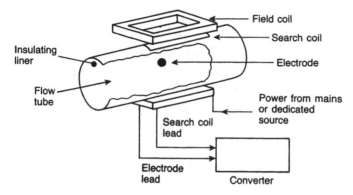

Figure 15.4 Electromagnetic flowmeter ('Magflow' meter).

more expensive than differential pressure meters as the pipe diameter increases. Frequent checking of the meter calibration is advisable if the highest accuracy is to be maintained. The principle of the electromagnetic meter is the same as that of the open channel gauge (Chapter 2).

In the turbine flowmeter (Fig. 15.5), a free spinning rotor is mounted axially in the pipe and a magnetic pick-up is employed to measure the speed of rotation which is proportional to flowrate except at very low flows. Turbine meters have a range of about 10:1 and have a much lower head loss than differential pressure meters but higher than electromagnetic meters.

Ultrasonic pipe flowmeters (Fig. 15.6) operate on the same principle as the open channel gauge (Chapter 12) and in recent years have become attractive as an option. They have no resistance to flow, are virtually independent of

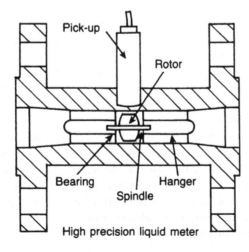

Figure 15.5 Turbine meter.

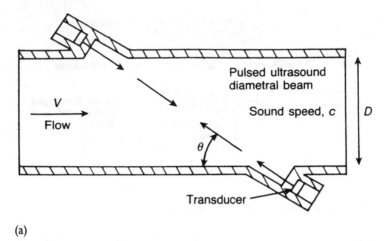

(a)

Figure 15.6(a) Ultrasonic pipe flowmeter.

viscosity and the cost for large pipes of 0.5 m and above is little different from that for small pipes and the accuracy improves as the pipe size increases. The configuration of the meter in its simplest form has one pair of transducers across a diameter but more sophisticated meters have four pairs of transducers located across four chords which do not pass through the centre of the pipe centre line. The transmissions in this case are averaged or weighted to give the mean velocity. Figure 15.6(b) shows a four-path system across four chords. More recent versions employ a clamp-on meter which has the advantage that flowrate can be measured in existing pipes without drilling or cutting but only one path having one pair of transducers is used, the path passing through the centre of the pipe.

The best ultrasonic meters use a 'time of flight' principle (Chapter 12) but some meters may employ a Doppler principle which relies on a single beam being reflected from particles moving with the flow (Fig. 15.7). The frequency with which these reflected signals are received is then a function of the frequency of the transmitted beam, the velocity of the water and the speed of the ultrasound. This system is less accurate than the 'time of flight' method which is mostly used in commercial meters.

The flowrate accuracy attainable with closed conduit meters is very much higher than that achieved in open channel gauges and meters manufactured and installed in accordance with national or international standards have an uncertainty of about 1.5% but a meter having an individual calibration by an absolute method can achieve an uncertainty of as low as 0.5%. It is essential, however, that the conditions under which the meter is used are identical to those under which it was calibrated.

(b)

Figure 15.6(b) Diagrammatic sketch of a four-path ultrasonic pipe flowmeter.

Example

An orifice plate meter has the following dimensions; calculate the flowrate

$$D = 0.229 \text{ m};$$

$$d = 0.121 \text{ m};$$

$$p_1 - p_2 = \text{differential pressure} = 0.08 \text{ m} = \text{Hg (mercury)};$$

$$C = 0.611 \text{ m};$$

$$h_w \text{ (water)} = 0.08 \left(\frac{13.6}{1} - 1\right) = 1.008 \text{ m}.$$

Now from equation (15.2)

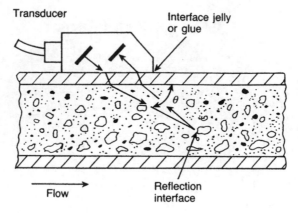

Figure 15.7 Ultrasonic Doppler flowmeter.

$$Q = CE\frac{\pi d^2}{4}\sqrt{(2gh_w)}\ \mathrm{m^3\ s^{-1}}$$

$$= \frac{C \times D^2}{\sqrt{(D^4 - d^4)}} \times \frac{\pi d^2}{4}\sqrt{(2 \times 1.008g)}\ \mathrm{m^3\ s^{-1}}$$

$$= 0.611 \times 1.032 \times 0.0115\sqrt{(19.777)}\ \mathrm{m^3\ s^{-1}}$$

$$= 0.033\ \mathrm{m^3\ s^{-1}}.$$

Using the ISO 5167/1 (BS 1042/1.1) equation

$$q_v = CE\frac{\pi d^2}{4}\frac{\sqrt{(2\Delta p \times \rho)}}{\rho}$$

where q_v is in $\mathrm{m^3\ s^{-1}}$;

   Δp is in Pascals ($\mathrm{kg\ m^{-1}\ s^{-2}}$)

   $\rho = 10^3\ \mathrm{kg\ m^{-3}}$.

Note: the conversion factor for mHg is taken as

   $\mathrm{mHg} \times 1.333 \times 10^5 = \mathrm{Pa}$.

Therefore

$$q_v = 0.611 \times 1.032 \times 0.0115\ \sqrt{(2 \times 1.333 \times 0.08 \times 10^8)}/10^3\ \mathrm{m^3\ s^{-1}}$$

$$= 0.033\ \mathrm{m^3\ s^{-1}}.$$

15.2 Discharge through unmetered pipes

Flowing full

The discharge of unmetered pipes flowing full is based on the head loss along the pipeline. The head loss is referred to as the 'head loss due to friction', h_f, and the energy gradient is given as $S = h_f/L$ where L is the length of pipeline. The flow through full pipes is a problem often encountered in hydrometry and several formulae are available for the estimation of discharge. The most common discharge equation is the Darcy–Weisbach equation for turbulent flow

$$h_f = \frac{4fLv^2}{2gD} \text{ (m)} \tag{15.4}$$

from which v can be determined. In equation (15.4) f is a non-dimensional coefficient dependent on the relative roughness and Reynolds number Re where

$$\text{Re} = \frac{vD}{v}. \tag{15.5}$$

In equations (15.4) and (15.5) D is the pipe diameter (m) and in equation (15.5) v is the kinematic viscosity ($\text{m}^2 \text{ s}^{-1}$). It should be noted that equation (15.4) is now usually replaced by

$$h_f = \frac{\lambda Lv^2}{2gD} \text{ (m)} \tag{15.6}$$

where $\lambda = 4f$. For laminar flow (Re $\leqslant$ 2000)

$$\lambda = \frac{64}{\text{Re}}. \tag{15.7}$$

Values of λ for steady uniform turbulent flow can be found from charts such as the *Moody diagram* or the *Charts for the hydraulic design of channels and pipes* by the UK Hydraulics Research Ltd. (Figs 15.8 and 15.9 respectively). Values of λ may also be determined from the Colebrook–White equation for full pipes

$$\frac{1}{\lambda} = -2 \log \left(\frac{k}{3.7D} + \frac{2.51}{\text{Re}\sqrt{\lambda}} \right), \tag{15.8}$$

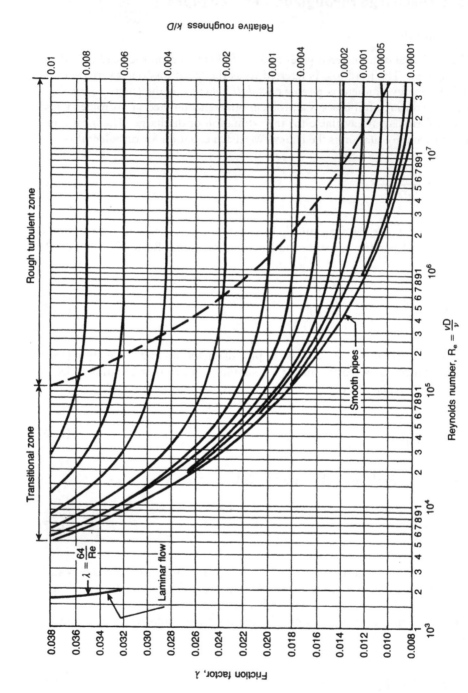

Figure 15.8 Moody diagram.

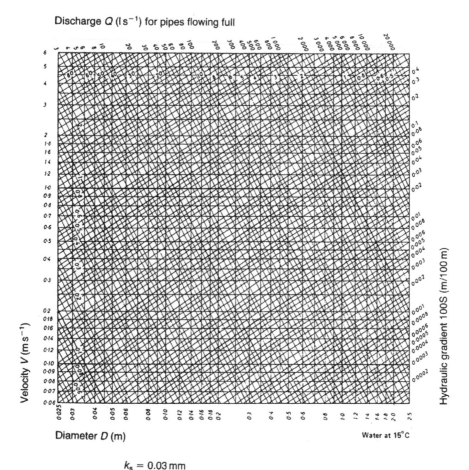

Discharge Q ($\mathrm{l\,s^{-1}}$) for pipes flowing full

Velocity V ($\mathrm{m\,s^{-1}}$)

Hydraulic gradient 100S (m/100 m)

Diameter D (m)   Water at 15°C

$k_s = 0.03\,\mathrm{mm}$

Figure 15.9 Extract from *Charts for the hydraulic design of channels and pipes* by Hydraulics Research Ltd.

where k is the effective roughness size of the pipe wall, and by combining equations (15.6) and (15.8) the following equation for v is obtained

$$v = -2\sqrt{(2gDS)} \log \left[\frac{k}{3.7D} + \frac{2.51v}{D\sqrt{(2gDS)}} \right].$$ (15.9)

Example

A pipeline 3000 m long and 300 mm in diameter carries water at 15°C between two reservoirs. The difference in water level between the reservoirs is 30 m. Neglecting entry losses, etc. and taking $\lambda = 0.015$, $v = 1.14 \times 10^{-6}$ and $k = 0.03$ mm, determine the discharge from

(a) equation (15.6);
(b) equation (15.9);
(c) Figure 15.8;
(d) Figure 15.9.

(a) From equation (15.6)

$$v = \left(\frac{19.62 \times 0.3 \times 30}{0.015 \times 3000}\right)^{1/2}$$

$$= 1.98 \text{ m s}^{-1}$$

and

$$Q = Av$$

$$= \pi \times \frac{0.3^2}{4} \times 1.98$$

$$= 0.140 \text{ m}^3 \text{ s}^{-1}.$$

(b) From equation (15.9)

$$v = -2\sqrt{(19.62 \times 0.3 \times 0.01)}$$

$$\times \log\left[\frac{0.03 \times 10^{-3}}{3.7 \times 0.3} + \frac{2.51 \times 1.14 \times 10^{-6}}{0.3 \sqrt{(19.62 \times 0.03 \times 0.01)}}\right]$$

$$= 2.03 \text{ m s}^{-1}$$

and $Q = Av = 0.143 \text{ m}^3 \text{ s}^{-1}$.

(c) From Fig. 15.8, with

$$\frac{k}{D} = \frac{0.03 \times 10^{-3}}{0.3} = 0.0001$$

and

$$\text{Re} = \frac{2 \times 0.3}{1.14 \times 10^{-6}} = 5 \times 10^5.$$

Fig. 15.8 gives $\lambda = 0.015$ and from (a) above this value of λ gives $Q = 0.140$ m^3 s^{-1}.

(d) From Fig. 15.9 and entering with the hydraulic gradient of

$$\frac{30}{3000} = 1.0 \text{ (m/100 m)} \quad \text{and} \quad D = 0.3 \text{ m}$$

v is found to be approximately 2.0 m s^{-1} and $Q = 0.15$ m^3 s^{-1}.

Partially full pipes

The Colebrook–White equation may be used to estimate flow in partially filled pipes. By replacing D in equation (15.9) by $4R$ the equation becomes

$$v = -\sqrt{(32gRS)}\log\left[\frac{k}{14.8R} + \frac{1.255v}{R\sqrt{(32gRS)}}\right] \qquad (15.10)$$

where R is the hydraulic radius $= A/P$ where A is the cross-sectional area of flow (m^2) and P is the wetted perimeter (m). It is evident that in sewers the value of k will have higher values than for clean pipes. Recommended values can be found in *Charts for the hydraulic design of channels and pipes* by UK Hydraulics Research Ltd.

Example

Calculate the flow in a sewer having $A = 0.75$ m^2, $P = 1.5$ m, $S = 0.001$, $k = 3$ mm, $v = 1.14 \times 10^{-6}$.
    From equation (15.10)

$$V = -\sqrt{(32\,g \times 0.5 \times 0.001)}$$

$$\times \log\left[\frac{0.003}{14.8 \times 0.5} + \frac{1.255 \times 1.14 \times 10^{-6}}{0.5\,\sqrt{(32g \times 0.5 \times 0.001)}}\right]$$

$$= 1.35 \text{ m s}^{-1}.$$

Therefore $Q = AV = 0.75 \times 1.35 = 1.01$ m^3 s^{-1}.

Methods of measuring flow in sewers, small channels and culverts

A number of instruments are now available for measuring flow in sewers, small channels and culverts by means of a sensor installed in the invert of the pipe or culvert (or on the bed or side of the channel). All of these instruments give velocity, depth of flow and discharge.
    Fig. 15.10 is a view of the SonTek acoustic Doppler monitor for pipes and culverts, (see also Chapter 6 ADCP).

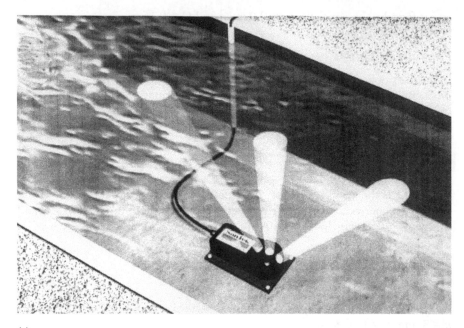

(a)

(b)

Figure 15.10 SonTek acoustic Doppler monitors: (a) acoustic Doppler monitor for pipe and culvert to measure velocity, water level and flow. (b) side looking acoustic Doppler monitor to measure velocity, water level and flow.

Further reading

Baker, R. C. *Flow Measurement – An Introductory Guide*. MEP, London 1989.

BS 5792. *Specification for Electromagnetic Flow Meters* (no ISO equivalent). HMSO, London 1980.

BS 7405. *Guide to the Selection and Application of Flowmeters* (no ISO equivalent). HMSO, London 1991.

Douglas, J. F., Gasiorek, J. M. and Swaffield, J. A. *Fluid Mechanics*. Pitman, London 1979.

Featherstone, R. E. and Nalluri, C. *Civil Engineering Hydraulics*. BSP Professional Books, Oxford 1988.

Hydraulics Research Ltd. *Charts for the Hydraulic Design of Channels and Pipes*, Fifth edn. Wallingford, Oxfordshire 1993.

ISO 5167/1 (BS 1042/1.1). *Square Edged Orifice Plates, Nozzles, and Venturi Tubes*. ISO, Geneva, Switzerland 1992.

ISO 9104 (BS 7526). *Evaluating the Performance of Electromagnetic Flow Meters*. ISO, Geneva, Switzerland 1991.

Lewit, E. H. *Hydraulics and the Mechanics of Fluids*. Pitman, London 1947.

Appendix

Conversion factors

Length			
	1 m	=	3.2808 ft
	1 ft	=	0.3048 m
	1 in	=	25.4 mm
	1 mile	=	1.6093 km
	1 km	=	0.6214 mile
	1 mm	=	0.0394 in
	1 cm	=	0.3937 in

Area			
	1 in^2	=	645.2 mm^2
	1 ft^2	=	0.0929 m^2
	1 sq mile	=	2.590 km^2 = 259 ha
	1 acre	=	4047 m^2
	1 acre	=	0.4047 ha
	1 ha	=	2.4710 acres
	1 km^2	=	0.3861 sq mile = 100 ha
	1 cm^2	=	0.1550 in^2
	1 m^2	=	10.763 ft^2

Volume			
	1 m^3	=	1000 l
	1 ft^3	=	28.32 l
	1 ft^3	=	0.02832 m^3
	1 ft^3	=	6.23 gal
	1 ft^3	=	7.4805 gal US
	1 gal	=	4.546 l
	1 gal	=	0.00455 m^3
	1 gal	=	0.1605 ft^3
	1 gal	=	1.201 gal US
	1 gal US	=	3.785 l
	1 gal US	=	0.003785 m^3
	1 cm^3	=	0.06102 in^3
	1 acre-ft	=	1233 m^3
	1 fl oz US	=	29.5 cc
	1 fl oz UK	=	28.4 cc

Velocity			
	1 m s^{-1}	=	3.2808 ft s^{-1}
	1 ft s^{-1}	=	0.3048 m s^{-1}
	mile h^{-1}	=	1.6093 km h^{-1}

Discharge (flow rate)	1 $m^3 s^{-1}$	=	86.400 Mld
	1 $m^3 s^{-1}$	=	35.3147 $ft^3 s^{-1}$
	1 $m^3 s^{-1}$	=	19.00526 Mgd
	1 Mld	=	0.0115 $m^3 s^{-1}$
	1 Mld	=	11.574 $1^3 s^{-1}$
	1 Mld	=	0.22 Mgd
	1 Mld	=	0.26 Mgd US
	1 ls^{-1}	=	0.019 Mgd
	1 Mgd	=	4.546 Mld
	1 Mgd	=	0.0526 $m^3 s^{-1}$
	1 Mgd	=	52.6161 s^{-1}
	1 Mgd	=	4546 $m^3 d^{-1}$
	1 $ft^3 s^{-1}$	=	0.0283 $m^3 s^{-1}$
	1 $ft^3 s^{-1}$	=	28.32 l s^{-1}
	1 Mgd US	=	3785 $m^3 d^{-1}$
	1 Mgd US	=	0.04381 $m^3 s^{-1}$
	1 $ft^3 s^{-1} mile^{-2}$	=	0.01094 $m^3 s^{-1} km^{-2}$
Density	Water	=	1000 kg m^{-3}
Mass	1 lb	=	453.6 g
	1 lb	=	0.4536 kg
	1000 kg	=	1 metric tonne
Pressure (head-water)	1 mm Hg	=	1.333×10^2 Pa
Gravitational acceleration (g)	9.807 m s^{-2}	=	32.175 ft s^{-2}

Index

Milton Keynes UK
Ingram Content Group UK Ltd.
UKHW021923071024
449327UK00022B/1699